Technological innovation and Third World multinationals

In recent years there has been a spectacular growth in the technological capabilities of developing countries and in outward direct investment from firms based in these countries. This book argues that these phenomena are related. The emerging technological capabilities of Third World firms are allied to the cumulative growth of their outward direct investment. The existence and accumulation of technological capabilities are a determinant as well as an effect of their international production activities.

Evidence is presented on the unique, innovative activities of MNEs from Asia and Latin America, which, although at an early state, have become increasingly important as international production evolves and as developing countries advance through higher stages of industrial development.

Paz Estrella E. Tolentino is a Transnational Corporations Affairs Officer at the United Nations Centre on Transnational Corporations, New York. The initial research for this book emerged from a doctoral thesis prepared by the author at the University of Reading where the author received the 1989 Academy of International Business Richard N. Farmer prize for the best Ph.D dissertation on international business.

Technological innovation and Third World multinationals

Paz Estrella E. Tolentino

London and New York

First published 1993
by Routledge
11 New Fetter Lane, London EC4P 4EE

Simultaneously published in the USA and Canada
by Routledge
29 West 35th Street, New York, NY 10001

© 1993 Paz Estrella E. Tolentino

Typeset in Garamond by Michael Mepham, Frome, Somerset
Printed and bound in Great Britain by
Mackays of Chatham PLC, Chatham, Kent.

British Library Cataloguing in Publication Data
A catalogue record for this book is available from the British Library
ISBN 0–415–04807–9

Library of Congress Cataloging in Publication Data
Tolentino, Paz Estrella E., 1960–
   Technological innovation and Third World
   multinationals / Paz Estrella E. Tolentino.
   p.   cm.
   Includes bibliographical references and index.
   1. Technological innovations—Developing countries.
   2. International business enterprises.   I. Title.
   T173.8.T65   1993
   338'.064'091724—dc20   92–5426
                                                         CIP

ISBN 0–415–04807–9

Contents

Figures

Tables

Foreword

It gives me great pleasure to write a foreword to this book. I have known the author since she came to Reading University in 1983 to pursue a new graduate course on *International Business and Economic Development* in the then Department of Economics. One of the intentions of offering the course was to encourage the better students to pursue a doctorate at the University and thus help maintain the reputation of the Department as one of the leading research centres in International Investment and Business Studies. Paz Tolentino not only distinguished herself in our MA programme but went on to complete an excellent Ph.D thesis, which has formed the basis for this monograph. Dr Tolentino was also of great assistance to John Cantwell and myself in our production of the first *Directory of Statistics on International Investment and Production*. This work has now been taken over by the Transnational Corporations and Management Division of the Department of Economic and Social development of the United Nations, with Paz Tolentino playing a leading role in the preparation of a five volume series. The first two volumes of the *World Investment Directory*, embracing Asia and the Pacific, and the OECD countries, have already been published. The remaining three volumes will be published in 1993.

There is another reason why I am delighted that the author asked me to contribute this foreword. For several years I have been interested in explaining the international direct investment transaction of countries. Why are some counties net outward investors, and other net inward investors? Why does the balance of a country's international investment balance change over time? Can one, indeed, trace a cycle or path which a country might follow as it moves through its various stages of development? In 1981, I first put forward (In a paper published in a volume edited by K. Khan on *Multinationals from the Developing Countries*) the idea of an investment development cycle (now renamed path) by which a country proceeded from engaging in very little inward or outward investment, to becoming a major inward investor, to becoming a net outward investor. Because of data limitations (especially of direct investment statistics from developing counties) I had to content myself with some cross-sectional statistical testing; which, when taking into account the different economic structures of countries, held up quite well.

In the last twelve years much has changed in the world economy. Several developing countries, and particularly those in East Asia, have begun to be major outward investors as their own industrial capabilities have become more competitive. This is entirely consistent with the predictions of the investment development path. At the same time, the relationship between a country's development and its international direct investment position is by no means as smooth or certain as my earlier analysis suggested. Neither does the behaviour of developed countries easily fit into the model. Why has the net outward investment position of the USA deteriorated so much since 1980? Why have France and Sweden become major outward investors? Why is Canada now a large foreign investor in the USA? Why is Japan's outward investment position so much out of line with the rest of the industrialised countries? What explains the dramatic growth of intra-Asian MNE activity since the mid-1980s?

These are some of the questions currently engaging the attention of several scholars. As one of these, in this volume, Paz Tolentino breaks new ground by examining the relevance of investment development path to explaining why some developing countries are important outward investors and other are not. In doing so she goes well beyond the idea behind my initial theorising to asking the question: Exactly what is the process by which a nation can create or acquire the kind of assets it needs to possess to be competitive in world markets; and why should it choose to engage in outward direct investment rather than penetrate the global economy?

To answer this question Paz Tolentino draws upon another thread of research, earlier undertaken by John Cantwell at the University of Reading, which has come to be known as the technological competence theory of international production. The author fully explains what she means by this in the early chapters of her book; but, perhaps, of greater interest to the reader is her application of the theory to examine the way in which developing counties might best organise their resources and capabilities so as to promote a technology growth trajectory which is best suited to their dynamic comparative advantage. *Inter alia*, such success is judged by its ability to attract the right kind of inward investment to promoting the capacity of its indigenous firms – or at least some of them – to compete efficiently in regional and global markets.

With its rich descriptive material, this monograph is fascinating to read. It also contains some astute policy recommendations with respect to national systems of innovation. I applaud these recommendations. However, I would add that the management of technology is only one instrument of government policy and that unless technology policy is systemically integrated into all other areas of macro-economic-organisational strategy, for example, education, competition, trade, the environment and fiscal strategy, even a satisfactory technology policy will be found wanting. I think the interaction between government and business in Japan in the 1960s, Korea and Taiwan in the 1970s and 1980s and Thailand, Malaysia and Mexico in the early 1990s provides ample evidence of this point.

John H. Dunning
Rutgers and Reading Universities

Preface

INTENTIONS

This book analyses the general trend towards internationalisation of business which in recent years has been common to firms of all countries, but whose growth has been most notable among firms from the developing countries.[1] The growth of newer multinationals from Japan, Germany and smaller developed countries and some of the more advanced developing countries is indicative of the capacity and incentive of their firms to follow the earlier outward multinational expansion of the traditional source countries for outward investment, the United States and the United Kingdom, at a much earlier stage of their national development.

The book focuses on the emergence and growth of Third World multinationals. There are few published books dealing with Third World MNEs *per se*. Some of these books are, Kumar and McLeod (eds) (1981), Chen (1983), Lall (ed.) (1983), Wells (1983), Khan (ed.) (1986) and recently, ESCAP/UNCTC (1988) and Bereznoi (1990). The two books by Wells and Lall deal with the theorectical aspects of the growth of Third World outward direct investment and offer empirical evidence that require updating and an analysis of broader magnitude in terms of the country coverage. Most of the edited books, book chapters, and journal articles deal with specific aspects of Third World international production activity such as MNEs from public-sector enterprises in developing countries (Kumar, 1981) or from a particular home country such as Argentina (Katz and Kosacoff, 1983), Brazil (Guimaraes, 1986; Villela, 1983; Wells, 1988), Hong Kong (Chen, 1983; ESCAP/UNCTC, 1986; Wells, 1978), India (Lall, 1982b, 1983b, 1985a; Morris, 1987), Singapore (Agarwal, 1987; Lecraw, 1985; Pang and Komaran, 1985), South Korea (Jo, 1981; Kumar and Kim, 1984, Koo, 1985; Euh and Min, 1986) and Taiwan (Ting and Schive, 1981; Chen, 1986). Other published studies deal with MNEs from groups of countries such as Hong Kong and Singapore (Pang and Komaran, 1984), Latin America (Diaz-Alejandro, 1977; White, 1981), Asia (ESCAP/UNCTC, 1988; Wells, 1984; World Bank, 1989) or ASEAN (Lecraw, 1981).

Yet, apart from Wells (1977, 1981, 1983), Dunning (1981b, 1986a), Lall

(1983a, 1983c, 1991) or Lecraw (1985) there is limited attempt to crystallise an integrated theory to explain the growth of Third World outward investment that is firmly grounded on empirical evidence drawn from an encompassing range of home developing countries and, more importantly, to assess how the innovative activities of firms have played an important role in their internationalisation. Some of the few exceptions to comparative analysis made between Third World MNEs from different countries or groups of countries are the studies of Wells (1980) and Agarwal (1985). In addition, the significance of Third World outward investment has increasingly become important, especially since the mid-1980s when many of the newly industrialising developing countries, particularly in Asia, widened the scope of their networks of international production in the face of balance-of-payment surpluses, appreciating currencies and protectionist pressures in their important export markets.

The present book attempts to fill this lacunae in the literature on Third World multinationals in at least three distinctive ways. First, the theorectical and empirical analysis of the emergence and growth of Third World multinational is based on a broad country coverage of home developing counties across Asia and Latin America and not just case studies of Third World MNEs from a single source country or groups of countries characteristic of existing literature dealing with this topic. As a special case study, this book also includes results of original empirical research on the emergence and growth of outward direct investment by Philippine firms. The widespread coverage enables a comprehensive comparative analysis to be made between MNEs from different developing countries that are at varying stages of development or belonging to particular regions upon which theoretical and policy implications can be derived. Second, the time-period covered by the research is timely and up-to-date which allows for cross-sectional and time-series analysis of the most recent developments in Third World outward direct investment. Third, and perhaps most importantly, the book seeks to document evidence not only of the growing significance of Third World outward investment, but how innovative activity among Third World MNES, although currently at an early stage, has tended to become increasingly important and more complex as their international production evolves and as their home countries advance through progressively higher stages of industrial development. These technological advantages are a means by which Third World MNEs have transformed the significance and character of their international production activities. However, over time, the evolution of international production dictates the accumulation of further and more complex technological advantages of firms.

Two major themes are proposed in explaining the development of Third World outward investment. The first theme pertains to the gradual process of expanding technological capabilities of selected Third World firms associated with the cumulative growth in their international direct investment. The existence and accumulation of technological capabilities are an important determinant as well as effect of the pattern and growth of their international production activities and are therefore a mutually reinforcing process. The pattern of technological capabilities of Third World firms fundamentally govern and is, in turn, governed by the achievement of international competitiveness and the extent of 'catching up' in international production. The second theme proposes that the emerging technological competence and growth of Third World firms is related to the pattern of their domestic industrial development and the cumulative process in which firms build upon their unique technological experience. Taken together, these propositions imply that the sectoral pattern of the outward direct investment of a developing country changes gradually over time, in a way that to some extent can be predicted.

The framework of analysis adopted in the core chapters of the book draws upon the evolutionary theories of the growth of the firm based on growth of production networks as opposed to the theories of the growth of the firm based on transaction and exchange in order to take account of dynamic elements of industrial change or evolution. The evolutionary stance taken of the analysis of the industrial and geographical development of Third World outward investment fundamentally requires an approach based on industrial dynamics that emphasises the evolution of the structure of an industry over time at the international, national and firm levels. The macroeconomic developmental theories of international production draw heavily from the industrial dynamics approach in describing the dynamic and developmental process of international production or the way in which stages of development or maturity of countries and firms affects the pattern of their international production activities. Such a framework of analysis is useful in explaining the differential growth rates of firms which, for the purposes of the book, pertain to how some Third World firms become MNEs relative to their domestic competitors in final product markets and how these firms achieve international competitive success and survival.

A major effort is undertaken in this book to explain not only why technological competence is significant in explaining the emergence and cumulative growth of Third World MNEs but also how such technological competence has been accumulated. In particular, the complementary role of foreign technology in strengthening the indigenous technological development and the growth of international production of indigenous firms in developing countries is carefully examined. The theoretical framework adopted is based on the Schumpeterian premise that economic growth and performance is dependent on the creation of new technology, the diffusion of technology and the efforts related to the economic exploitation of innovation and diffusion.

The concept of technological competence has been previously used to explain the growth of MNEs based in industrialised countries (see for example, Cantwell, 1989a, 1991a). This book attempts to show how the concept is equally relevant to explain the growth of MNEs based in developing countries, taking into account the distinctive, path-dependent and differentiated nature of innovative activities of firms associated with the level of industrial development of their home countries. The lower levels of industrial development of developing countries, however, implies that there are liable to be important differences in the significance and form of the technological accumulation of Third World MNEs as opposed to that of MNEs from the more developed countries. The MNEs based in the United States, United Kingdom, Japan, Germany and other industrialised countries are frequently technological leaders, unlike Third World MNEs which rely to a greater extent upon simpler and less research-intensive forms of technology creation. Hence, although an evolutionary trend in the sectoral and geographical pattern of outward investment may also be evident among firms in the Third World, these firms do not exhibit the same rapid pace of technological advancement as firms from the more developed countries. Third World multinationals have emerged, developed and 'caught up' not through the process of advancing the frontiers of technological development but by a process of technological learning as new technologies discovered in the more industrialised countries find applications in developing countries in the course of technological diffusion.

The distinctive innovative activities of MNEs based in Third World countries provide evidence that these firms have followed a technological course that is to some extent independent and a function of their own unique learning experience. The unique technological course of their firms is not necessarily limited to descaling and increasing the labour intensity of sophisticated technologies or to specialisation in the production of lower skill and lower technologically intensive sectors. Rather, their innovations are based on lower levels of research, size, technological experience and skills and, in some cases, the modernisation of an older technique, including foreign outdated technology. Without having yet developed a strong reliance on R & D, a greater part of their innovatory capacity has been given to production engineering, learning by doing and using, and organisational capabilities, and, as will be shown in the book, the access to more advanced but complementary forms of foreign technology. However, there are a steadily rising number of examples of Third World MNEs that have become genuinely innovative even though such activities have tended to be less scientifically refined and have not generally involved frontier technologies.

The proposition therefore adopted in this book relates to the changing forms of innovation as opposed to their existence as a variable factor in stages of development. Hence, although the forms of innovation may be different among firms from the Third World and those in industrialised counties, the existence of innovation within the firm determines the growth of their international production, especially in manufacturing. Hence although the technological activities of Third World MNEs requires an empirical investigation of technology creation beyond R&D and international patenting activities, the concept of rapidly emerging technological capabilities or technological competence is a useful means of analysing the international growth of manufacturing firms from quite difference environments, and at different stages of technological development and capacity (Cantwell and Tolentino, 1990).

The process of independent technological accumulation on the part of

Third World MNEs requires the formulation of theories of the growth of Third World MNEs which of necessity recognises that the role of innovation within firms has been increasing, and has become more sophisticated as international production has developed. In particular, in the formulation of a more dynamic evolutionary theory of technological change to explain the growth of new sources of outward direct investment from the Third World, the book attempts to assess the significance of the two current main schools of thought on the growth of Third World MNEs, i.e. between the framework of the product cycle model articulated by Wells (1983, 1986) and that of the model of localised technological change advanced by Lall (1983a).

The concept of stages of development in the international production of Third World MNEs developed in the book also represents a contribution to the advancement of the theories of international production based on macroeconomic development. The analysis of the empirical evidence will attempt to show that MNEs from different developing countries belong to different stages in their international production based on the extent of their technological competence allied partly to the upgrading of the domestic industrial structure of their home countries. However, since countries can and do deviate from a general trend, there may be a case for the formulation of a multi-modular framework to explain the growth of MNEs from different developing countries that are at different stages of development in their international production activities and whose home countries have pursued particular industrial development patterns.

Developing countries may therefore be categorised in groups according to the increasing complexity in the pattern of their outward investment as opposed to the levels of their outward investment. The newly industrialising developing counties (NICs), hereto defined as including South Korea, Taiwan, Singapore and Hong Kong in Asia and Mexico, Brazil and Argentina in Latin America, are regarded to have more complex forms of outward investment because of their greater capacity for technological accumulation. They are therefore steadily moving into a higher stage of development in international production. By comparison, the international production activities of lower-income developing countries are at a relatively early stage of development. Although some attempts are made to describe differences in the general characteristics among firms and countries belonging to a particular group (as for example, the differences in the general characteristics within the group of the higher income newly industrialising countries as the NICs as opposed to the lower-income developing countries), there is no attempt at this stage to describe more specific or detailed idiosyncratic characteristics or deviations from the general characteristics of firms and countries belonging to a particular group. For purposes of the present analysis, the emphasis is macro-economic oriented in as much as comparison is made *between* groups of firms, i.e. between MNEs from the NICs on the one hand and the lower income developing countries as opposed to a comparison *within* these groups.

The theoretical framework adopted to explain the phenomenon of Third

World Multinationals is multi-disciplinary in nature and traverses the fields of innovation and technological transfer, economic development and foreign direct investment and the MNE. Although the narrow focus of the book is centred on Third World multinationals, the concept of stages of development in international production elaborated here has general relevance to the analysis of the changing character and composition of outward direct investment of any country or firm as development proceeds. The emergence and growth of Third World MNEs is but one manifestation of the more recent fundamental changes in the pattern of international production associated with the general trend towards internationisation of business.

Given the fundamental importance of technological accumulation in the process of 'catching up' and maintaining international competitiveness in the face of increasing competition, some implications for policy formulation in developing countries are examined. In particular, emphasis is placed on the development of national systems of innovation in supporting the dynamic growth path of comparative advantage at the international, national and firm levels.

BACKGROUND

There have been two decisive factors which have greatly influenced the direction taken by the book. First, I have a genuine interest in economic development in general and the impact of the multinational enterprise and international production on developing countries in particular, which was nurtured throughout my graduate degree courses at the University of Reading in the United Kingdom during the period 1983-7. While being a graduate student at the University, I was privileged to work under the joint supervision of Dr John A. Cantwell and Professor John H. Dunning, two leading researchers in the field of international business. The econometric tests of the concept of the investment development cycle advanced by Professor Dunning have provided the foundation for the further development of research on the growth of new sources of outward direct investment, particularly from the Third World. I am greatly indebted to the insights of Dr John A. Cantwell on the areas of classical economic theory which helped shape the dynamic approach of my work. The broader scope of technological innovation of firms from the newly industrialising countries (NICs) has enabled me to adopt his theory of technological accumulation (later broadened to the concept of technological competence) in explaining their increased capacity to engage in more complex forms of international production and their greater scope for geographical diversification.

This book emerged from the dissertation that I submitted to the University of Reading in October 1987 in fulfillment of the requirements of the degree of Doctor of Philosophy in economics. Over the course of more than four years, however, the book has acquired a distinctive character of its own as my insights on the phenomenon of Third World multinations have gradually evolved.

There have been several factors providing the motivation to pursue further research which have led to the publication of this book, First, during the course of my thesis defence, I obtained invaluable insights from my external examiner, Professor Sanjaya Lall, one of the more renowened researchers in the area of Third World multinationals, upon whose theory of localised technological change I have elaborated as a vital component of the dynamic theory of Third World outward investment that I had developed. While nurturing a great respect for Professor Lall's work, I also derived encouragement from his genuine interest in the future direction of my post-Ph. D research.

Second, there have been two occasions when the main findings of the research were presented in international conferences which have served as fora to test new ideas and gather constructive comments and suggestions that have undoubtedly enriched the quality of the research, particularly in the fine tuning of theoretical constructs and in the search for more empirical evidence. The first international conference in which the research findings were presented coincided with the Annual Meeting of the European International Business Association in Antwerp, Belgium, in December 1987, shortly before I had even defended the thesis in a *viva voce* examination at the University of Reading. The second international conference entitled: 'World Trade and MNE in the 21st Century' was sponsored by the Association of International Business Studies of the Chinese Culture University and was held in Taiwan, Republic of China in May 1992. These conferences have provided the opportunity for Dr John Cantwell and myself to write collaborative papers on the area of technological accumulation and Third World multinationals.

Third, and perhaps more important, in 1989 the thesis distinguished itself with the award of the Academy of International Business Richard N. Farmer Prize for the best thesis in international business. The dissertation competition committee comprised Professors Stephen Kobrin, Donald Lessard, Susan Douglas and Robert Grosse who described the thesis as 'interesting, very thorough and a significant contribution to the literature'. Such honour and recognition has not only represented one of the most important prizes for academic achievement bestowed upon me which I will treasure *ad infinitum* but, more importantly, has become an important source of personal motivation, determination and inspiration over time.

Fourth, since graduation from the University of Reading, I have had the excellent opportunity to be connected with the Transnational corporations and Management Division (formerly centre on Transnational Corporations) of the United Nations. The execution of my work responsibilities at the United Nations requires that I keep up-to-date with recent trends in international business and, in the course of doing so, to gather more recent empirical evidence of the emergence and growth of Third World outward investment. Although I necessarily have to pursue my personal research interests on Third World multinationals outside my full-time employment at the United Nations, I nevertheless stand to gain from being part of one of the more important research centres on transnational corporate activity in the world. This has been particularly important at a time when the growth of Third World outward direct investment has assumed increasing relevance.

This book is therefore a product of considerable personal sacrifice, but nevertheless pursued with great motivation and dedication as a humble contribution to the advancement of current economic thought. It is with great pride, once again, that I dedicate this book to my parents, Dr and Mrs (Dr) Paulino D. Tolentino for their steadfast support to my career advancement and to the Philippines for which I constantly nurture great hopes.

Pas Estrella E. Tolentino
New York City

NOTE

1 Some remarkable evidence of the general trend towards internationalisation of business since the mid-1970s, owing in part to the rise of outward direct investment from the Third World and other new source countries, became apparent with the econometric testing of the concept of an investment development path undertaken in the book. The growth of international production activities by firms from newer source countries, including the Third World, that are at intermediate stages of national development, has resulted in a structural change in the previously established relationship between net outward investment and a country's relative stage of development.

1 Introduction

THE GROWTH OF THIRD WORLD MULTINATIONALS

The post-war period has witnessed the sustained and unprecedented growth of the multinational enterprise (MNE). The growth of international business financed by foreign direct investment has been accompanied by increasing diversity in its geographical source distribution. In the period since 1960 and most notably since the mid-1970s the emergence and growth of newer home countries from both developed and developing countries has provided additional competitive elements in the general trend towards internationalisation. These newer home countries acquired the capacities and the incentive to catch up in the process of international production and in some cases their relatively faster rate of growth and increasing competitiveness have served to challenge the expansion of the more traditional investor countries, the United States and United Kingdom.

This book reports on the findings of extensive theoretical and empirical research on the emergence and growth of multinationalisation of firms from developing countries based in Asia and Latin America. As a special case study, Chapters 7 and 8 report on the findings of original empirical research on the emergence and growth of multinational investment activities by Philippine firms. The early chapters of the book set the scene for the analysis of Third World multinationals. Chapter 2 attempts to examine the emergence and growth of outward direct investment from developing countries in the post-war period within the overall context of worldwide developments in international production in the immediate pre-war and post-war periods. A broad analysis of the historical evolution of the pattern of international production activities is undertaken to determine the extent to which the more recent growth of newer source countries from the Third World can be compared with the historical source countries in terms of industrial and geographical patterns at a specific point in time as well as over time.

The analysis of the trends in foreign direct investment in Chapter 2 will show that prior to the Second World War, a majority of international production activities, apart from those organised by individual entrepreneurs in the period prior to 1914, were resource based and directed towards the colonies

(Svedberg, 1981). Developing countries received two-thirds of the total stock of accumulated inward direct investment in 1938, originating mainly from the UK and USA. These two countries accounted for 40 per cent and 28 per cent respectively of the total stock of accumulated outward direct investment in that year (Dunning, 1983a). The increasing significance and complexity of the innovative activities of MNEs as a result of greater experience in international production has led in part to more sophisticated forms of outward investment in the post-war period. In particular, there has been a shift in the industrial and consequently the geographical distribution of foreign direct investment (FDI) from resource-related investments in developing countries towards import-substituting manufacturing investments in the developed countries.

Such increasing sophistication in the sectoral and geographical pattern of outward direct investment of UK and US firms has also been observed in the dynamic analysis of the development of outward direct investment activities of more recent investors from the newer source countries such as Japan and Germany. The examination of the historical patterns of international production provides a framework within which to examine the distinctive nature of the sectoral and geographical structure of Third World outward investment and how this structure might be expected to evolve across countries at different stages of development as well as within countries over time.

THE THEORETICAL FOUNDATIONS

The theoretical framework for the study of Third World MNEs is provided by Chapters 3 and 4. Since one of the main aims of this book is to re-examine the relevance of existing theories of Third World multinationals, an understanding of the major strands in the development of the theory of international production since 1960 is demanded. The emergence and development of a distinct theory of international production and the MNE from 1960 can be traced to changes in the pattern of international production during the post-war period. In part, the declining dominance of the USA as a major source of FDI in the 1950s and 1960s and the increasing importance of Germany, Japan, Switzerland and other developed countries, as well as developing countries, as sources of FDI in the 1970s and 1980s necessitated the adaptation of existing theories to take account of these new sources. Efforts in this direction are pursued in Chapter 3, which provides valuable insights into modern theories of international production based on opposing theories of the growth of the firm and industrial organisation. The framework of analysis adopted in the core chapters leans heavily on the evolutionary theories of the growth of the firm based on growth of production networks, as opposed to the theories of the growth of the firm based on transaction and exchange. The latter set of theories are essentially static and fail to take account of dynamic elements of industrial change or evolution (Carlsson, 1987).

Taking the perspective of growth of production networks in the theory of the growth of the firm necessitates a view of the MNE as an active firm or an agent that exercises control over ownership advantages and not as a passive product of market structure as described by the theories of the growth of the firm based on transaction and exchange. The adoption of the more dynamic elements of theories of the growth of the firm based on growth of production networks requires that ownership advantages are regarded as necessary for the existence and growth of MNEs and international production. Such a framework of analysis is useful in explaining the differential growth rates of firms: how some Third World firms become MNEs relative to their domestic competitors in final product markets and how these firms achieve international competitive success and survival.

The concept of stages of development in the international production of Third World MNEs elaborated in this book represents a contribution to the advancement of the theories of international production based on macroeconomic development. In Chapter 3, some of the major country- and industry-based theories that pertain to the macroeconomic development aspects of the growth of international production of the industrialised countries are reviewed, most notably Kojima's and Ozawa's theory of trade and direct investment and Dunning's concept of the investment development path. These macroeconomic development theories explain the growth of international production of countries and firms based on their stage of development. The Kojima theory provides a useful foundation for examining the general way in which the stage of national development helps to explain the industrial structure of international production by indigenous firms. Evidence for this is provided by Ozawa with reference to the evolutionary trend in the pattern of Japanese outward investment from 1950. Kojima's and Ozawa's more general, comparative advantage based models present a sound explanatory framework within which to view the growth of international production of countries undergoing rapid growth such as Japan, Germany and the newly industrialising countries. In the early stages of international production, the models can explain the process of relocation of production of mature or technologically standardised industries, because locational advantages favour countries at an earlier stage of development. Such relocation of production is undertaken while domestic firms still have the technological and organisational advantages associated with lower technology and more labour-intensive production activities which can be more profitably exploited in a foreign environment that has lower levels of technological capacities and production costs. Although most Third World MNEs are unlikely to develop as fast as their Japanese counterparts, who can today draw on frontier technologies, the trend is towards a persistent upgrading of activities for which these firms are responsible. The phases of development suggested by Ozawa for Japanese MNEs therefore become relevant to the study of Third World multinationals.

The underlying developmental processes that underlie the Kojima–Ozawa theory may be explained by the more general concept of an investment development path advanced by Dunning. The concept of the investment development path shows in more general terms the impact of the national stage of development on both the level of net outward investment and the character and composition of outward direct investment. The early versions of the concept of an investment development path advanced by Dunning (1981a, 1981b, 1982, 1986a, 1986b, 1988a) are based on the proposition that the *level* of inward and outward investment of different countries, and the balance between the two, is a function of their stage of development as measured by gross national product (GNP) per capita. The econometric testing of the concept of an investment development path in Chapter 5 attempts to show how the general trend towards the internationalisation of business, owing in part to the rise of outward direct investment from the Third World and other new source countries, has affected the relationship between the investment position of countries and their national level of development. Some interesting findings become evident in the examination of empirical evidence for the period since the mid-1970s when a structural change occurs in the relationship between net outward investment and a country's relative stage of development as a result of the general rise in the internationalisation of firms from countries at lower stages of development. The growth of newer multinationals from Japan, Germany and smaller developed countries, as well as some of the richer developing countries, is indicative of the capacity and incentive of firms to follow the earlier outward multinational expansion of the traditional source countries, the USA and the UK, at a much earlier stage of their national development.

More recent versions of the investment development cycle have argued that apart from the level of net outward investment, the *character* and *composition* of outward direct investment of a country's firms vary with the national stage of development (Dunning, 1986a, c). The early forms of foreign investment are frequently resource based or sometimes import substituting, but in each case have a quite specific location associated with a particular type of activity. However, as firms mature, their outward investments evolve from a single activity or product in a particular location and they adopt a more international perspective on the location of their different types of production activities. Such MNEs are responsible for directly organising an international division of labour within the firm in place of markets. At this stage, the character of their international production activities and their ownership advantages are also determined by firm-specific factors in addition to country-specific factors.

THE STAGES OF DEVELOPMENT IN INTERNATIONAL PRODUCTION

Although the narrow focus of this book is centred on Third World multinationals, the concept of stages of development in international production elaborated here has general relevance to the analysis of the changing character and composition of outward direct investment of any country or firm as development proceeds. The concept can be thought of in the context of Johanson's and Vahlne's (1977) *Model of Knowledge Development and In-*

creasing Foreign Commitments. In their work, the internationalisation process of the firm is regarded as following a sequential pattern in which initial foreign investments are directed to those markets with the greatest proximity in terms of psychic distance. The psychic distance at which foreign ventures are located is likely to increase directly with the complexity of the sectoral pattern of outward direct investments. Hence the sectoral pattern of outward direct investment has important implications for the geographical pattern of such investments.

The outward direct investment of countries generally follows a developmental or evolutionary course over time which is initially predominant in resource-based or simple forms of manufacturing production in the earlier stages of development and then evolves towards more sophisticated and complex forms of manufacturing investments. Such changing sectoral composition of investment is associated with changing geographical composition. The simpler forms of investment in resource-based and manufacturing sectors are frequently undertaken in resource-rich or labour-abundant LDCs while the more complex research-intensive investments are undertaken in industrialised countries.

The industrial and geographical distribution of Third World outward investment examined in this book is analysed within a stages of development context that is determined largely by the pattern of domestic industrial development of home countries and the emergence of ownership advantages by local firms. The survey encompasses the primary, secondary and construction sectors whose growth may be explained by the application of the concept of technological competence.[1] The developmental course of these most recent outward investors from the Third World has been faster and has a distinctive technological nature compared to the more mature multinationals from Europe, the USA and Japan, owing to the different stages of their national development.

The increasing diversification in the geographical pattern of Third World outward direct investment, as a result of the increasing complexity of its sectoral pattern, may be described as following three sequential stages.

The first stage, which is considered the most important for new international investors involved largely in resource-based and simpler manufacturing and service investments, is directed towards investments in neighbouring and/or ethnically related territories. Evidence for this is shown in the relatively high share of intra-regional direct investment by firms in Asia and Latin America, most of which, as predicted by the product cycle hypothesis, is located in other developing countries that are further down the pecking order than the home country. However, the evidence suggests that the importance of investments in these neighbouring and/or ethnically related territories declines as investing firms gain greater experience in international production and as their home countries progress to higher stages of industrial development.

The increased sectoral sophistication of international production activities from purely extractive resource-based investments directed largely to the export of primary products to the home country, towards the downstream processing of these primary products in the host country in the first instance, or alternatively the relocation of labour-intensive manufacturing abroad, may increase the geographical scope of these investments. At this stage, investments directed towards more sophisticated markets in other countries with increasingly further psychic distance may become increasingly more important. The second stage may therefore entail international production in several neighbouring and/or ethnically related territories or in other, non-ethnically related developing countries, of simple forms of manufacturing that are either resource based or labour intensive. Lower production and transport costs, close psychic distance as well as the presence of favourable investment opportunities and the desire to make fuller use of regional economic integration, seem likely to explain significant direct investments within distinct regional groupings at this stage of development. The ownership advantages of Third World MNEs in these first and second stages are based on an adapted product or process innovation obtained through experience in developing countries and which is applicable to other developing countries at a lower stage of development.

In the third stage, the sectoral evolution of international production progresses to more sophisticated and complex manufacturing investments. Firms with more complex capabilities become better able to conduct international production activities further away from their home base and in countries further away in terms of psychic distance, possibly including investments in developed countries.

In this book, developing countries are categorised in three groups according to the increasing complexity in the pattern of their outward investment as opposed to the levels of their outward investment. The newly industrialising developing countries (NICs) are regarded as having more complex forms of outward investment because of their greater capacity for technological accumulation compared to lower-income developing countries. Although some attempts are made to describe differences in the general characteristics among firms and countries belonging to a particular group, for example, differences in general characteristics *within* the group of higher-income newly industrialising countries as opposed to lower-income developing countries, there is no attempt at this stage to describe more specific or detailed idiosyncratic characteristics or deviations. For the purposes of the present analysis, the emphasis is macroeconomic oriented, in that comparison is made *between* groups of firms, i.e. between MNEs from the NICs and those from the lower-income developing countries, as opposed to a comparison *within* these groups.

The general hypothesis to be tested is that the industrial and geographical development of MNEs from the more advanced developing countries such as South Korea, Taiwan, Singapore and Hong Kong in Asia, and Mexico, Brazil and Argentina in Latin America, is steadily moving into the third stage of development described above. By comparison, the industrial and geographical development of MNEs from lower-income developing countries, whose international investment is at a relatively early stage of development, is still largely concentrated within neighbouring ethnically related developing countries and other non-ethnically related developing countries.

TECHNOLOGICAL CONSIDERATIONS IN THE GROWTH OF THIRD WORLD MULTINATIONALS

In the context of explaining the stages of development of Third World outward investment, this book proposes two major themes. The first theme pertains to the gradual process of expanding technological capabilities of selected Third World firms associated with the cumulative growth in their international direct investment. The existence and accumulation of technological capabilities is an important determinant as well as an effect of the pattern and growth of their international production activities and is therefore a mutually reinforcing process. The pattern of technological capabilities of Third World firms fundamentally governs and is in turn governed by the achievement of international competitiveness and the extent of catching up in international production. The second theme proposes that the emerging technological competence and growth of Third World firms are related to the pattern of their domestic industrial development and the cumulative process by which firms build upon their unique technological experience. Taken together, these propositions imply that the sectoral distribution of the outward direct investment of a developing country changes gradually over time, in a way that to some extent can be predicted.

An important connection is evident between the evolutionary pattern of domestic industrial development, the emergence of ownership advantages by local firms and the growth of outward investment. This framework of analysis requires an approach based on industrial dynamics most elaborately proposed by Carlsson (1987) that emphasises the evolution of the structure of an industry over time at the international, national and firm level. The macro-economic developmental theories of international production draw heavily on the industrial dynamics approach in describing the dynamic and developmental process of international production or the way in which stages of development or maturity of countries and firms affect the pattern of their international production activities. Such a link between industrial development and international production finds a theoretical antecedent in the classical theory of David Ricardo (1772–1823) which established the link between international trade (and production) and industrial restructuring in the course of economic development. Based on the classical growth model and the law of diminishing returns, Ricardo drew attention to the necessity for a structural shift in the path of development of an economy from agriculture to manufacturing. The sectoral shift in the pattern of both domestic and international production of firms from resource-abundant developing countries from an initial concentration in the primary sector towards the secondary sector may be regarded in the context of the classical theory of Ricardo. Such a path of sectoral development, away from agricultural production which over time results in diminishing returns towards the manufacturing sector which offers greater opportunities for long-term growth, ensures a more favourable path of national development. In the case of countries that are resource poor, or the developed countries, the sectoral shift envisioned by Ricardo may occur within the manufacturing sector as simple types of activities are structurally shifted and replaced by more complex and higher valued ones.

In the early stages of national development, technological advantages tend to be more country specific and, in the case of resource-based activities, associated in part with the abundance or scarcity of natural resources in the home country. In either case, firms gain accumulated expertise in the exploitation and processing of natural resources present in the home or host country. The early forms of technological ownership advantages of US and UK firms in wood processing and metal and coal processing respectively were acquired because of the abundant availability of timber and coal in the USA and the UK (Rosenberg, 1976). This book will show that this general trend also holds true in the early stages of technological development of Third World MNEs.

However, as home countries advance through progressively higher stages of industrial development and as firms accumulate greater experience in international investment, the technological embodiment of their outward direct investment activity becomes more significant and more complex. This is seen in part in more complex technologically intensive investments as well as research-intensive investments in centres of innovation in the developed countries. These research-intensive investments enable Third World firms with narrower technological capabilities to gain access to more advanced, complementary forms of foreign technology which can be adapted and integrated with their indigenously generated technology.

Chapters 8, 9 and 10 offer evidence for the growing significance of Third World outward investment, and for how innovative activity among Third World MNEs, although currently at an early stage, has become increasingly important and more complex as expansion towards international production develops and as their home countries advance through progressively higher stages of industrial development. The trend in the industrial and geographical composition of Third World outward investment is explained in the context of an underlying accumulation of technological advantages. These advantages are a means by which Third World MNEs have transformed the significance and character of their international production activity. Over time, the evolution of international production dictates the accumulation of further and more complex technological advantages.

Owing to the different stages of national development, there are liable to be important differences in the significance and form of the technological accumulation of Third World MNEs as opposed to that of firms from the more developed countries. The MNEs based in the USA, UK, Japan, Germany and other industrialised countries are frequently technological leaders, unlike Third World MNEs which rely to a greater extent upon simpler and less research-intensive forms of technology creation. Although an evolutionary trend in the sectoral and geographical pattern of outward investment may also be evident among firms in the Third World, they do not exhibit the same rapid pace of technological advancement as those in the more developed countries. Third World multinationals have emerged, developed and caught up not through the process of advancing the frontiers of technological advancement but due to a process of technological learning as new technologies discovered in the more industrialised countries find applications in developing countries.

The distinctive innovative activities of firms based in Third World countries provide evidence that these firms have followed a technological course that is to some extent independent and a function of their own unique learning experience. However, the unique technological course of Third World firms is not necessarily limited to descaling and increasing the labour intensity of sophisticated technologies or in specialisation in the production of lower-skill and lower technologically intensive sectors but rather in innovations that are based on lower levels of research, size, technological experience and skills, and that achieve improvements by modernising an older technique, including outdated foreign technology. Without a strong reliance on research and development (R&D), a greater part of their innovatory capacity has been given to production engineering, learning by doing and using, and organisational capabilities and, as will be shown in Chapter 6, access to more advanced but complementary forms of foreign technology. There is a steadily rising number of examples of Third World MNEs that have become genuinely innovative even though such activities have tended to be less scientifically refined and have not generally involved frontier technologies.

The premise adopted in this book relates to the changing forms of innovation as opposed to their existence as a variable factor in stages of development. Although the forms of innovation may be different among firms from the Third World from those in industrialised countries, the existence of innovation determines the growth of international production, especially in manufacturing. Although the technological activities of Third World MNEs requires an empirical investigation of technology creation beyond R&D and international patenting activities, the concept of rapidly emerging technological capabilities or technological competence is a useful means of analysing the international growth of manufacturing firms from quite different environments and at different stages of technological development and capacity (Cantwell and Tolentino, 1990).

The nature of technological accumulation as firm specific, cumulative and differentiated was originally advanced by Pavitt (1988) and has previously been applied to explain the historical evolution of the sectoral pattern of comparative advantage in innovative activity of national groups of firms in industrialised countries (Cantwell, 1990) and in work on competition between rival technologies (Arthur, 1988, 1989). These innovative activities, especially in R&D, have been found to be statistically significant determinants of differences in export and productivity performance (Soete, 1981b; Fagerberg,

1988) and also of internationalisation of business among the major OECD countries (Vernon, 1974, 1979; Cantwell, 1989). The concept of technological accumulation is consistent with the evolutionary theories of technological change of Usher (1929), Marx (1976), Atkinson and Stiglitz (1969), Rosenberg (1976, 1982), Nelson and Winter (1977), Stiglitz (1987) and Cantwell (1989a). The dynamic theories of technological change are evolutionary in their treatment of technological change as a gradual and progressive process from crude designs to more refined manufacturing systems that benefit from scientific discipline and cumulative experience. More important, in the Marxist context, is the transformation of technology resulting from adaptation according to the requirements of particular applications and users.

THE THEORIES OF THIRD WORLD MULTINATIONALS

The process of independent technological accumulation on the part of Third World MNEs requires the formulation of theories on the growth of Third World MNEs that of necessity recognise that the role of innovation within firms has been increasing, and has become more sophisticated as international production has developed. As will be shown in Chapter 4, the scope of innovation of Third World MNEs in the two main schools of thought differs between the framework of the product cycle model articulated by Wells (1983, 1986) and that of the model of localised technological change advanced by Lall (1983a).

In the product cycle model, the expansion of Third World MNEs is associated with a Schumpeterian process within which a competitive fringe of firms emerging from the developing countries are catching up and imitating the innovative activities of the technologically leading firms during the tail end of the product and industry life cycle, in accordance with the requirements of Third World markets and production conditions (Wells, 1986). However, some evidence is presented in this book to support the model of localised technological change and the more general theory of technological accumulation and competence. Unlike the predictions of the product cycle model, the proposition is that indigenous firms in developing countries have the capability to follow an independent or localised technological trajectory in which imported technology is integrated with the development of indigenous technology in a manner consistent with the dynamic growth path of changing comparative advantage. As will be evident in the analysis of the role of foreign technology in the technological development of Third World MNEs in Chapter 6, the integration process is more successful when foreign generated technology is complementary to the path taken by indigenous innovation, and can therefore be beneficially integrated and adapted within a unique programme of technology generation.

These views of the technological advantages of Third World firms, which encompass not only the simple imitation and adaptation, by a competitive fringe of Third World firms, of innovations from more advanced countries, but also a further set of related, although differentiated, innovations as a result of the unique technological experience of Third World firms, are consistent with the evolutionary theories of technological change. The innovative activities are triggered and guided not only by evolving market signals as predicted by the product cycle model but also by evolving technologies (Atkinson and Stiglitz, 1969; Rosenberg, 1982; Pavitt, 1988).

Indeed, there are new multinationals operating in the Third World, and particularly in the newly industrialised countries, that are genuinely and uniquely innovative, and in some cases these firms have progressed to the early stages of R&D activities, in part through investments in research facilities in the developed countries. Although these firms are still at a relatively early stage of development, their abilities in technological accumulation continue to expand. Some Third World MNEs now operate in sectors the products of which are far from standardised, and their activities are helping to generate a fresh pulse of innovation. Due to the localised and irreversible nature of technological change, their distinctive innovative capacities cannot be readily replicated by the affiliates of industrialised-country MNEs located in developing countries that have progressed to new technologies, since there are high costs in efficiently reproducing or transferring older technologies. For example, if MNEs from industrialised countries attempt to replicate the technological advantages of Third World MNEs in small-scale manufacture, they lose the advantages of accumulated experience in large-scale manufacture (Wells, 1978).

THE ROLE OF FOREIGN TECHNOLOGY IN THE DEVELOPMENT OF THIRD WORLD OUTWARD INVESTMENT

This book attempts to explain not only *why* technological competence is significant in explaining the emergence and cumulative growth of Third World MNEs but also *how* such technological competence has been accumulated. In particular, Chapter 6 is devoted to examining the role of foreign technology in strengthening indigenous technological development and the growth of international production of indigenous firms in developing countries.[2] In addition, Chapter 7 analyses the particular case of dynamic interdependence between inward and outward direct investment in the Philippines. The theoretical framework adopted in these chapters is based on the Schumpeterian premise that economic growth and performance are dependent on the creation of new technology, diffusion of technology and efforts related to the economic exploitation of innovation and diffusion (see, for example, Fagerberg, 1988; Badulescu, 1991; and applications for developing countries in Kim, 1980). This idea is implicit in the theory of technological competence as an important determinant of international competitiveness and the differential growth rates of firms (for further elucidation of the theory, see Cantwell, 1991a).

The argument is advanced here that even in the Asian NICs that have developed some national innovative capabilities, the main sources of growth of developing countries in the catch-up phase of technological development are their ability to access technology from abroad through diffusion and their indigenous capacity to exploit the benefits of technology created abroad. The various forms of modality of foreign technology transfer that have helped to strengthen indigenous technological development and growth of production of Third World MNEs are examined. The analysis of the industrialisation and growth of developing countries in the late twentieth century on the basis of international diffusion of foreign technology has historical antecedents in the industrialisation of Europe and the USA in the nineteenth century and Japan earlier in the twentieth century.

The book's main thesis is that more advanced foreign technology transfer has acted as a trigger mechanism for modern economic growth in some developing countries which are on a lower level of economic and social development, in the manner predicted by Kuznets (1968). Since developing countries have limited technological and capital abilities, the transfer and implementation of packaged technology from abroad provide the initial basis for their technological development. Foreign technology helps to develop local technical and entrepreneurial capabilities which provide the major sources of innovation at a more intermediate and advanced stage of technological development, where research and development within the firm, in addition to engineering results obtained in the early stages, provide the basis for technological change and growth of production in new technologies (Kim, 1980; Kwon and Ryans, 1987; Lee, Bae and Choi, 1988).

The focus of this book is to show how in certain cases foreign technology has been an agent in supporting dynamic changes in comparative advantages of firms and industries in developing countries towards higher value-added and higher technology-intensive industries, in accordance with policies implemented by governments in these countries. In particular, the growth of more complex sectors of industry associated with more advanced stages of development is reflected in the growth of international production of local firms in these countries. The technology transferred from abroad has helped some local firms in developing countries to strengthen their technological capabilities and to generate competitive forms of ownership advantages necessary in carrying out their international production activities.

This work concludes by assessing the implications of empirical evidence about Third World MNEs for the formulation of current theories and policies. The emergence and growth of international production by Third World firms is one aspect of the dynamic changes in the geographical and industrial pattern of international production that have been occurring in the post-war period to which the theories of foreign direct investment are being adapted to provide an explanation. Building upon the theoretical framework established by the development of various strands in the modern and distinct theories of international production from 1960, including the two basic theories formalised to explain the growth of Third World MNEs, combined with documented empirical evidence on the pattern of their emergence and growth, the main objective of this book is to provide a dynamic evolutionary approach to the analysis of Third World outward investment. The essence of the dynamic theory of Third World outward investment proposed here is the relative shift in importance of the models of the product cycle, localised technological change and technological accumulation in explaining the outward investment of firms, as developing countries advance through progressively higher stages of domestic industrial development and their firms acquire greater experience in international investment. The notion of a cumulative, firm-specific, distinctive and path-dependent process of technological accumulation embodied in the theories of localised technological change and technological accumulation increasingly assumes greater significance in explaining the sustained growth of Third World MNEs over time.

Given the fundamental importance of technological accumulation in the process of catching up and maintaining international competitiveness in the face of increasing competition, some implications for policy formulation in developing countries are examined. In particular, crucial emphasis is placed on the development of national systems of innovation in supporting the dynamic growth path of comparative advantage at the international, national and firm levels. Government intervention has a central role to play in the technological learning and catching up by Third World firms in the dynamic process of industrial change towards more complex technologically intensive sectors that offer greater promise for long-term growth and advancement.

2 Trends in foreign direct investment

INTRODUCTION

This chapter attempts to analyse the increasing significance of outward direct investment from developing countries in the post-war period within the overall context of developments in the geographical and industrial pattern of foreign direct investment in both the immediate pre-war and post-war periods. The broad analysis of foreign production activities of firms in the post-war period reveals the increasing industrial and geographical diversity of multinational investment activity.

Prior to the Second World War, a majority of these international production activities, apart from those organised by individual entrepreneurs in the period prior to 1914, were resource based and directed towards the colonies (Svedberg, 1981). Developing countries received two-thirds of the total stock of accumulated inward direct investment in 1938 originating mainly from the UK and USA. These two countries accounted for 40 per cent and 28 per cent respectively of the total stock of accumulated outward direct investment in 1938 (Dunning, 1983a).

The increasing significance and complexity of the innovative activities of MNEs as a result of greater experience in international production has led in part to more sophisticated forms of outward investment in the post-war period. In particular, there has been a shift in the industrial and geographical distribution of FDI from resource-related investments in developing countries, which embody production experience and simpler forms of technological accumulation, towards import-substituting manufacturing investments in the developed countries, which embody more important and complex forms of technological accumulation. Such increasing sophistication in the sectoral and geographical pattern of outward direct investment of UK and US firms has also been observed in the dynamic analysis of the development of outward direct investment activities of more recent investors from Japan, Germany and the developing countries.

However, although an underlying pattern exists between this increasing sophistication, associated with firms' increasing importance and complexity of technological advantages as greater international production experience is acquired, there are liable to be important differences in the significance and form of the technological accumulation of Third World MNEs as opposed to that of MNEs from the more developed countries. MNEs based in the USA, UK, Japan, Germany and other industrialised countries are frequently technological leaders, unlike Third World MNEs which rely to a greater extent upon simpler and less research-intensive forms of technology creation. Although an evolutionary trend in the sectoral pattern of outward investment is also evident among firms in the Third World, they do not exhibit the same rapid pace of technological advancement as firms from the more developed countries.

The growth of outward direct investment from newer source countries such as Germany, France, Italy and, more recently, the developing countries can *inter alia* be viewed within an interpretation of the concept of an investment development path in which the dynamic interaction between technologically innovative indigenous firms and foreign technology, introduced in some cases through inward foreign direct investment, has strengthened the technological capabilities and growth of production of indigenous firms. Such continuous assimilation and adaptation of foreign technology, channelled primarily through licensing agreements, have resulted in Japan's export of technology particularly in connection with, and in support of, their overseas direct investments (Ozawa, 1974). The role of foreign technology in strengthening indigenous technological development and growth of international production of indigenous firms based in the developing countries is further elaborated in Chapter 6.

The analysis of the evolving patterns of international production is divided into three distinct periods: from 1939 to 1960, from 1961 to 1975 and from 1976 to the present. Considerable emphasis is placed on the central theme of the book which is the shift in the structure of FDI towards newer Third World MNEs which are increasingly gaining the capacity to engage in international production activities and catch up with older and more established MNEs from the developed countries.

1939 TO 1960

Prior to 1939, the majority of direct capital exports consisted primarily of resource-based investments in developing countries. The primary-product sectors and railroads accounted for 55 per cent and 20 per cent respectively of FDI in 1914. Manufacturing, however, accounted for 15 per cent of foreign direct investment abroad in 1914 while trade and distribution activities, public utilities and banking accounted for 10 per cent of the global stock of direct investment abroad. This sectoral distribution remained essentially unchanged by 1938.

These early forms of multinational investment activity were fundamentally motivated by the need to integrate backwardly into the extraction of raw materials or commodities which did not exist in Europe and other developed countries but existed in resource-rich developing countries in great abundance. These vertically integrated resource-based investments provide early evidence of the presence of an international hierarchical organisation, and were more significant for UK and European investors than for US investors in the pre-First World War period, *inter alia* reflecting trans-Atlantic differences in domestic resource availability. The availability of capital coupled with the absence of favourable alternative investment opportunities in the limited domestic market in several European home countries, as well as the opportunity for supply-oriented investments abroad, are other important reasons (Wilkins, 1974).

The limited foreign activity of the US MNEs in the period prior to 1914 was mainly market oriented rather than resource oriented and consisted of the establishment of sales branches to stimulate sales within markets. Factories were only built, when necessary, to maintain or expand markets or when good markets could not be reached, enlarged or maintained through exports. However, there were a few supply-related investments undertaken by US MNEs in neighbouring countries, chiefly to obtain primary products that were not available or to supplement inadequate supplies in the USA.

The UK was the largest foreign capital stake holder prior to 1939 with shares of 45 per cent and 40 per cent of the world stock of direct investment abroad in 1914 and 1938 respectively (see Table 2.1). The USA was a poor second with an estimated share of 18 per cent of the world stock in 1914, but this had increased to 28 per cent by 1938. France, Germany, Italy, the Netherlands, Sweden and Switzerland were the other significant outward investors prior to 1939.

Developing countries accounted for 63 per cent of the accumulated stock of inward foreign direct investment in 1914, a share that further increased to 66 per cent in 1938 (see Table 2.2). The significant recipients within the developing countries were those in Latin America and the Caribbean, followed by Asia and the Pacific. In general, developing countries that had colonial ties with the UK, France and Belgium received considerable amounts of direct investment from these countries (Svedberg, 1981).

However, between 1870 and 1900, resource-based investments grew more slowly as a new type of industry, more closely related to new consumer needs, began to emerge. Direct capital exports were concentrated on the establishment of foreign branches by enterprises already operating in their home countries, essentially the kind of foreign activity of the modern MNE which mainly predominates at the present time. Such new forms of MNE activity, which rapidly developed in 1875 and became firmly established in 1914, were directed primarily to more developed countries.

The shift in the sectoral distribution of FDI from resource-based investments towards secondary-sector investments, particularly in consumer goods industries, occurred *inter alia* in response to policies of economic nationalism in the developed countries. The introduction of tariffs in Germany, France, Scandinavia, the USA and in the British colonies and dominions from 1873 to

Table 2.1 Estimated global stock of outward foreign direct investment by country of origin, selected years 1914–89, US $ million and percentage

Country	1914 Value	1914 Share	1938 Value	1938 Share	1960 Value	1960 Share	1975 Value	1975 Share	1989 Value	1989 Share
Developed countries	14,302.0	100.0	26,350.0	100.0	67,000.0	99.2	272,035.6	98.8	1,308,600.0	95.5
Western Europe	11,000.0	76.9	16,850.0	63.9	30,600.0	45.3	117,687.3	42.7	673,800.0	49.2
United Kingdom	6,500.0	45.4	10,500.0	39.8	12,400.0	18.4	37,001.7	13.4	224,100.0	16.4
West Germany	1,500.0	10.5	350.0	1.3	800.0	1.2	14,353.8	5.2	121,700.0	8.9
France	1,750.0	12.2	2,500.0	9.5	4,100.0	6.1	10,607.5	3.9	70,500.0	5.1
Belgium	NA	NA	NA	NA	1,300.0	1.9	3,038.4	1.1	15,800.0	1.2
Italy	1,250.0 }	8.7 }	3,500.0 }	13.3 }	1,100.0	1.6	3,298.7	1.2	50,900.0	3.7
Netherlands					7,000.0	10.4	19,922.3	7.2	81,600.0	6.0
Sweden					400.0	0.6	4,670.4	1.7	33,500.0	2.4
Switzerland					2,300.0	3.4	22,442.7	8.2	48,300.0	3.5
Other Europe	NA	NA	NA	NA	1,200.0	1.8	2,351.8	0.9	27,400.0	2.0
North America	2,802.0	19.6	8,000.0	30.4	34,400.0	50.9	134,406.2	48.8	443,900.0	32.4
Canada	150.0	1.0	700.0	2.7	2,500.0	3.7	10,356.2	3.8	63,900.0	4.7
USA	2,652.0	18.5	7,300.0	27.7	31,900.0	47.2	124,050.1	45.1	380,000.0	27.7
Other developed	500.0	3.5	1,500.0	5.7	2,000.0	3.0	19,942.1	7.2	190,900.0	13.90
Australia	180.0 }	1.3 }	300.0 }	1.1 }	200.0	0.3	1,108.8	0.4	28,900.0 }	2.1 }
New Zealand							25.0	0.01		
South Africa					1,300.0	1.9	2,867.3	1.0	7,600.0	0.6
Japan	20.0	0.1	750.0	2.8	500.0	0.7	15,941.0	5.8	154,400.0	11.3
USSR	300.0	2.1	450.0	1.7	NA	NA	NA	NA	NA	NA
Developing countries	neg.	neg.	neg.	neg.	540.0	0.8	3,304.1	1.2	61,662.0	4.5
Total	14,302.0	100.0	26,350.0	100.0	67,540.0	100.0	275,339.7	100.0	1,370,262.0	100.0

Sources: Dunning, 1983a, for data for the years 1914 and 1938. Dunning and Cantwell, 1987, for data for the year 1975. Various national sources for data for the years 1960 and 1989. The outward FDI stock of developing countries for the years 1960, 1975 and 1989 represent author's estimate.

Notes: NA not available.
neg. negligible.
Data on the USA exclude direct investments in the finance, insurance and real estate sectors in the Netherlands Antilles.

Table 2.2 Estimated global stock of inward foreign direct investment in major host countries and regions, selected years 1914–89, US $ million and percentage

Country	1914 Value	1914 Share	1938 Value	1938 Share	1960 Value	1960 Share	1975 Value	1975 Share	1989 Value	1989 Share
Developed countries	5,235.0	37.2	8,346.0	34.3	36,700.0	67.6	183,399.0	75.5	1,133,300.0	80.8
Europe	2,500.0	17.7	2,200.0	9.0	12,500.0	23.0	98,710.6	40.6		
Western Europe	1,100.0	7.8	1,800.0	7.4	12,500.0	23.0	97,978.1	40.3	539,900.0	38.5
United Kingdom	200.0	1.4	700.0	2.9	5,000.0	9.2	24,490.4	10.1	138,800.0	9.9
Other Europe	1,400.0	NA	400.0	16.4	NA	NA	732.5	0.3	NSA	NSA
USSR	1,000.0	7.1	NA	NA	NA	NA	NA	NA	NA	NA
North America	2,250.0	16.0	4,096.0	16.8	20,500.0	37.8	64,447.7	26.5	503,800.0	35.9
Canada	800.0	5.7	2,296.0	9.4	12,900.0	23.8	36,785.7	15.1	103,000.0	7.3
USA	1,450.0	10.3	1,800.0	7.4	7,600.0	14.0	27,662.0	11.4	400,800.0	28.6
Other developed	485.0	3.4	2,050.0	8.4	3,700.0	6.8	20,241.1	8.3	89,600.0	6.4
Australia ⎱	450.0	3.2	1,950.0	8.0	3,600.0	6.6	18,744.1	7.7	69,300.0	4.9
New Zealand ⎰										
South Africa									11,100.0	0.8
Japan	35.0	0.2	100.0	0.4	100.0	0.2	1,497.0	0.6	9,200.0	0.7
Developing countries	8,850.0	62.8	15,969.0	65.7	17,600.0	32.4	59,439.5	24.5	269,600.0	19.2
Africa	900.0	6.4	1,799.0	7.4	3,000.0	5.5	15,445.2	6.4	29,800.0	2.1
Asia and Pacific	2,950.0	20.9	6,068.0	25.0	4,100.0	7.6	13,039.0	5.4	123,800.0	8.8
China	1,100.0	7.8	1,400.0	5.8	neg.	neg.	NA	NA	NSA	NSA
India and Ceylon	450.0	3.2	1,359.0	5.6	1,100.0	2.0	2,223.9	0.9	NSA	NSA
Latin America and Caribbean	4,600.0	32.7	7,481.0	30.8	8,500.0	15.7	29,386.6	12.1	103,900.0	7.4
Southern Europe ⎱	400.0	2.8	621.0	2.6	500.0	0.9	NA	NA	NA	NA
Middle East ⎰					1,500.0	2.8	650.0	0.3	12,100.0	0.9
Australasia	neg.	neg.	neg.	neg.	neg.	neg.	918.7	0.4	NSA	NSA
Total	14,085.0	100.0	24,315.0	100.0	54,300.0	100.0	242,838.9	100.0	1,402,900.0	100.0

Sources: Dunning, 1983a, for data for the years 1914 and 1938. Dunning and Cantwell, 1987, for data for the year 1975. Various national sources for data for the years 1960 and 1989.

Notes: Tables 2.1 and 2.2 are intended to show the geographical patterns of inward and outward direct investment. The total estimated stock of inward and outward direct investment are not equal due to differences in the definitions and methodologies used in the collection of FDI data both within and between countries and possible statistical errors. *neg.* negligible. *NA* not available. *NSA* not separately available.

1896 created opportunities for the establishment of a whole range of consumer goods and the industries that produced them. These industries overtook railways and mining both domestically and internationally and became typical representatives of the new kind of multinational enterprise. The growth of many British, Dutch, Belgian and Swiss firms was constrained by relatively small domestic markets and high tariff barriers in their export markets, a feature of economic-nationalist policies in the age of combined depression and attempted growth. These firms were compelled to turn from exports based on their home factories to the establishment of factories in important export markets in foreign countries in order to maintain and expand their business. The production and consumption of these goods in the host countries could be adapted with relative ease with the use of new concepts in advertising.

The shift that occurred in the sectoral pattern of FDI from resource-based to import-substituting manufacturing investment brought about a shift in the geographical pattern of FDI from resource-rich developing countries to market-rich developed countries in Western Europe and the USA. However, although these investments were beginning to become more significant, accounting for 25 per cent of total FDI in 1938, resource-based investments in developing countries still accounted for two-thirds of the total stock of inward FDI, a situation very little changed since 1914. Import-substituting manufacturing investments in developed countries in particular constituted only 34 per cent of the estimated stock of accumulated FDI in the pre-Second World War period.

There are several factors that explain the geographical distribution of import-substituting manufacturing investments in the developed countries, of which geographical distance, language, cultural, political and trading ties are the most significant. The product cycle model is a useful analytical framework within which to view the shift of location of production from the home country in the innovative new stage (or innovation-based oligopoly) into other relatively advanced countries and previous export markets in the maturing product stage (or mature oligopoly). The important stimulus to the initial import-substituting investment of these firms was the high tariff barriers imposed in their important export markets in other advanced industrialised countries against products that had a high income elasticity of demand.

The period from 1945 to the end of the 1960s was the best-known phase of multinational enterprise growth, characterised by the generation of major technological advances and an economic and political climate particularly conducive to the growth of international business. Such growth was largely accounted for by American firms (see Servan-Schreiber, 1967). US firms established a total of 2,181 manufacturing subsidiaries during the period from 1939 to 1961, which represented 60 per cent of the total new manufacturing subsidiaries established abroad in that period by US, UK, Continental European and Japanese parent firms combined (Vaupel and Curhan, 1974). By 1960, 47 per cent of total foreign direct investment stock originated from the USA, a phenomenal increase from a share of 28 per cent in the period prior to the

Second World War. US firms became the leading manufacturers and exporters of products of superior technology, primarily to Europe and Canada, but by the 1950s and 1960s, with increased duties and transport costs in these countries, the direct investment route became a more important modality of servicing these markets. Latin America and other developing countries became less important recipients of US direct investment despite cheaper labour and prospects of higher profitability in these areas.

A developmental linkage was then established between US MNEs and some technologically innovative indigenous firms in the UK and the advanced Continental European nations as local firms in these host countries enhanced their technological ownership advantages either through the development of their own indigenous technology or the importation of American technology (Dunning and Cantwell, 1982). The accumulation of technological advantages by European firms is reflected partly in the value of their direct investment which reached US $30,600 million by 1960, representing a global share of 45 per cent of which the UK, the Netherlands and France together accounted for about 35 per cent. A total of 718 and 653 new manufacturing subsidiaries were established between 1939 and 1961 by UK and Continental European MNEs respectively, accounting for 20 per cent and 18 per cent of the total manufacturing subsidiaries established by US, UK, Continental European and Japanese MNEs during the period. Around 73 per cent of UK manufacturing subsidiaries and 71 per cent of Continental European manufacturing subsidiaries were established in developed countries by 1960. In particular, Australia, South Africa, North America and other EC countries were the main recipients of direct manufacturing investments by UK MNEs. On the other hand, Continental European MNEs still tended to locate their manufacturing subsidiaries mostly in EC and other European countries.

By contrast, the early growth of Japanese investments in resource-related and import-substituting activities from the 1950s until the early 1970s was directed towards neighbouring Southeast Asian countries. These resource-based investments, which have expanded with the support of the Japanese government, were geared to looking abroad for sources of raw materials and natural resources with which Japan is poorly endowed (Ozawa, 1977, 1982). On the other hand, a total of 105 Japanese overseas manufacturing plant were established in the immediate post-war period, of which 7 were located in developed countries and 98 in developing countries, mainly in the Asian and Pacific regions. The growth of Japanese manufacturing investments in developing countries represents a sharp reversal of the product cycle trend demonstrated by other developed country MNEs which located their manufacturing subsidiaries in other developed countries. This may be largely explained by the earlier stage of development of Japanese outward investments. The early forms of direct investments in the manufacturing sector of developing countries by Japanese MNEs have been concentrated in a large number of small- to medium-scale labour-intensive industries in response to the rise of domestic wages and other production costs in Japan (Kojima, 1978;

Ozawa, 1979a). The relatively few manufacturing subsidiaries established in developed countries such as the USA and Europe were initially marketing outlets representing trade investments as well as banking sector investments geared to gain access to international capital markets and to promote exports from Japanese firms located in the Southeast Asia region (Tsurumi, 1976; Sekiguchi, 1979).

The structural shifts in the sectoral distribution of FDI resulted in the dominance of market-oriented, import-substituting manufacturing investments in the post-war period. By 1960 the share of the manufacturing sector in the total foreign direct investment stock of UK and US firms had grown to 35 per cent from a share of 25 per cent in 1938. Foreign interests in supply-oriented investments such as agriculture and public utility activities decreased significantly, although mining investments grew at an average rate and some mining investments by UK and US MNEs in non-ferrous metals grew at a very fast rate. Apart from state-owned oil MNEs, the major European capital exporting firms in the 1950s such as France, the Netherlands and Switzerland directed their interests mainly to manufacturing, trade and service activities.

In consequence, the share of accumulated FDI directed to developed countries doubled from 34 per cent in 1938 to 68 per cent in 1960 as a result of the structural shift of FDI towards import-substituting manufacturing investments associated partly with the imposition of trade barriers in these countries. Green-field ventures accounted for 48 per cent and 53 per cent respectively of manufacturing investments of non-US and US parent firms during this period (Vaupel and Curhan, 1974).

1960 TO 1975

An important indication of the sustained expansion of multinational enterprises during the 1960s and the first half of the 1970s is the 10 per cent annual average increase in the value of the worldwide stock of outward foreign direct investment from US $67,540 million in 1960 to US $275,340 million in 1975. Such high rates of growth reached a peak in the late 1960s but decelerated in the early and mid-1970s.

A significant development during the period is the increased diversity of the source countries of FDI. Table 2.1 suggests that the share of multinational companies based in the USA, UK, France and the Netherlands declined to 45 per cent, 13 per cent, 4 per cent and 7 per cent respectively in 1975 compared to their corresponding shares of 47 per cent, 18 per cent, 6 per cent and 10 per cent in 1960. Other countries, notably West Germany, Switzerland and Japan had increased their share of outward direct investments to 5 per cent, 8 per cent and 6 per cent respectively in 1975 from corresponding shares of 1 per cent, 3 per cent and 1 per cent in 1960.

The large increase in direct investments originating from West Germany and Japan is a reflection of the developmental industrial upgrading of these countries associated with stronger rates of technological innovation of indigenous firms complemented by access to foreign technology. Inward import-substituting FDI acted as a stimulus to indigenous technological innovation in West Germany while licensed foreign technology acted as a corresponding stimulus in Japan. Japanese firms initially responded to competition in international markets through export growth, which in turn acted as a spur to further international competitive performance in the form of FDI.

Although some outward direct investments were sourced from some other developed countries, such as Australia and Canada, and some developing countries such as Bermuda and Hong Kong during this period, most of them actually represented direct investments by foreign affiliates in these countries undertaken for administrative or tax convenience. In some cases, direct investments are undertaken because the affiliate is better able to serve and/or produce in some foreign markets than the parent company (Dunning, 1983a).

Increasing diversification in the geographical distribution of outward direct investment is confirmed by an examination of the growth of 483 of the world's largest industrial enterprises. In 1962, 292 of these enterprises originated from the USA, 128 from the EC, 14 from other parts of Europe, 29 from Japan and 17 from other developed countries. Only 3 of the 483 originated from developing countries. By 1977, the USA and the EC had become less important home countries of direct investment. The number of firms originating in these countries declined to 238 and 109 respectively in 1977, while those originating in other European countries, Japan and developing countries increased to 33, 69 and 17 respectively (Dunning and Pearce, 1985).

The increasing concentration of direct investments directed towards developed market economies during the period from 1960 to 1975 is reflected in Table 2.2. Developed countries received 75 per cent of the total stock of direct investment in 1975 compared to 68 per cent in 1960. Europe, in particular, received 41 per cent of direct investments in 1975 compared to 23 per cent in 1960. Australia, New Zealand and South Africa marginally increased in significance as host countries with collective shares of 7 per cent in 1960 and 8 per cent in 1975. The share of North America, on the other hand, declined from 38 per cent in 1960 to 27 per cent in 1975.

The share of developing countries as recipients of foreign direct investment further declined to 25 per cent in 1975 from a share of 32 per cent in 1960. The diminution in the relative importance of foreign direct investment in developing countries is further reflected when compared with the faster growth of trade and domestic investment in these countries during the 1970s. Foreign direct investment in developing countries was shifted from the lower-income developing countries towards a few of the larger, more industrialised countries, otherwise known as newly industrialised countries (NICs), as well as those that had natural resources of vital interest to multinational enterprises. While many of the lower-income developing countries received little or no direct investment in 1975, oil-producing countries received more than 20 per cent of the direct investment stock in developing countries and there have been significant amounts of direct investment held in tax havens. About ten other countries accounted for 40 per cent of the FDI stock in developing countries.

1975 TO THE PRESENT

In contrast with the immediately preceding period, the rate of growth of international production during the late 1970s and 1980s accelerated faster than the pace of growth of trade or output. In the 1960s, FDI grew at twice the rate of GNP; since 1983 foreign direct investment outflows have increased at an unprecedented annual rate of 29 per cent, three times faster than the growth rate of exports and four times faster than the growth rate of world output (Julius, 1990; UNCTC, 1991). By 1989, the stock of accumulated foreign direct investment by country of origin had reached approximately US $1.4 trillion, almost five times that in 1975, representing an annual average increase of 12 per cent since 1975. The swift pace in the growth of FDI has also been associated with its rapidly evolving character and composition.

In part, the late 1970s and 1980s witnessed a greater diversification in both the origin and destination of foreign direct investment and as a result the world economy rapidly achieved a new instrument of international economic integration (Julius, 1990). There have been major shifts in the geographical pattern of foreign direct investment flows during this period, as will be confirmed in Chapter 5 where it is shown that the previously significant relationship between net outward investment (NOI) flows and stage of development of countries in the period prior to 1975 had undergone some structural change in the period from the mid-1970s.

Increasing diversity in the origin of foreign direct investment is shown in an examination of an equalised sample of 483 of the world's largest industrial enterprises in 1982. The number of firms originating from the USA and non-EC European countries declined to 213 and 29 respectively from a corresponding level of 238 and 33 in 1977. On the other hand, the number of firms originating from EC countries, Japan and other developed as well as developing countries increased to 118, 79, 20 and 24 respectively from corresponding levels of 109, 69, 17 and 17 in 1977 (Dunning and Pearce, 1985). In terms of value of foreign direct investment, while the USA accounted for about two-thirds of the total outflows and about 10 per cent of inflows of FDI in the 1960s, its share in total outflows had declined to less than 50 per cent by the early 1980s while its share of total inflows had increased to more than 30 per cent, with most of the growth taking place between 1977 and 1980 and since 1986. This may reflect the erosion of the technological and managerial superiority with which US industry emerged from the Second World War, which made US firms effective competitors abroad and, even more so, in their home market (Graham and Krugman, 1989). By 1989, the significance of the USA as a source country of outward direct investments declined to 28 per cent from a share of 45 per cent in 1975 while its significance as a host country increased to 29 per cent from a corresponding share of 11 per cent in 1975. The leading role of the USA in the internationalisation of economic activity has therefore changed significantly since the late 1970s from being a main source to a main recipient of foreign direct investment.

The increasing diversity in the source countries of FDI is also reflected in the list of the fifteen largest outward investors by size of foreign direct capital stock in 1975 and 1983 as presented in Table 2.3. While the USA, UK, West Germany and Australia maintained their positions at first, second, sixth and thirteenth place respectively, Switzerland, the Netherlands, France, Sweden and Belgium moved down one place and occupied fourth, fifth, eighth, tenth and twelfth positions respectively in 1983. Canada, Italy and South Africa, on the other hand, improved their positions one place and respectively occupied seventh, ninth and eleventh place in 1983. Japan's position improved tremendously, moving up to third place in 1983. Hong Kong and Brazil, both newly industrialised developing countries, occupied the fourteenth and fifteenth places among the world's largest outward investors in 1983.

Chapter 5 of this book provides further evidence of the significant increase in the number of MNEs originating in newer source countries such as Japan and other developed countries, as well as the developing countries, in the period from 1975. The share of other developed countries in the total stock of outward FDI doubled from 7 per cent in 1975 to 14 per cent in 1989 with Japan, in particular, responsible for 11 per cent of total outward foreign direct investment stock in 1989 compared to a 6 per cent share in 1975.

Table 2.3 The fifteen largest outward direct investors by size of foreign direct capital stock, 1975 and 1983, US $ million and percentage

Country	1975		1983	
	Value	*% Share*	*Value*	*% Share*
USA	124,050.0	45.1	226,962.0	38.3
UK	37,001.7	13.4	88,543.2	14.9
Japan	15,941.0	5.8	61,276.0	10.3
Switzerland	22,442.7	8.1	40,532.2	6.8
Netherlands	19,922.3	7.2	39,120.9	6.6
West Germany	14,353.8	5.2	38,934.6	6.6
Canada	10,356.2	3.8	28,795.4	4.9
France	10,607.5	3.9	17,242.3	2.9
Italy	3,298.7	1.2	8,495.3	1.4
Sweden	4,670.4	1.7	6,761.4	1.1
South Africa	2,867.3	1.0	5,679.7	1.0
Belgium	3,038.4	1.1	4,211.0	0.7
Australia	1,108.8	0.4	2,997.9	0.5
Hong Kong	*NSA*	*NSA*	2,540.0	0.4
Brazil	279.0	0.1	2,010.0	0.3
Other countries	5,401.9	2.0	19,300.1	3.3
Total	275,339.7	100.0	593,402.0	100.0

Source: Dunning and Cantwell, 1987. The figures for 'other countries' for 1975 and 1983 represent author's estimate.

Since 1983, Japanese FDI outflows, as measured in local currency terms, have doubled every two years, promoting Japan to third place in the league of international investors after the USA and the UK (Julius, 1990). Such growth has also been associated with a rapidly evolving character and composition (Ozawa, 1990). The concentration of Japanese FDI in resource-based and labour-intensive or technologically standardised industries in developing countries from the 1950s to the early 1970s has evolved towards the mass production of assembly-based consumer durables in high-income countries such as the USA and Europe, supported by a network of subcontractors since the late 1960s. By the mid-1970s the transformation of Japanese outward FDI towards more complex manufacturing activities in industrialised countries such as the USA and Europe had been well established (Franko, 1983). Since the early 1980s Japanese FDI has also been characterised by the flexible manufacturing of highly differentiated goods, involving the application of computer-aided designing (CAD), computer-aided engineering (CAE) and computer-aided manufacturing (CAM).

The rapidly evolving comparative advantage of Japanese MNEs has been assisted by a continuous process of industrial restructuring and economic growth sustained largely through the emphasis of the Japanese government on upgrading their industrial structure towards the production of more technology-intensive products, in part through the introduction of the most advanced foreign technologies in the form of licensing (Ozawa, 1985). As a result, Japanese MNEs such as Sony, Matshushita and Sanyo, as well as other electrical/electronic firms, have successfully penetrated markets for microelectronics in the developed countries at an early stage, in order to fulfil their objective of participating in oligopolistic competition on the basis of well-differentiated products and good marketing techniques (Baba, 1987).

Direct investments from developing countries also increased in importance. Firms in some of the larger developing countries began to expand their business abroad through direct investment from the mid-1970s, mainly within their geographical regions. Of the 24 MNEs known to originate in developing countries in the period from the mid-1970s, two MNEs each emerged from Brazil, India and Taiwan while nine MNEs emerged from South Korea. In addition, one MNE originated from each of the following countries: the Netherlands Antilles, Israel, Argentina, Chile, Colombia, Kuwait, Mexico, the Philippines and Venezuela (Dunning and Pearce, 1985). As of 1975, the stock of foreign direct investment originating from developing countries was estimated to be around US $3,304 million, representing a 1.2 per cent share in the world stock of outward direct investment, which reflects a 13 per cent annual average growth from an estimated level of US $540 million in 1960, or a 0.8 per cent share. These estimates of direct investment originating from developing countries exclude investments made as part of tax minimisation or related programmes, or investments by countries in which the ultimate parent company could be traced to MNEs from developed countries. Hence, direct investments from countries such as Panama, the Netherlands Antilles, Bermuda, the Bahamas, the Caribbean Islands, Haiti and other tax haven countries are omitted. Similarly, oil investments from the Middle East are also excluded since these investments are more properly described as portfolio rather than direct investments.

By 1989, direct investments from developing countries increased almost twentyfold to a conservative estimate of US $61,662 million, representing an annual average increase of 23 per cent from the 1975 level. Such an annual average growth rate of outward FDI from developing countries is almost twice as fast as that of developed countries, at 12 per cent. As a result of the high growth rates of their outward FDI, the share of developing countries in the world stock of outward direct investment was estimated to have increased to at least 4.5 per cent by the end of the 1980s from a share of 1.2 per cent in 1975 and 0.8 per cent in 1960.

In terms of inward direct investment, the share of FDI directed to developed countries further increased to 81 per cent in 1989 from 75 per cent in 1975. Table 2.2 reveals that Western Europe became a slightly less important recipient of foreign direct investment, with a declined share of 39 per cent in 1989 compared to 41 per cent in 1975 and 23 per cent in 1960. The share of the North American continent, however, increased significantly from 27 per cent in 1975 to 36 per cent in 1989. The share of the USA, in particular, more than doubled, to 29 per cent in 1989 from an 11 per cent share in 1975 and 14 per cent in 1960. On the other hand, developing countries received 19 per cent of the estimated stock of accumulated foreign direct investment in 1989, a significant decline from their 24 per cent share in 1975 and 32 per cent share in 1960. This shows the declining attraction of FDI in these countries, due partly to the increased importance of technologically intensive investments that necessitate location in high income, developed markets. Moreover, the debt problems of Latin American countries, the declines in per capita income in African countries, the general excess capacity in world commodity markets and weak commodity prices may have made developing countries less attractive as locations for foreign direct investment (see Riedel, Büttner and Ernst, 1988, and also Julius, 1990). The majority of the investments directed to the developing countries are concentrated in the NICs as well as developing countries with large domestic markets such as Nigeria, India, Zimbabwe, Malaysia and Indonesia. These latter countries are considered to occupy the tier immediately below the NICs and count among the fifteen largest net debtor nations of foreign direct investment in 1975 and 1983 as well as being the new source countries of outward FDI.

Another indication of the changing composition of foreign direct investment is shown in its sectoral pattern. Foreign direct investment in the late 1970s and 1980s was increasingly concentrated in the services sector and to a lesser extent in the manufacturing sector of both developed and developing countries. Nevertheless, the manufacturing sector accounted for an increasing share of FDI in developing countries during the latter part of the 1970s and early 1980s in contrast to the period up to the 1960s and the first half of the 1970s, when direct investments were usually made in the extractive and primary sectors. Such manufacturing foreign direct investments took place both in import-substituting and export-promotion industries and some of them exerted a large developmental impact on the economies of these countries and, in particular, on the internationalisation of their industries. Indeed, a common feature of the internationalisation of firms from these countries is, *inter alia*, the concentration of import-substituting FDI at an earlier period. This may have perpetuated a virtuous cycle by generating development spill-over effects to technologically innovative indigenous firms that have developed the capacity to respond independently to the technological competition offered by foreign MNEs.

The capacity of these indigenous firms for localised technological innovation enabled them to catch up with the older and more established MNEs from the developed countries in the process of international economic expansion through trade and foreign direct investment. Inward investment, along with other modalities of foreign technology transfer, fulfilled a similar developmental upgrading role in the internationalisation of firms from Germany, Italy and, to a lesser extent, Japan, in the immediate post-war period. Chapter 6 analyses the role of foreign technology in encouraging the accumulation of indigenous technological capacity and growth of production of Third World MNEs. The role of inward foreign direct investment in particular is studied partly through an analysis of the industrial distribution of inward and outward direct investment in a number of developing countries, which attempts to show in concrete terms the particular sectors where inward foreign direct investment and other modalities of technology transfer from the more developed countries may have fulfilled a developmental role in the growth of production of indigenous firms. In sectors where foreign technology has fulfilled a developmental role, local firms in developing countries may have initially caught up with the more established investors through export growth, which in turn acted as a spur to further competitive performance in the form of FDI. As happened in the earlier stages of international production of firms in the more industrialised countries, the imposition of tariff barriers in export markets provided opportunities for outward FDI by developing countries – especially in the case of Asian NICs – which enabled firms to expand their network of productive activities across national boundaries.

The increasing technological capacities of some local firms in developing countries enhanced by the developmental role of FDI and other modalities of foreign technology transfer are reflected not only in the increasing degree of internationalisation from exports to direct investment but also in changing patterns in the composition of such international economic activities. In the case of exports, the importance of manufactured products has surpassed that of primary goods, especially for the larger of the developing countries such as the NICs. In particular, the Southeast Asian NICs such as Hong Kong, South Korea, Taiwan and Singapore doubled their share of total world exports of manufactured products from 3.7 per cent in 1980 to 7.2 per cent in 1987. These

Asian NICs accounted for 65 per cent of the total exports of manufactured products from developing countries as early as 1976 compared to a corresponding level of 42 per cent in 1965. Such rapid growth in the manufactured exports of Asian NICs has been accompanied by significant shifts in the composition of exports from labour-intensive manufactured products towards manufactured products that embody greater skill, capital and technological intensity (Park, 1989). In fact, these countries have become more competitive than Japan in the export of such products as iron and steel, ships, metals and some transport machinery, and are equally competitive in the export of textiles, organic and inorganic chemicals, ceramics, clay and non-ferrous metals (Japan Economic Planning Agency, 1987). The pattern of foreign economic involvement – from exports of certain manufactured goods at the initial phase of international expansion followed by, or accompanied by, outward direct investment – which characterised the growth of foreign investment by firms from Germany and Japan in the immediate post-War period can therefore be applied to the more recent growth of FDI by firms from the NICs.

In contrast to the general sectoral trend towards the increasing importance of services, from an earlier emphasis on resource-based and manufacturing as an area for foreign direct investment, the data in Table 2.4 show that the manufacturing sector is proportionately more important for the newer outward investors in the Third World which are catching up with the more traditional investors. This may be explained by the close link between trade and investment in the early phases of international production as demonstrated by Cantwell (1989a) in a study of the sectoral distribution of the growth of exports and outward investment in the developed countries. Such findings show that the sectoral distribution of the growth of exports and outward investment are more correlated for the relatively newer investors such as Germany and Japan and, by extension, the present NICs. By comparison, there is a much greater discrepancy between the sectoral distribution of growth of exports and outward investment and hence greater scope for substitution between trade and investment for the more established international investors such as the UK and the USA.

However, although the evolving sectoral pattern of Third World outward investment bears a certain resemblance to the changing sectoral composition of the international production of firms from the developed countries during an earlier period, a distinction has to be made between the significance and form of their technological accumulation, owing to the different stages of their national development and the nature of technology as firm specific, cumulative and differentiated. The MNEs of the USA, UK, Japan and Germany are frequently technological leaders, unlike Third World MNEs which rely to a greater extent upon simpler and less research-intensive forms of technology creation. Although a quicker pace is observed in the evolution of the sectoral pattern of Third World outward investment from resource based towards manufacturing and, to some extent, services, as developing countries advance

Table 2.4 Sectoral distribution of outward direct investment, 1982, US $ million

Country	Primary	Secondary	Tertiary
Developed countries	132,332.1	214,163.3	163,886.2
Europe	50,678.7	92,877.0	61,564.1
EEC	49,574.3	88,427.0	56,694.6
France	5,270.8	6,718.9	7,317.4
W. Germany	1,754.3	23,295.0	15,093.8
Italy	1,803.6	3,694.2	2,617.5
Netherlands	13,900.5	17,187.8	8,649.2
UK	26,845.1	37,531.1	23,016.7
Other Europe	1,104.4	4,450.0	4,869.5
Austria	*neg.*	320.4	355.5
Finland	11.7	405.6	474.6
Norway	*neg.*	533.3	815.0
Spain	136.1	365.8	674.8
Sweden	956.6	2,824.9	2,549.6
North America	70,848.9	103,449.9	75,090.2
Canada	7,716.9	12,840.9	6,988.2
USA	63,132.0	90,609.0	68,102.0
Other developed	10,804.5	17,836.4	27,231.9
Australia	581.5	892.4	1,267.9
Japan	10,223.0	16,944.0	25,964.0
Developing countries	571.4	1,247.2	811.9
Asia and Pacific (except			
Japan and M. East)	522.8	1,163.1	699.7
Hong Kong	342.2	963.0	514.8
India	2.4	103.2	20.4
S. Korea	149.4	33.5	106.7
Philippines	28.8	63.4	57.8
Latin America and			
Caribbean	48.6	84.1	112.2
Argentina	48.6	75.3	20.1
Chile	*neg.*	8.8	92.1
Total	132,903.5	215,410.5	164,698.1

Source: Dunning and Cantwell, 1987.

through higher stages of development, and as the innovative activities of indigenous firms become increasingly more significant and complex with greater experience in international production, these firms do not exhibit the same rapid pace of technological advancement as firms from the more developed countries. Such different forms of technological innovation between developed and developing countries and between the NICs and the lower-income developing countries are further discussed in the following chapters.

CONCLUSION

This chapter traced post-war developments in the geographical and industrial distribution of foreign direct investment and, in particular, the increasing significance of outward FDI from developing countries, within the overall context of developments prior to the Second World War in which foreign investments were primarily resource based and directed towards the colonies. The increasing significance and complexity of the innovative activities of MNEs as a result of greater experience in international production have led to more sophisticated forms of outward investment, from resource-based investments in developing countries towards import-substituting manufacturing investments in the developed countries.

The growth of MNEs can be analysed *inter alia* within an interpretation of the concept of an investment development path in which indigenous firms in particular sectors in the host countries strengthened their capacities for technological innovation through foreign technology accessed in some cases through inward foreign direct investment. Such a development link is found, for example, between US inward investments and the outward investments of technologically innovative firms from the UK and the advanced Continental European nations that have enhanced their technological competence by combining US technology with their indigenously generated technologies. The strengthened technological capabilities of these firms have been associated with their increasingly competitive international expansion in the form of exports and outward investments.

The increasing sophistication in the sectoral and geographical pattern of outward direct investments of traditional investor countries based in the UK and USA, as a result of the increasing significance and complexity of their technological accumulation, has also been observed in the dynamic analysis of the evolutionary pattern of outward direct investment of newer investors from Germany and Japan in the period since 1960. This book aims to show that while a similar evolutionary trend is evident in the development of Third World outward investment, the form of technological accumulation varies among the leading technological firms from the USA, UK, Japan and Germany as opposed to the simpler, less research-intensive technological creation of Third World MNEs.

3 The theories of international production

INTRODUCTION

This chapter aims to analyse the different strands in the emergence of the modern and distinct theories of international production from 1960. There was no distinct theory of international production prior to 1960 because, apart from that organised by individual entrepreneurs in the period prior to 1914, it consisted largely of resource-based investments undertaken by firms from developed countries in developing countries. These resource-based direct investments, as well as portfolio investments, were largely explained by locational theories of international trade and capital movements. The relevance of historical antecedents to the emergence of the distinct theory of international production was first analysed by Cantwell, Corley and Dunning (1986). The modern theory of international production has drawn from the theories of capital movements, trade, location, industrial organisation, innovation and the firm.

The emergence of a distinct theory of international production and the MNE from 1960 can be traced to changes in the pattern of international production during the post-war period. Such changes are manifested *inter alia* in terms of the increasing dominance of import-substituting manufacturing investments in high technology-intensive sectors in developed countries; the rapid growth of cross-investments and intra-industry trade and production between developed countries; and the emergence of globally integrated MNEs. In particular, the occurrence in the post-war period of cross-investments and intra-industry trade and production between countries with similar factor proportions necessitated industry-based explanations. Such explanations are opposed to a framework based on locational theories of trade and capital movements which suggest that capital moves in response to profit or interest rate differentials across countries. In addition, the declining dominance of the USA as a major source of FDI in the 1950s and 1960s, and the increasing importance of Germany, Japan, Switzerland and other developed countries, as well as developing countries, as sources of FDI in the 1970s and 1980s, necessitated the adaptation of the existing theories.

The analysis of the different theories of international production starts with the pioneering contributions of Hymer (1960, 1972) which draw on major developments of the theories of Bain (1956) and Penrose (1959) of the growth of the firm and also marginally on the theories of Williams (1929), Robinson (1933) and Schumpeter (1934). The analysis then traces the development of the industrial organisation approach to international production with the contributions of Kindleberger, Caves and Johnson. These writers focused on the increasing importance of ownership advantages of firms in order to explain the emergence of intra-industry investment.

The importance of internalisation advantages in the growth of the firm were first incorporated in the analysis of international production during the 1970s with the growth of FDI in industries characterised as having a high technological intensity, the competitiveness of which greatly depended on the integration of R&D, production and marketing activities within the firm. The internalisation theories of Casson drew essentially from Coase (1937).

The macroeconomic development theories of international production represent an attempt to explain the growth of international production of countries and firms based on their stage of development. Some of these theories are evident in Vernon's product cycle model, Kojima's and Ozawa's theory of trade and direct investment, Dunning's concept of investment development path and the concept of stages of development elaborated by Tolentino (1987) and Cantwell and Tolentino (1990). The increased possibilities for competitive interaction between firms in international industries brought about by the emergence of competing sources of FDI have led to the development of the theories of the MNE based on oligopolistic competition.

The crucial development in the area of location theories in the period since 1960 is the incorporation of oligopoly to spatial decision-making which reflects the interdependence in the activities of economic agents on location. However, the models of the choice of modality of foreign market servicing between exports and foreign production are increasingly inadequate in the light of the development of rationalised international investments and global competition in the 1980s. The eclectic paradigm of international production is an attempt to synthesise the major strands in the theory of international production.

THE THEORY OF DIRECT INVESTMENT: HYMER

Post-war developments in international production necessitated a shift in emphasis to the analysis of ownership advantages based on the theories of the firm and the growth of the firm. The previous emphasis placed on location advantages based on orthodox theories of international trade and capital movements failed to explain the increasing significance of cross-investments and intra-industry trade and investment especially between countries that have similar factor proportions. The increasing importance of cross-investments between countries required industry-based explanations as opposed to locational explanations. The doctoral dissertation of Hymer (1960, published

1976) is regarded as the precedent of the modern theory of the MNE based on theories of the firm and industrial organisation.

Hymer's major contribution was in advancing a unique and distinct theory of the MNE *per se* on the basis of capital movements and intermediate products associated with the international operations of firms. The analysis therefore shifted from the neoclassical trade and investment theory and focused on the unique features of FDI as a modality within which MNEs maintain control over non-financial and ownership-specific intangible assets. The distinctive contribution of Hymer over the neoclassical trade and investment theory is in considering the MNE as an institution for international production rather than international exchange.

The element of control over the use of assets transferred abroad is considered important by the MNE for three reasons. The two major reasons are to obtain some form of monopoly power through the elimination of competition between the foreign enterprise and enterprises in other countries, and to appropriate fully the returns on certain skills and abilities. The minor reason is to minimise risks through diversification in which control is not necessarily involved.

Hymer's main thesis is his explanation of the process of FDI as an international extension of industrial organisation theory. In addition to recognising the importance of control to direct investment, he distinctly viewed FDI as a modality for transferring intermediate products such as technology, management and organisational and marketing skills. The expected return on the transfer of these intermediate products, not the expected return on capital, prompts firms to become MNEs. Capital is regarded as a modality for the transfer of these intermediate products.

The view is therefore one of an active firm that erects barriers to entry, colludes with other firms within the industry and exercises control over intermediate product markets and creates imperfections. The major cause of firms' international operations is seen to derive from the possession of oligopolistic forms of ownership advantages that arise from the presence of firms of unequal abilities. The analysis of these advantages is based largely on the advantages of established firms that create entry barriers for firms of a different nationality (Bain, 1956). These advantages, which arise from scale economies, knowledge advantages, distribution networks, product diversification and credit advantages, enhance the ability of the MNE to restrict competition and therefore increase market power.

However, the presence of unequal abilities of firms is a necessary but not sufficient condition for international operations. The second major cause of international operations is the exploitation of an advantage. The extent to which a firm engages in direct investment depends largely on whether that firm licenses the advantage or exploits the advantage within the firm through international production. The firm would only choose to exploit such advantages because of the imperfect nature of the market for the advantage in an oligopolistic market structure.

In the final part of his theoretical exposition, Hymer identified the three main characteristics of the sectors in which FDI is more likely to concentrate: the prevalence of imperfect (and noticeably) oligopolistic competition; the presence of firms of unequal ability; and, finally, the greater scope for economies of joint operation, i.e. product or process diversification. The absence of these characteristics is seen to be consistent with the absence of FDI. Hymer also pointed to two important implications of the presence of oligopolistic elements in international production. The first implication is associated with the indeterminacy of international operations. The second implication is policy oriented, whereby host governments can increase their bargaining power and tax the higher than opportunity costs return of firms without affecting their behaviour.

LIMITATIONS OF THE HYMER THEORY OF DIRECT INVESTMENT: THE VIEW OF ALTERNATIVE THEORIES OF THE GROWTH OF THE MNE

The oligopolistic theory of the firm adopted by Hymer which allows for active firms and interactive processes in which the conduct of firms determines the market structure is distinct from the theory of monopolistic competition adopted by later theorists more entrenched in the industrial organisation tradition, based on a structure–conduct–performance model. These theorists, such as Kindleberger, Hirsch, Lall, Helpman and Krugman, who regarded market structure described by monopolistic competition between differentiated products as determining the conduct and performance of firms, have a narrower and more restricted view of ownership advantages of firms. Their theories are essentially static and fail to take account of industrial dynamics or evolution (Carlsson, 1987).

These two opposing views of the firm can be analysed within the underlying framework of the growth of production networks *vis-à-vis* the framework of transaction and exchange. The framework of growth of production networks adopted by Hymer, Schumpeter and Vernon, and those in the technological accumulation tradition such as Pavitt (1988) and Cantwell (1989a), consider the active role of the firm in initiating new forms of oligopolistic ownership advantages hand-in-hand with growth of production as necessary for the long-term survival of the firm. In their view, the growth of the firm is the essential cause of market imperfections and failure.

By contrast, Coase's (1937) theory of monopolistic competition based on the framework of transaction and exchange, adopted by Buckley and Casson *et al.*, views the hierarchical organisational structure of the firm as a reactive or passive product of the transaction costs of markets, since there is no independent scope for the managerial strategy of the firm. The hierarchical organisation of the firm is seen as an efficiency device for organising transactions and decreasing transaction costs with respect to the market. The firm grows through the displacement of markets which operate in a costly, imperfect and inefficient way. Apart from the advantages accruing through the process of internalisation, ownership advantages are not necessary for the existence of the MNE and international production in the standard theory of transaction costs and vertical integration, as the existence of transaction costs suffices.

Hymer adopted the framework of growth of production in his analysis which sees the growth of the firm as resulting from the motive of firms to seek new means of collusion and market power by reducing competition and increasing industrial concentration. However, the process of growth of firms through cooperative arrangements finds application only in circumstances where relatively weak firms within the industry seek to strengthen their ownership advantages and where there is intense competition. In any case, collusion between MNEs is generally undertaken to improve the relationship between international production networks rather than being an explanation of their existence and growth (Cantwell, 1991b).

This theory was applied in the work of Hymer and Rowthorn (1970), Sanna Randaccio (1980) and more recently by Cowling and Sugden (1987) who argued that monopolisation and collusion are important means by which firms attain profitability and security. However, Graham (1975, 1985) argues that although collusive agreements and cartels were an important means of achieving industrial stability and avoiding price warfare in world markets in the period prior to 1914, exchange of threats or intra-industry production by MNEs in the major markets of their rival firms became more typical of the strategy adopted by firms in the period from 1960 as a means of attaining security. Assuming that no merger among major rival firms occurs, the modality of achieving stability in international markets through cross-investments reduces the likelihood that collusion can be achieved successfully in world markets. This exchange of threat view of oligopolistic interaction as a means to preserve price stability has recently been integrated within the internalisation theory by Casson (1987). Similarly, the idea of competitive action and reaction has also been central to the work of Hamel and Prahalad (1988) in explaining the process of cross subsidisation of national market share battles as an important feature of global strategy adopted by MNEs.

The recent theories of technological accumulation proposed by Pavitt (1988) and Cantwell (1989a) are also opposed to the Hymer market power view of the firm. In their view, the cumulative process of generating technological ownership advantages is associated with rivalry and competition between firms in an international oligopoly rather than serving as a barrier to entry against potential new entrants. In some cases, the emergence of new and interrelated technologies since the 1980s may result in the emergence of clusters or galaxies of firms which are strategically linked, not necessarily by ownership but by contractual arrangements which often take the form of quasi-integration (Dunning, 1988c). These organisational networks or corporate family relationships between firms have resulted in declining industrial concentration and greater competition between corporate family constellations (Karlsson, 1988). Increasing competition and declining industrial concentration act as an incentive to further advance the accumulation of technology, especially of the more research-intensive kind that leads to the creation of new products and processes.

The antecedents of the Hymer view of the firm can be traced to Adam Smith, who viewed the motive of firms as seeking new means of collusion and market power in order to increase profits. The major difference between Smith and Hymer lies in the role of investment. While Smith viewed competition between firms as an incentive to further investment, which in turn would discourage collusive behaviour, Hymer viewed investment as a means of encouraging and reinforcing collusive behaviour (Cantwell, 1985).

The emphasis placed by Hymer on the impact of the conduct of foreign enterprises on market structure laid the foundations for the defensive oligopolistic theory of the MNE formalised by Vernon, Knickerbocker and Graham. Moreover, his unprecedented emphasis on the importance of ownership advantages inspired economists such as Kindleberger (1969) and Caves (1971). Kindleberger approached Hymer's work from the standpoint of the theory of trade and capital movements and developed his analysis of ownership advantages in terms of the theory of the firm and industrial organisation. On the other hand, Caves approached Hymer from the standpoint of industrial economics. The contributions of each of these economists, as well as others, to the development of the modern theory of international production based on industrial organisation is examined in turn.

THE INDUSTRIAL ORGANISATION APPROACH TO THE THEORY OF INTERNATIONAL PRODUCTION

Charles Kindleberger, who had been the supervisor of Hymer's dissertation, also elaborated on the industry-specific nature of the growth of FDI, which is distinct from the locational theories of international trade and capital movements adopted prior to 1960 as well as the reformulated theory of international capital movements adopted by Aliber (1970, 1971). However, in contrast to Hymer who had an oligopolistic theory of the growth of production networks across national boundaries, Kindleberger had a monopolistic theory of international production. He viewed the firm that undertakes direct investment as having monopolistic advantages over existing or potentially competitive firms in the country. Direct investment is seen as stemming from imperfections in the markets for goods and factors, including technology or some interference in competition by government or by firms which separates markets (Kindleberger, 1969).

The distinctive contribution of Kindleberger is in proposing a useful categorisation of the nature of monopolistic advantages which lead to direct investment. First, departures from perfect competition in goods markets, including product differentiation, special marketing skills, retail price maintenance, administered pricing, and so forth. Second, departures from perfect competition in factor markets, including the existence of patented or unavailable technology, of discrimination in access to capital, of differences in skills of managers organised into firms rather than hired in competitive markets. Third, internal and external economies of scale, the latter being taken advantage of by vertical integration. And fourth, government limitations on output or entry.

Kindleberger's contribution is distinct from that of Hymer in one major respect, which pertains to the nature of the firm and its impact on competition. With regard to the former, Kindleberger misunderstood Hymer as he spoke of monopolistic advantages and monopolistic competition as a form of market structure whereas Hymer referred to oligopolistic competition. It is Kindleberger's focus on monopolistic advantages associated with the market power view of the firm and the existence of monopolistic market structure that lead Casson *et al.* to claim that ownership advantages are not a necessary condition for international production. But although monopolistic advantages are not a necessary condition for FDI, the kind of oligopolistic advantages Hymer referred to are necessary.

As a result of his emphasis on monopolistic competition, Kindleberger characterised a passive rather than an active firm. The MNE is reinterpreted as a *product* of market structure characterised by monopolistic competition between differentiated products and not as an *agent* that is involved in oligopolistic competition with other firms in order to exercise control over their ownership advantages. While Hymer analysed the impact of the conduct of firms, and specifically their attempts at achieving market power, on market structure, writers in the industrial economics tradition, such as Kindleberger, Hirsch, Lall, Helpman and Krugman, analysed the impact of a determinate monopolistic market structure on the conduct and performance of firms.

Caves (1971) similarly identified and explained the nature of ownership advantages and indicated that the fundamental element of FDI is a particular trait of market structures in both lending (home) and borrowing (host) countries. The precise nature of ownership advantages varies according to the three forms of firm expansion in a new, geographically segregated production facility. These forms are horizontal extension (producing the same goods elsewhere); vertical extension (adding a stage in the production process that comes earlier or later than the firm's principal processing activity); and conglomerate diversification. Oligopoly with product differentiation predominates where MNEs engage in horizontal direct investment. Oligopoly without necessary differentiation predominates where MNEs engage in vertical direct investment. A *differentiated product* is defined here as a collection of functionally similar goods produced by competing sellers, but with each seller's product distinguishable from its rivals' by minor physical variations, brand name and subjective distinctions created by advertising or differences in the ancillary terms and conditions of sale.

Johnson (1970) asserts that such ownership advantages as managerial and technical knowledge are the prime determinants of direct investment. But as, in his view, the process of knowledge creation requires an investment of resources but once created assumes the character of a public good, there must be some incentive for private investment to engage in the costly process of knowledge creation. These incentives may be in the form of granting the producer of that knowledge a temporary monopoly in the use of the knowledge created through public tolerance and legal protection of commercial secrecy. These forms of achieving monopoly have increasingly become more important than the patent system. However, this practice of rewarding the producer of commercially useful knowledge with the right to charge a monopoly price for the product leads to the unfortunate consequence of economic inefficiency.

The framework of transaction and exchange which regards technology as akin to information and as having public good characteristics that are costly to create but costless to transfer is common not only among earlier works on industrial organisation but also among economists, not least in the internalisation literature. However, Teece (1976, 1977) has reformulated the framework in a more dynamic context by referring to the declining value of information upon dissemination and the concomitant high cost of technology transfer. Such a view of technology is fundamental in the framework of growth of production which perceives the distinctive firm-specific and cumulative technological characteristics of firms as difficult to transfer to and be implemented by another firm with different technological characteristics. Such transfer of technology is even more difficult between firms in countries that are in different stages of economic development (Rosenberg, 1976).

Through an analysis of the main features of US investment in Canada, Horst (1972) found that most, if not all, of the principal ownership advantages necessary for foreign production can be subsumed under the heading of the size of the firm. On the other hand, Mason, Miller and Weigel (1975) pointed out that the sources of ownership advantages which enable foreign firms to engage in international business include superior product and production technology, superior management skills and preferential access to production inputs such as capital.

Hirsch (1976) referred to firm-specific know-how and other intangible income-producing proprietary assets in developing a model of choosing an optimal modality of foreign production. These assets are a result of past investments in product or process R&D and/or investments in advertising and other promotional techniques which enable the firm to engage in some form of product differentiation. This is broadly similar to the work of Johnson (1970). The intangible assets, of which one component is managerial know-how, are analysed within a static neoclassical viewpoint and taken to have the characteristic of a public good that is subject to obsolescence. The possession of these assets constitutes an effective barrier to entry to rival firms and renders a temporary monopoly power which yields a monopoly rent over and above the profit rates prevailing in the industry.

In analysing the empirical evidence of industrial diversification and internalisation of US manufacturing firms, Wolf (1977) generalised that the important ownership advantages of these firms are those that stem from size and technical know-how. Foreign investment is therefore perceived as a modality for large firms that have a high level of technical expertise to maximise the use of their under-utilised resources. On the other hand, Lall (1980a), in broad conformity with the views of Kindleberger and Hirsch, maintained that the extent of foreign involvement in each industry is determined by the possession of a distinctive combination of monopolistic advantages. By focusing his analysis on the foreign involvement of US manufacturing, Lall identified the main monopolistic advantages that determine the extent of foreign involvement: technology, product differentiation, scale economies and skills. With the use of a simple model that ignores the differences in production costs abroad, Lall confirmed that significant relationships exist between internal barriers to entry and foreign involvement and between the mobility or transferability of monopolistic advantages and the export–foreign production choice.

With respect to the nature of monopolistic advantages, Lall regards the combination of technological and product differentiation factors as exerting the largest influence on all forms of foreign involvement and constituting the most important ownership advantage of US manufacturing industries. Scale economies and the possession of production and skill advantages have the same, although less statistically significant, effect as technological intensity. Finally, capital intensity does not indicate a statistically significant influence on either exports or foreign production.

THE ROLE OF FINANCIAL FACTORS IN INTERNATIONAL PRODUCTION: ALIBER

Unlike other economists who focused on the importance of industry- and firm-specific ownership advantages, Aliber (1970, 1971) focused on the importance of country-specific advantages and specifically on those ownership advantages that accrue to firms located in a particular currency area. Aliber departs from the analysis of the theories of international production derived from the theory of industrial organisation by indicating that the post-Hymer theories of industrial organisation, through their emphasis on the importance of real factors over financial factors in explaining direct versus portfolio investments, have limited power in explaining the advantages of source country firms; the geographical and industrial pattern of foreign investment; foreign investments that result from take-overs; other forms of foreign economic involvement such as exports and licensing that source country firms might exploit in the host country; and the factors that distinguish national economies.

Following on from these perceived limitations of the industrial organisation theory of international production, Aliber proposes that the fundamental factor in explaining the pattern of foreign direct investment which helps to finance international production centres on financial factors such as capital market relationships, exchange risks and the preferences of the market for holding assets denominated in selected currencies. In general, Aliber explains that the pattern of FDI is determined by the size of the host country market; the value of patents; the height of tariffs; the costs of doing business abroad in a particular industry; and the dispersion of capitalisation rates by country and industry. As a result of these determinants, the prediction is that foreign investment will predominate in the more capital- and research-intensive industries.

Aliber's assertion of the dominance of financial over real factors suggests that there is no need for a separate theory of FDI and the MNE *vis-à-vis* portfolio investment. He argues that the earlier general theory of international capital movements, in which currency valuation and relative interest rates are important determinants of foreign investment flows, can be restored with only minor qualifications. He dismisses the factors related to the growth of firms and industries which provide a more important explanation of the general trend towards internationalisation of business characterised in part by intra-industry production and changing sectoral composition of international production over time.

In contrast to the theories of industrial organisation which deal with the industrial pattern of international production, Aliber's theory is constrained to a particular geographical pattern of investment and in particular the impact of currency valuation on the flow of capital movements. While his theory may explain the capital movements into Europe by US firms and individuals in the 1960s with the overvaluation of the US dollar, such a theory weakened even on a macroeconomic level in the 1980s when the overvaluation of the dollar actually led to European and Japanese investment in the USA rather than the reverse. The growth of inward direct investment into the USA was sustained both in periods of US dollar strength in the period 1982–5 and in periods of US dollar weakness in the early and late 1980s. However, the more accentuated growth during periods of US dollar weakness in the period 1979–82 and since 1986 leads to the conclusion that trends in the international currency markets may affect the timing but not the long-term trend of FDI (Dunning, 1988b).

Despite their limitations, the role of financial factors may be incorporated in some theories of international production and in particular the product cycle model. The model deals with the process by which technologically innovative firms become very significant exporters in the first stage of foreign market servicing and then become outward direct investors at a later stage when locational advantages helped by a strong local currency favour international production. The presence of a strong local currency as a result of a strong export position and overall balance of payments surplus may provide the impetus for the growth of outward direct investment by locally based innovative firms (Cantwell, 1991b). This process has effectively characterised the growth of UK firms in the nineteenth century, US MNEs in the early post-war period and, more recently, of firms from Japan and Germany (Cantwell, 1989b). This growth process also holds true for some Third World firms based in the Asian NICs in the period since the mid-1980s.

THE THEORIES OF INTERNALISATION

Internalisation advantages were first incorporated in the analysis of international production during the 1970s with the growth of FDI in industries characterised as having a high technological intensity whose competitiveness greatly depended on the integration of R&D, production and marketing activities within the firm. In addition, the increased importance of intra-industry investment, the formation of customs unions both in Western Europe and North America and the growth of rationalised investment which led to global integration of subsidiaries reinforced the significance of transaction power of the firm as against asset power. Hence, an alternative theory of FDI, apart from the Hymer-Kindleberger-Caves industrial organisation approach, developed. Its main premise is that MNEs use hierarchical structures as a substitute for inefficient market systems. The reorganisation of firm networks helps to generate a shift towards internalisation factors. However, the theories of the firm based on internalisation are similar to the conventional industrial organisation theory based on a structure–conduct–performance model to the extent that the firm is seen as a passive reactor to markets.

The new trends towards the importance of internalisation factors to the growth of the MNE enabled McManus (1972) and Buckley and Casson (1976) to formulate their theories of international production based on an extension of the Coase (1937) view of the firm. Similarly, the emphasis placed by Williamson (1975) on the transaction cost theory was constructed on the basis of foundations laid down by Commons (1934) and Coase (1937), and was applied and extended to an international context by Teece (1981a, 1981b, 1982, 1983, 1985). Williamson (1975) is regarded as the father of modern transaction cost economics (Rugman, 1985a). The emphasis of these economists on the impact of transaction costs on the growth of the MNE led Dunning to refer to the presence of transactional market imperfections. A review of some of the theories of the MNE based on internalisation within a framework of transaction costs and exchange is in order.

The main thesis of the markets and hierarchies approach of Williamson (1975) is that markets and firms are alternative routes for completing a related set of transactions. The particular modality chosen depends upon relative efficiency. The relative inefficiency of markets over the hierarchical organisation fundamentally stems from the presence of transaction costs. In the absence of transaction costs, the form of economic organisation is regarded as irrelevant since any advantage a firm holds over another is effectively eliminated by costless contracting (Williamson, 1979). He has made no attempt to extend his markets and hierarchies approach to the MNE, although some indications in this direction were pointed out (Williamson, 1981).

Another distinct application of the Coasian theory of the firm is given by McManus (1972) who developed a theory of FDI based on the existence of economic interdependence among certain activities conducted in various countries. The international firm and the price mechanisms, perhaps supplemented by contractual arrangements, are seen as the methods by which these interdependent activities in different countries are co-ordinated. In general, the interdependent producers will choose to centralise control if the international firm yields the highest level of efficiency for a given cost of co-ordinating their joint activity. The market is considered an expensive way to effect some forms of exchange or to constrain some forms of interdependence because of high transaction costs. The principal conclusion of the theory is that the international firm is chosen as the means of organising production internationally in those industries in which both the interdependence among producers in various countries and the costs of co-ordinating their activities are relatively high.

The long-run theory of the MNE formulated by Buckley and Casson (1976) represents another extension of the Kaldor (1934), Coase (1937) and Penrose (1959) analyses of the firm. The basic premise of the long-run theory is that the growth of the MNE is one aspect of a radical change in business organisation which results from very general forms of imperfection in intermediate product markets and the costs of organising markets. The integration of R&D, production and marketing activities and the internalisation of knowledge form the basis for a simple theory of the growth of the MNE. The long-run theory is based on the premise that firms maximise profit in a world of imperfect markets by bringing under common ownership and control activities which are linked by the market. Such internalisation of markets across national boundaries generates MNEs.

The markets where internalisation is likely to be advantageous are perishable agricultural products, intermediate products in capital-intensive manufacturing processes and raw materials whose deposits are geographically concentrated. For example, Buckley and Casson explain the pattern of growth of the MNEs in resource-based and agricultural industries as resulting from the internalisation of intermediate product markets in multi-stage production processes. In addition, the growth of the more modern MNE in the post-war period and the growth of cross-investments or intra-industry production among developed countries, as well as research-intensive MNEs, is regarded as resulting from the internalisation of the markets in knowledge, resulting in the increased potential profitability of R&D activities. The long-run theory of the MNE has been further extended and clarified by Casson (1979, 1986). Consistent with the predications of the long-run theory, Casson (1979) suggests that internalisation will predominate in industries that rely heavily on proprietary information and in industries that operate multi-stage processes under increasing returns to scale or with capital-intensive techniques.

A possible limitation of the internalisation theory of Buckley and Casson stems from their narrow emphasis on economies of scale in production which are also important determinants of the gains of internalisation (Kojima, 1989). In any case, as will be shown later on, the use of a framework of transaction and exchange in explaining the pattern of growth of MNEs is a limited view that concentrates only on the impact of markets on the growth of the firm. This view has led Dunning (1983a, 1983b, 1988a) and Teece (1983) to refer to the presence of transaction ownership advantages (Ot) of MNEs that arise specifically from the capacity of MNE hierarchies *vis-à-vis* external markets to recoup the transactional benefits or lessen the transaction costs of the common governance of separate but interrelated activities located in different countries. However, in addition to transaction ownership advantages there are also asset ownership advantages (Oa) analysed in the framework of growth of production, which are also important aspects of the growth of firms. These ownership advantages of firms arise from the exclusive or privileged possession of individual assets. In the technological accumulation view, such asset ownership advantages are manifested in terms of gradual accumulation of technology and experience in production as well as methods of organising production.

Rugman (1980, 1981a, 1982) stressed the importance of internalisation as a general theory of FDI and a unifying paradigm for the theory of the MNE. Broadly similar to the views of Buckley and Casson, Rugman's theory of internalisation is taken to mean the creation of an internal market within the firm that functions as efficiently as a potential but unrealised external market because of internal transfer prices. The MNE is considered as a modality for retaining firm-specific advantages in knowledge on a worldwide scale since transaction costs in non-standardised innovative technology may be potentially high in licensing or joint venture arrangements. This theory of internalisation is regarded by Rugman as a theory of centralisation in decision-making following the hierarchical and centralised analysis of the firm by Williamson (1975).

The major limitation of Rugman's theory of internalisation stems from the restrictive interpretation of the internalisation of a market. The creation of internal markets leads to a centralisation in decision-making only in special circumstances as in the case of highly synergistic systems in which the decentralisation of units is difficult or impossible to achieve (Kay, 1983). When ownership is unified, the creation of internal markets requires decentralised control using transfer prices or other flexible budgetary control within the unit (Parry, 1985). Alternatively, the operation of internal markets involves decentralised profit centres transmitting shadow price signals to other decision makers within the organisation (Buckley, 1983). Unlike Williamson, who stresses the centralisation of decision-making as the determinant of internalisation, Rugman indirectly stresses decentralisation of decision-making, in tandem with McManus (1972) and Buckley and Casson (1976).

A deviation from the pure transaction cost features of internalisation theory provided by Coase, McManus, Williamson, Teece and Rugman is the *appropriability theory* of Magee (1977a, 1977b, 1981). The theory stems from the industrial organisation theory of international production developed by Arrow (1962), Demsetz (1969), Johnson (1970), and Caves (1971) on the creation and appropriability of the private returns from investment in information. The appropriability theory can therefore be considered as belonging to the framework of transaction and exchange although some attempts have been made to introduce dynamic elements with the concept of the industry technology cycle.

The analysis of appropriability is based on the static neoclassical premise that the creation of new information is considered to have the nature of a public good in that once it is created, the use of that information by the creator does not preclude its use by second parties (Johnson, 1970). The appropriability problem arises when second parties reduce the private return on privately created information by the first party and firms face difficulties in protecting their proprietary rights to information and in securing rents on know-how over time (Arrow, 1962). Appropriability is taken to be largely determined by the efficiency of the legal system and the industry structure. For example, the presence of oligopoly or monopoly *ceteris paribus* would promote investments in R&D because appropriability costs are low for these industry structures. Major innovations, in turn, promote an increase in optimum firm size leading to a more concentrated industry structure. The appropriability theory therefore suggests that greater efficiency results when high technology is transferred worldwide within the firm because the probability of the technology being copied or stolen by outsiders is less than if the technology is transferred through the market. Hence, MNEs specialise in skilled labour training and sophisticated technologies because there is greater difficulty associated with the imitation of these technologies, which result in higher private returns and therefore higher appropriability than simple technologies.

The major contribution of Magee is in introducing dynamic elements in the appropriability theory through the development of the *industry technology cycle*. Following the analysis of Williamson (1979), the main premise is that industry moves over time from less competitive to more competitive market structures and from less standardised to more standardised trade in intermediate goods as a result of diffusion and changes in product technology. Although the appropriability theory shows a close relationship to the internalisation approach to the MNE adopted by Coase, McManus, Buckley and Casson, Williamson and Teece, the theories of this latter set of economists show in more general terms the way in which MNEs respond to a variety of market imperfections. The main thesis of internalisation as applied to the MNE is the replacement of the external market by the internal organisational structure wherever and whenever such internal operations are more efficient than the external market or where appropriate markets do not exist.

However, where market exchange is characterised by monopolistic or monopsonistic elements, a firm or an MNE may attain both market and administrative co-ordination since the firm may exercise control without internalisation. The firm or the MNE becomes both a controller and coordinator of a network of production or income-generating assets (Cowling and Sugden, 1987). This occurs, for example, where the MNE is the monopsonistic buyer of the products whose international production is subcontracted to another firm in a foreign location.

As was pointed out earlier, the major limitation of modern theories of international production based on the internalisation of markets whose analysis is based on a neoclassical framework of exchange and the transaction costs of such exchange stems from their neglect of the role of ownership advantages and their limited emphasis on locational dispersion of activity as important influences on transaction costs and the pattern of MNE growth. Some implications of the process of technological change for transaction costs is evident in the work of Armour and Teece (1980).

These theories of the growth of the firm based on the internalisation of intermediate product markets that link firms, as opposed to trade between independent parties, provide a good analytical framework to explain the international expansion of firms through vertical integration. Perhaps the most comprehensive empirical application of the theory of the growth of the firm through managerial hierarchies is provided by Chandler (1980, 1986). The development of modern industrial firms in the USA, Europe and Japan emerged from administrative hierarchies. These firms have largely grown through forward and backward integration followed less often by investment in R&D. The development of the first modern industrial MNEs may be traced to these integrated industrial enterprises that initiated direct investment abroad in marketing in the first instance and then in production facilities and personnel.

Although internalisation may explain the growth of firms from national to multinational, the theory has limited explanatory power in analysing the continued and sustained growth of these multinational firms in international markets or the way in which they sustain a process of competition *vis-à-vis* other firms in world markets. The absence of dynamic elements related to the growth of firms and industries has resulted in a static theory which explains the growth of the firm as a tautology. For example, Casson (1983) postulated that MNEs grow because transaction costs within the firm relative to the market fall and the reason transaction costs fall is attributed to improvement in the organisation of the firm. Teece has attempted to circumvent the tautological nature of transaction cost by showing that the public good characteristics of technology and transaction costs are not necessary in a framework of transaction and exchange (Teece 1976, 1977, 1981a, 1981b, 1982, 1983, 1985). He reformulated the transaction cost framework in a dynamic context by referring to the high costs of technology transfer as a result of the declining value of disseminated information.

The high governance costs of licensing brought about by opportunism, asset specificity, asymmetries in information and other contingencies encourage the MNE to substitute internal market transactions for licensing agreements (or exporting, which is ruled out by locational cost factors) (Teece,

1985). The markets that are internalised by the MNE depend on whether the investment abroad is horizontal or vertical (Teece, 1981a, 1981b, 1983). A horizontally integrated MNE internalises markets for proprietary and non-proprietary know-how such as technology, managerial know-how and goodwill, while a vertically integrated MNE internalises intermediate product markets.

The difficulty of transferring know-how associated with horizontal FDI through market transactions is the major drive for internalisation. In the face of this, Teece concludes that intra-firm transfer to a foreign subsidiary has advantages over autonomous trading in terms of better disclosure, easier agreement, better governance and more efficient transfer. On the other hand, the substitution of internal governance structure by vertically integrated MNEs circumvents the problems of strategic manoeuvring and costly haggling associated with bilateral dependence and therefore makes a substantial contribution to economic efficiency.

Such a view of internalisation is fundamental to the framework of growth of production in which the firm is taken to possess distinct technological characteristics which are difficult to transfer to and even more difficult to implement by another firm with different technological characteristics. The difficulty and high cost of a technological exchange between firms is seen in terms of the nature of technology as cumulative, firm-specific and differentiated.

LOCATIONAL THEORIES OF THE DIVISION OF LABOUR

Developments in international production in the period since 1960 have exerted a major influence on the emergence of a distinct theory of international production. The origins of location advantages of firms which have been based on the Hecksher-Ohlin-Stolper-Samuelson model of trade and international capital movements, which suggests that capital moves in response to interest rate differentials, do not prove acceptable in explaining intra-industry FDI. In general, neoclassical trade and location theories have focused solely on the locational explanations and have not considered the more important determinants of the growth of the firm based on ownership advantages. The importance attached to location factors in more recent theories of the MNE generally lies in two distinct but interrelated areas: first, location factors determine why and where MNEs produce in foreign markets; second, location factors interact with ownership and internalisation factors.

With regard to the importance of location factors in determining why and where MNEs produce in foreign markets, both early and recent location theories stress that market size and growth are the major locational factors. However, tariffs and other barriers to trade, transport costs, host government foreign investment policy, availability, quality and price of inputs, infrastructure, psychic proximity, stability, strength of indigenous competition, opportunities for economies of scale, etc., are also identified as important in the more recent location theories of the MNE (Svedberg, 1981).

Hymer's emphasis on ownership advantages as a necessary but not sufficient condition for FDI diverted his attention from the analysis of location advantages in the internationalisation of production, since the exploitation of ownership advantages in a foreign *vis-à-vis* domestic location was analysed in a market power context as opposed to a locational theory context. However, in his later work, Hymer suggested that firms locate activities in accordance with a hierarchy, with high grade activities in industrialised countries and low grade activities in developing countries (Hymer, 1972), Such a view of dependent development and the presence of a world hierarchical system has recent applications in the work of Newfarmer (ed., 1985) and Bornschier and Chase-Dunn (1985).

Aliber (1970, 1971) stresses that the market servicing decision, i.e. licensing, exports and FDI, fundamentally depends on several locational factors. First, the relative costs of production in source and host countries; second, transport costs and tariffs; third, the costs of doing business abroad, an extra cost relevant to the FDI modality; fourth, the different rates of capitalisation of income streams depending on the currency of denomination; and fifth, market size and market growth in the host country.

The theories of internalisation have analysed the growth of the firm in terms of intermediate product markets that link firms and have taken ownership and location advantages of firms as given and exogenous. However, some locational implications of the internalisation of intermediate product markets are evident in the work of Buckley and Casson (1976), Casson (1979, 1986), Casson *et al.* (1986) and Buckley (1988). Casson *et al.*, in particular, analysed some implications of internalisation for vertical integration and the international division of labour. In their work, an MNE is created whenever intermediate product markets are internalised across national boundaries. A vertically integrated firm becomes multinational when the location strategy of the firm, which is determined primarily by the interplay of comparative advantage, barriers to trade and regional incentives to internalise, makes the location of different stages of production in different countries optimal. On the other hand, the optimal location strategy of a firm with integrated R&D, production and marketing activities would be to service the world market through horizontal integration, i.e. multi-plant production, taking account of the impact of trade barriers such as tariffs and transport costs (Casson, 1979).

Locational factors therefore influence the incentive to internalise intermediate product markets among horizontally and vertically integrated MNEs. The prediction is that internalisation would predominate in markets where there are greater opportunities for transfer pricing; where there are higher tariffs; where there are large tax differentials between fiscal areas; and where there are severe restrictions on capital movements between currency areas. Conversely, internalisation would be least in markets where there are significant differences in social structure, culture and language, and in politically hostile countries (Casson, 1986; Buckley, 1988).

The *rational global planning model* of Robock and Simonds (1983) represents another attempt to analyse the relevance of internalisation on location. The basic premise of the model is that firms develop logistic world-scale models that represent rational patterns for supplying selected markets on the basis of market objectives and location economics. These logistic models incorporate sources of raw materials such as labour, management and capital and production sites, service, marketing and research facilities and form the basis for firms' optimal business strategies. The global planning approach assumes the same relevance for expansion in domestic and international markets although additional variables and risks result from international expansion, namely currency risks, tariffs, local content requirements, etc. The optimal business strategy for international expansion covers a wide range of options: traditional trade, licensing, loan-purchase agreements for resource seekers, establishment of foreign procurement or subcontracting offices, marketing and purchasing facilities, assembly and re-packing operations and full-scale foreign production, either through acquisition or green-field investment. The precise mode chosen depends on the internalisation advantages. The absence of imperfections in intermediate product and factor markets would enable the international trade route to be most feasible; otherwise, location factors modified by government intervention determine the location of production for either horizontal or vertical integration. However, the internalisation strategy adopted may be such that initial exporting would lead to FDI at some later stage.

The theory of internalisation of Rugman (1981a) similarly stresses that the location of production follows the dictates of relative costs and therefore the locational strategy of the firm is largely determined by the interaction of comparative advantage and barriers to trade. Several other models designed to predict alternative modes of market servicing were presented by Hirsch (1976), Weinblatt and Lipsey (1980), Grosse (1985) and Cantwell (1989a), but apart from Cantwell all these models are comparatively static. Cantwell's findings show that the locational choice between exports and foreign production is dependent on the stage of international growth or maturity of firms. The sectoral correlation of exports and foreign production is higher for newer international investors such as Germany, Japan and the NICs than for established investors such as the USA and the UK. As a result, the export versus foreign production issue becomes a far more important location decision for more established and mature investors that already have a network of overseas subsidiaries than for countries that are at an early stage of industrial expansion and whose firms have strong technological advantages.

In any case, models of the choice of the modality of foreign market servicing, even when placed in a dynamic context, are to a large extent outdated when considering the increasing significance of global competition since the late 1970s (see also Vahlne and Nordstrom, 1988, and Johanson and Matson,

1988). Nevertheless, locational factors are still fundamental to the growth of globally integrated MNEs which are engaged in rationalised (efficiency seeking) investment. However, these firms are more concerned with issues pertaining to the location of particular products and processes across their international networks, which reflect different preferred locations for different activities, than with the modality of foreign market servicing. These firms derive competitive advantages more from the integration of their activities on a worldwide basis through investments in worldwide distribution networks and global brands and their ability to cross-subsidise across product segments and national borders. These competitive advantages are based less on the location of their activities due to country differences in factor costs or extra-national scale economies (Hamel and Prahalad, 1988). The opportunities for reducing risks through cross-border arbitrage and the enhanced leverage produced by global co-ordination of activities provide the MNE with unique economic and rent-earning capacities (Kogut, 1984).

Apart from the interrelation between internalisation and location factors, recent theories of international production have focused on the interrelation between ownership and location factors. The analyses of Rugman (1975, 1976, 1977, 1979, 1980, 1981a), Lessard (1976, 1977, 1979), Agmon and Lessard (1977), Kogut (1983) and Doz (1986) on the generation of the transaction ownership advantages have shown the importance of several locational factors, namely the ability to exploit differences in capital costs, exchange rates and government policies that determine factor costs, or the specific costs of doing business for the firm in the host country; the ability to integrate and specialise production across national boundaries in order to take advantage of international differences in factor endowments and costs; the ability to exploit differences in national tax rates and structures through transfer pricing; the ability to reduce risks and achieve greater stability through the international diversification of factor and product markets which are not perfectly correlated; and the ability to gain more favoured access to and/or better knowledge about information, inputs and markets. However, against these transactional benefits are set the costs of greater communication and administration, greater exposure to risk of loss of fixed assets and greater possibility of national discrimination.

Other studies show that location exerts an important influence on firms' international competitiveness at all stages of technological development (Hirsch, 1967; Cantwell, 1989a). These and other studies explain the growth and increasing complexity of international production in terms of developmental factors are further analysed in the macroeconomic development theories of international production.

MACROECONOMIC DEVELOPMENT THEORIES OF INTERNATIONAL PRODUCTION

The macroeconomic developmental theories of international production describe the dynamic and developmental process or the way in which stages of development or maturity of countries and firms affect their international production activities. The major strands in macroeconomic development theories are Vernon's product cycle model, Kojima's and Ozawa's models of direct investment, Dunning's investment development cycle and the stages of development concept elaborated below.

The product cycle model

An innovation theory that links firms' ownership advantages to countries' locational advantages is exemplified in Vernon's (1966) product cycle model (PCM). While the full elaboration of the PCM is discussed with the theories of Third World MNEs in the next chapter, it is worth noting the context in which the model belongs with other macroeconomic developmental theories of international production. The model emerged as a result of the limitations provided by the traditional theory of international trade in explaining the growth of trade and international production between the US and Europe, whose factor endowments are similar. In particular, the model sought to formulate an effective theoretical tool for explaining the composition of US trade and the existence of a persistent trade surplus in the 1950s, and also for predicting the likely patterns of foreign direct investment by US firms in Europe in the 1960s. The major contribution of the PCM to the theory of FDI is in analysing trade and investment as part of the dynamic process of exploiting foreign markets. In particular, the model emphasised the role of market demand and relative factor prices in the extent and form of innovation and product development and as important determinants of the location of production. The initial version of the PCM, which was used to explain patterns of international trade and balance of payments, provided a useful theoretical tool for analysing firms' initial foreign economic involvement in final products.

The theory of direct investment: the contributions of Kojima and Ozawa

Kojima's integrated theory of international trade and direct investment also analyses the interaction between ownership advantages and the changing location of production. The theory is based on the Hecksher-Ohlin principle of comparative advantage (or costs) (Kojima 1973, 1975, 1978, 1982, 1990). Kojima's basic theorem is that foreign direct investment should complement comparative advantage patterns in different countries. Such investment must therefore originate from the comparatively disadvantaged (or marginal) industry of the source country which leads to lower-cost and expanded volume of exports from the host country. This type of FDI is referred to as *pro-trade, Japanese-type FDI*. The non-equity forms of Japanese resource-based investment in resource-rich countries are regarded as trade-oriented investment because of the assurance of a supply quota or production sharing arrangements with indigenous enterprises in the host countries.

Such investments contrast with the wholly owned vertically integrated resource-based production activities of firms from the USA and major European countries which originate from the comparatively advantaged industries of the source country and lead to misallocation of resources and a decreased volume of exports from the host country. This type of FDI is referred to as *anti-trade, American-type FDI*. The oligopolistic and technologically advanced US firms that engage in FDI are seen to be motivated by the defence of their oligopolistic position, the exploitation of factor markets and the presence of tariff barriers in the developed countries.

An important criticism of Kojima's theory is the way in which import-substituting investments are referred to as anti-trade oriented. While import-substituting investments may be considered anti-trade oriented at the microeconomic level, they are not anti-trade oriented at the macroeconomic level unless very restrictive assumptions are introduced in the model. The growth of FDI from the USA, Germany and Japan is often accompanied by an increasing level of exports. There is also evidence to suggest that export-oriented investments may play a less significant role in industrial adjustment or in increasing the welfare of the host country since these investments are likely to be an enclave kind (Dunning and Cantwell, 1990).

Kojima's theory is essentially industry based as opposed to country based, as he claims, in that Japanese and American MNEs were concentrated in different industries which helps to explain why their overseas production was aimed principally at serving either export or local markets. The extent to which the theory is country based derives essentially from the general way in which the stage of national development helps to explain the industrial structure of indigenous firms. Although further support for Kojima's macroeconomic approach to FDI was provided by Ozawa (1971, 1977, 1979a, 1979b, 1985), he acknowledges in his later work that the distinctive characteristics of Japanese FDI from 1950 to the early 1970s in resource-based and labour-intensive or technologically standardised industries in developing countries belonged to the first two phases of Japanese overseas activity. Since the late 1960s, as a result of industrial restructuring in Japan, overseas investment has evolved to the third, import-substituting or recycling, phase of international production. This is centred on the mass production of assembly-based, consumer durables in high-income countries such as the USA and Europe, supported by a network of subcontractors. Moreover, since the early 1980s, a fourth phase of Japanese FDI can be discerned. This is characterised as the flexible manufacturing of highly differentiated goods, involving the application of computer-aided designing (CAD), computer-aided engineering (CAE) and computer-aided manufacturing (CAM) (Ozawa, 1990).

The growth of Japanese overseas investment has therefore followed an evolutionary course from resource-based and simple manufacturing towards more technologically sophisticated forms of outward investment. The difference from the evolutionary path taken by US overseas expansion lies essentially in the swiftness with which Japanese overseas activities have made the transition. Hence, although US and Japanese firms and industries are at a different stage of evolution, Japanese firms have demonstrated their ability to catch up quickly and keep pace with advances in technological development in the West.

Kojima (1990) himself acknowledges that recent US FDI in Japan is largely akin to a Japanese type FDI while Japanese FDI in the USA is like the American-type FDI identified previously. Such a rapid pace in the evolution of Japanese direct investment may be attributed in part to the efforts of the Japanese government systematically to change the composition of the country's comparative advantage, aided in part by licensed foreign technology (Ozawa, 1974). The view of pro-trade Japanese FDI now appears to be a reflection of the early stages of development of Japanese multinationals. The history of Japan's post-war trade development does not lend support to this form of application of the static theory of comparative advantage which thirty years or so ago would have suggested that Japan should concentrate on the production of labour-intensive goods and import capital- and technology-intensive goods. Indeed, the increasing technological competitiveness and trade surplus of Japan in technologically intensive products provide support for the development of future-oriented technologies, so that the sectors in which a country enjoys the greatest potential for innovation and in which investment may be most beneficial are not necessarily those in which it currently has a comparative advantage (Blumenthal and Teubal, 1975; Pasinett, 1981). Major new investments by Japanese companies in the electronics components industry and the diversification into new product lines of those in the consumer electronics sector show that Japan is continually seeking to build an industrial structure based on sectors in which the country currently has no comparative advantage (Dunning, 1986b). As a necessary extension of the argument, Kojima and Ozawa (1985) argue that global welfare is increased where international production helps to restructure industries in line with dynamic comparative advantage.

Although most Third World MNEs are unlikely to develop as fast as their Japanese counterparts, who can today draw on frontier technologies, the trend is towards a persistent upgrading of activities for which these firms are responsible. The phases of development suggested by Ozawa for Japanese MNEs are thus relevant to the study of Third World multinationals. The first two phases of Japanese outward investment from the 1950s to the early 1970s in resource-based and labour-intensive manufacturing activities in developing countries are relevant in explaining the early stages of the growth of outward investment from the newer investors based in the developing countries. In the early stages, as domestic firms in the developing countries upgrade their domestic production activities, resource-based and less technologically sophisticated manufacturing activities are transferred to countries at an earlier stage of development with lower levels of technological capacities and production costs. In the later stages, more sophisticated manufacturing activities are transferred abroad, even to the developed countries. The present overseas investment activities of the newly industrialising countries (NICs) in the developed countries, although based on a less research-intensive form of technological innovation than the Japanese MNEs, can nevertheless be described as being in the early phases of the import-substituting or surplus-recycling stage of international production.

The underlying developmental processes that underlie the Kojima-Ozawa theory may be explained by Vernon's product cycle model and Dunning's investment development path. The underlying issues of the product cycle model of Vernon and Hirsch in particular and Kojima's integrated theory of international trade and investment are very similar despite the different theories of trade upon which these two models are based (Mason, 1980). Kojima's and Ozawa's more general comparative advantage-based model presents a useful explanatory framework within which to view the growth of international production of countries undergoing rapid growth, such as Japan, Germany and the newly industrialising countries. Their models explain the process of relocation of production of mature or technologically standardised industries as locational advantages favour countries at an earlier stage of development. Such relocation of production is undertaken while the domestic firms still have the technological and organisational advantages associated with lower technology and more labour-intensive production activities which can be more profitably exploited in a foreign environment with lower levels of technological capacities and production costs. The similarity in the concept between the two models is obscured by the misleading framework adopted by Kojima in his analysis.

The concept of the investment development path advanced by Dunning shows in more general terms the impact of the national stage of development on both the level and the character and composition of outward direct investment.

The concept of an investment development path

Early versions of the concept of an investment development path advanced by Dunning (1981a, 1981b, 1982, 1986a, 1986b, 1988a) are based on the proposition that the level of inward and outward investment of different countries, and the balance between the two, is a function of their stage of development as measured by gross national product (GNP) per capita. The concept is further elaborated in Chapter 5, but the main propositions are that the countries at the lowest stage of development have very little inward or outward investment, and consequently a level of net outward investment that is close to zero. Countries at somewhat higher levels of development attract significant amounts of inward investment, but as the outward investment of their own firms is still limited, their net outward investment (NOI) is negative. The argument then is that, past some threshold stage of development, outward investment increases for countries at yet higher levels of development. The balance between inward and outward investment in developed countries results in the return of their net outward investment to zero. The continued growth of their outward investment at a later stage results in a positive net outward investment.

Empirical evidence provided in this book for the period since the mid-1970s suggests the existence of a structural change in the relationship between NOI and the country's relative stage of development as a result of the general rise in the internationalisation of firms from countries at lower stages of development. The growth of newer multinationals from Japan, Germany and smaller developed countries, as well as some of the richer developing countries, suggests their firms' capacity to follow the earlier outward multinational expansion of the traditional source countries, the USA and the UK, at a much earlier stage of their national development. The increased significance of outward investments from these newer source countries provides first-hand evidence of the general trend towards internationalisation so that the national stage of development no longer becomes a good predictor of a country's overall net outward investment position.

More recent versions of the investment development cycle have argued that, apart from the level of net outward investment, the character and composition of such investment vary with the national stage of development (Dunning, 1986a, c). The early forms of foreign investment are frequently resource based or sometimes import substituting, but in each case have a quite specific location associated with a particular type of activity. However, as firms mature their outward investments evolve from a single activity or product in a particular location and they adopt a more international perspective on the location of their different types of production activities. Such MNEs are responsible for directly organising an international division of labour within the firm in place of markets. At this stage, the character of their international production activities and their ownership advantages are also determined by firm-specific in addition to country-specific factors. However, home country-specific factors still dominate over firm-specific ones in determining national technological performance (Patel and Pavitt, 1991).

The concept of stages of development in international production

This book will elaborate on the proposition that the character and composition of outward direct investment change as development proceeds. Countries' outward direct investment generally follows a developmental or evolutionary course over time which is initially predominant in resource-based or simple forms of manufacturing production which embody limited technological requirements in the earlier stages of development and then evolve towards more technologically sophisticated forms of manufacturing investments. The developmental course of the most recent outward investors from the Third World has been faster and has a distinctive technological nature compared to the more mature multinationals from Europe, the USA and Japan, owing to the different stages of their national development. Such changing sectoral composition is associated with changing geographical composition. The simpler forms of investment in resource-based activity are frequently undertaken in resource-rich developing countries while more complex research-intensive investments are undertaken in industrialised countries.

Countries' changing sectoral and geographical composition is explained by the theory of technological accumulation in terms of a firm's ability to develop more sophisticated forms of technological ownership advantages. In the early stages of technological development, advantages tend to be more country specific and associated in part with the abundance or scarcity of natural resources in the home country. In either case, firms gain accumulated expertise in the exploitation and processing of natural resources present in the home or host country. The early forms of technological ownership advantages of US and British firms in wood processing and metal and coal processing respectively were acquired because of the abundant availability of timber and coal in the USA and the UK (Rosenberg, 1976). This book hopes to show that the general trend also holds true in the early stages of technological development of Third World MNEs.

As home countries advance through progressively higher stages of industrial development and as firms accumulate greater experience in international investment, the technological embodiment of their outward direct investment activity becomes more significant and more complex. This is seen in part in research-intensive investments in centres of innovation in the developed countries through which firms gain access to more advanced, complementary forms of foreign technology which can be adapted and integrated with their indigenously generated technology.

THE THEORIES OF INTERNATIONAL PRODUCTION BASED ON OLIGOPOLISTIC COMPETITION

Oligopolistic theories of international production provide better explanations of the growth of globally integrated MNEs and the process of competition by which these firms maintain stability in international industries. These theories have taken into account the more balanced technological competition occurring among MNEs in the USA, Europe and Japan in contrast to the period during the 1950s and early 1960s when the USA fulfilled a technological hegemonic role. The presence of oligopolistic market structures associated with the progressive internationalisation of industries in the 1970s led to subsequent modification of the basic theory of the PCM which was elaborated comprehensively in 1966. The limitations of the basic theory mainly centre on the failure of the early model to explain firms' increasing global strategy; their ability to develop, mature and standardise products almost simultaneously; the foreign economic involvement of established MNEs; and the diminished dominance of the USA as a source country for FDI. Subsequent work by

Vernon (1971, 1974, 1977, 1979) led to the Mark II version of the product cycle model which represents one of the earliest analyses of the oligopolistic nature of international production and, in particular, the impact of risk-minimising strategies on the relocation of production abroad. The PCM Mark II model, which allows for different kinds of barriers to entry in the different stages of the product cycle in order to overcome some of the limitations of the model's initial version, is further discussed in the next chapter.

The exchange-of-threats behaviour described in PCM Mark II is different from the follow-the-leader behaviour of oligopolistic firms although both forms have the objective of risk minimisation. Knickerbocker (1973, 1976) explains that the follow-the-leader behaviour of firms of one nationality, which undertake new product lines or penetrate new foreign markets in response to initial foreign investment by a leading firm of the same nationality, also generally occurs in oligopolistic industries. These forms of behaviour, as well as the imposition of industry-wide standards for enterprises bidding on new concessions, pricing conventions, the establishment of joint subsidiaries among the leader firms or, alternatively, long-term bulk purchase-and-sale contracts that amount to quasi-partnerships, are some of the strategies that can be adopted in an oligopoly that may lead to a declining rather than increasing concentration of investment and the maintenance of stability in the industry.

The concept of intra-industry production and the exchange-of-threat behaviour of firms in oligopolistic industries has similarly been used by Graham (1975, 1978, 1985). Graham conforms with Vernon in arguing that the growth of a firm entails greater oligopolistic interaction between firms especially since the increasing maturity of products leads to greater capital intensity and economies of scale in production. This results in increasing threats in the form of price-cutting strategies by rival firms and therefore considerations of risk minimisation and profit maximisation become important. However, the firm may have to accept a trade-off between risk minimisation and profit maximisation (Rothschild, 1947).

An important feature of PCM Mark II and Graham's theory is the emphasis placed on the global strategy of the firm in response to world-wide competition, but in these theories strategies are still determined by the maturity of the industry when technology creation is diffused abroad. Competition and intra-industry trade and production only occur at the mature oligopoly stage when there is a threat to international market equilibrium. In practice, however, MNEs compete in industries that are often in the forefront of technological innovation and that are sustained in part through intra-industry investment to gain access to complementary forms of foreign innovation which can be used to upgrade and extend indigenously generated innovation (Cantwell, 1987b).

Hymer's and Rowthorn's (1970) explanations of the rapid growth of European firms in the USA also embodied oligopolistic considerations based on the maintenance of security, i.e. a response to earlier US expansion in Europe. In their view, leading firms in each industry seek to have a similar geographical distribution of sales or production through the formation of mergers which increase the financial strength required for international production.

Related ideas of oligopolistic interaction are seen in the model of Sanna Randaccio (1980). This stems from the limitations in the explanations offered by the product life cycle model implicit in the work of Graham (1975, 1978) and Franko (1976), and Hymer's and Rowthorn's (1970) analysis of firm behaviour, in accounting for the increase in the share of total European direct investment in US manufacturing since the mid-1960s as well as changes in the pattern of such expansion since the 1960s. The model explains how, at a particular stage in the development of the new oligopolistic firm, the costs of growth associated with continued expansion in the domestic market increase as a direct result of the firm's increased market share and the increased industrial concentration. The search for security may make the internationalisation alternative more important, especially when international and national competitors exist. The particular strategy adopted in international expansion is based on Hymer's theory of the growth of the firm through collusion and take-overs. The particular method of international expansion chosen is dependent on the stage of firm development.

The initial international expansion of the new oligopolistic firm in a foreign country at a similar stage of development and pattern of demand is associated with the establishment of cartels or the formation of collusive arrangements with major rival firms. However, as the firm matures and becomes an established oligopolist, its international expansion is associated with investments in the home market of major rivals. At this stage, the greater number of large independent producers in all the major markets and the highly diversified nature of established oligopolistic firms with considerable international experience make the probability of collusive arrangements less likely. The mode of international expansion is frequently undertaken through the acquisition of a small firm as this represents the lowest cost of growth in the new market with fewer risks of provoking damaging competitive warfare. Higher costs of growth are associated with the acquisition of a firm with a larger market share and industrial concentration.

The increased share of the firm in the world market and the increasing cost advantage relative to other established oligopolists in the industry prompts the latter to increase their total volume of sales in the home market of the firm. These intra-industry investments, which represent a mutual penetration of home markets of rival firms, are also undertaken through the acquisition of small indigenous firms as this is associated with the lowest cost of growth and minimum disruption of the market equilibrium. In the later stages, the profitability of continued expansion of established firms in rival markets declines as acquisition becomes more difficult and the costs of promotion increase.

Recent work on competitive theories of international production has focused on explaining its growth in terms of the process of technological as opposed to price competition between rival firms in an oligopolistic industry. For example, the theory of technological accumulation explains the increasing internationalisation of manufacturing production in terms of the process of technological accumulation organised within the internationally integrated networks of the MNE, which helps to sustain technological competition between MNEs. Such a process arises because internationalisation has helped sustain the growth of complementary or interrelated technologies which are in competition with one another (Pavitt, 1988; Cantwell, 1991b). Similar themes of firms sustaining the process of competition through innovation are evident in Jenkins' (1984, 1987) concept of internationalisation of capital. In his view, MNEs functioning in an integrated world economy described by increasing standardisation of products and process in each industry aim to preserve their competitive advantages through the continual process of differentiation of their products and technology.

These competitive theories of international production and technological competition are helpful in explaining the expansion of international production in the industrialised countries up to the early 1970s, when the process of technological accumulation and competition provided the strongest driving force. However, the slow down in the growth of markets, innovation and productivity in the period since the mid-1970s has meant that intra-industry trade and production are increasingly made in response to the global strategy of MNEs characterised by the rationalisation of international production and not to threats of instability made by rivals. MNEs have progressed from being suppliers of technology and finance for scattered international production to becoming global organisers of economic systems, including systems for allied technological development in different parts of the world (Cantwell, 1987a, 1991b).

THE ECLECTIC PARADIGM OF INTERNATIONAL PRODUCTION

The eclectic paradigm of international production is an attempt to synthesise the different theories of international production in a general framework of analysis consistent with both the neoclassical and dynamic theories of international investment. It is an organising framework in which competing theories of international production can be analysed both from the viewpoint of MNEs as a response to imperfect markets and as a particular feature of the growth of firms and the change in ownership advantages over time. The organising framework brought to bear by the eclectic paradigm therefore helps to identify, relate and compare different strands in the theories of international production.

The foundation of the eclectic paradigm is provided by three sets of conditions which determine the extent, form and pattern of international production. First, firms must possess net ownership advantages *vis-à-vis* other firms in serving particular markets. There are three kinds of ownership-specific advantages identified: those that stem from the ownership of proprietary or intangible assets and that need not arise due to multinationality; those that branch plants of established enterprises have over *de novo* enterprises producing in the same location; and those that specifically arise because of multinationality, which is an extension of the first two (Dunning, 1981a).

An analysis of the factors that generate and sustain ownership advantages must depend on firm-specific, industry-specific and country-specific characteristics. In general, the greater these ownership advantages and/or the lower the barriers to entry into a foreign market, the greater the likelihood that foreign affiliates of MNEs will supply a large share of a foreign market. A distinction has recently been made between asset (Oa) advantages, which arise from the exclusive or privileged possession of individual assets by MNEs *vis-à-vis* other enterprises, and transaction (Ot) advantages, which reflect the ability of the MNE hierarchy *vis-à-vis* external markets to reclaim the transactional benefits (or reduce the transactional costs) of the common governance of separate but interrelated activities located in different countries (Dunning, 1983a, 1983b; Teece, 1983). These two types of ownership advantages are regarded as a function of structural and transactional market imperfections respectively (Dunning and Rugman, 1985).

Second, firms that possess ownership advantages must find it in their best interests to utilise these advantages themselves, i.e. to internalise the ownership advantages within their own organisations through an extension of their activities across national boundaries, rather than sell these advantages, or the rights to these advantages, to independent firms through licensing and similar contracts. Internalisation advantages may arise both from appropriability and co-ordination. Appropriability arises when the integrated firm is able to capture a full return on the exclusive or privileged ownership of individual assets such as technology, while co-ordination ensues when the firm is able to capture transactional benefits from the common governance of a network of complementary assets located in different countries (Dunning, 1988a).

This distinction between asset and transaction advantages as well as appropriability and co-ordination has taken into account the Coasian framework of transaction and exchange which focuses on markets as an integral explanation of the growth of the firm. Such a view is opposed to the classical framework of analysis which focuses on accumulation of technology and experience in production as an important determinant of such growth. However, although the distinctions remain important in internalisation theory, such divisions are increasingly blurred as explanations of the greater internationalisation of industries associated with the global nature of firms (Teece, 1990). These firms increasingly gain both ownership and internalisation advantages from their multinationality, i.e. from the control and co-ordination of a whole range of complementary assets located in different countries rather than from the possession of distinctive forms of ownership advantages.

Third, given that firms possess ownership and internalisation advantages, foreign direct investment will take place only when it becomes profitable to transfer intermediate products originating in the home country with at least some immobile factor endowments, or other intermediate products (including natural resources), outside a firm's home country, i.e. there must be locational advantages in producing in a foreign country. The absence of these locational advantages means that foreign markets are serviced entirely through exports and domestic markets are served by domestic production.

Although ownership, internalisation and locational advantages are considered separately, the major strength of the eclectic paradigm is in determining the interrelationship between these advantages. For example, locational advantages may be ownership specific, in which case they become exploitable only through international production. In addition, locational advantages such as tariff barriers, investment incentives, transfer costs, formation of free trade zones or regional trading blocs in several host countries may have important implications for internalisation advantages.

CONCLUSION

This chapter has attempted to analyse the major strands in the modern and distinct theories of international production from 1960. The development of a distinct theory of international production and the MNE from 1960 can be traced to the growth of market-oriented and import-substituting direct investments; the increasing importance of cross- and intra-industry investments; and the emergence of the globally integrated MNE during the post-war period. Prior to 1960, the theory of international production was taken within the context of the Ricardian theory of international trade, i.e. that international production is an extension of domestic economic activity and an instrument of colonial policies. Such a Ricardian view was reinforced because a majority of these investments, apart from those organised by individual entrepreneurs in the period prior to 1914, were resource based and directed towards the colonies.

The analysis of the major stands in the modern theory of international production from 1960, as distinct from the theory of international trade and capital movements, has shown the development of both competing and related theories that aim to explain the growth of the MNE under changing patterns of international production in the post-war period. The conceptual diversity and conflict, as well as interrelations in the development of the theory of international production from the theories of the growth of the firm, industrial organisation, trade, location and innovation, are indicative of the complex and dynamic nature of international production. The eclectic paradigm represents an attempt to construct a general framework within which to analyse the different theories.

4 The theories of Third World multinationals

INTRODUCTION

This chapter seeks to extend the overall theoretical framework developed in the previous one in order to compare and contrast the theories of Third World multinational enterprises. Two basic theories have been formulated to explain the growth of Third World MNEs: the *product cycle model (PCM)* originally formulated by Vernon (1966, 1971, 1974, 1979) and applied to Third World MNEs by Wells (1977, 1981, 1983, 1986) and the *theory of localised technological change* formulated by Lall (1981, 1982b, 1983a, 1983c). This chapter develops each of these theories and suggests a more general framework to explain the growth of Third World MNEs in the theory of technological accumulation and competence.

THE PRODUCT CYCLE MODEL (PCM)

The PCM was, at one time, referred to as the only dynamic theory of international trade and investment. The model, which emerged as a criticism of the traditional and neoclassical Hecksher-Ohlin-Samuelson theory of trade, was previously used to address microeconomic issues but was applied by Vernon (1966) to explain patterns of international trade and investments and balance of payments. The model emphasised the role of market demand and relative factor prices in the extent and form of innovation and product development and as important determinants of the location of production.

The PCM is a significant exemplification of the *Schumpeterian theory of economic development* in the area of international trade. Development, in the Schumpeterian tradition, is a spontaneous and discontinuous change in the channel of the circular flow which forever alters and displaces the previously existing equilibrium state. These spontaneous and discontinuous changes are manifested in the implementation of new combinations and cover the following five cases: first, the introduction of a new good or of a new quality of a good; second, the introduction of a new method of production; third, the opening of a new market; fourth, the conquest of a new source of supply of raw materials or half-manufactured goods; and fifth, the formation of a new organisation of any industry, as in the creation or dissolution of a monopoly position (Schumpeter, 1934).

The PCM analyses, in particular, the impact of discontinuous changes, brought about by innovation and the introduction of entirely new products, on international trade and the location of production as a new cost-determined equilibrium is achieved. The model then analyses the Schumpeterian process by which a disequilibrium position brought about by the innovative activity of a single entrepreneur or technologically leading firm is subsequently imitated by a competitive fringe of firms which catch up with the leading firm and, by increasing productivity growth and output, remove its monopoly rent in order to return to an equilibrium position. The innovative activity of firms and the resultant temporary monopoly that they enjoy increase the rate of growth of the firm. Writers in the modern Schumpeterian tradition, such as Freeman, Clark and Soete (1982), who analyse the swarm of imitator firms responsible for increasing productivity growth and output in the process of catching up with the technologically leading firms have not made much use of the product cycle model.

The Schumpeterian analysis of the product cycle model certainly has applications for explaining the major technological developments in the pre-1914 period when European firms were the technology leaders but competitive firms from the USA caught up rapidly with their technological pace and were subsequently the major sources of innovative activity. The model also has applications for explaining the second wave of major technological innovations that led to the creation of new industries in the aftermath of the Second World War. Firms from the USA maintained the leading position until 1960 when competitive firms from Europe and Japan began to catch up with them. However, these European and Japanese firms not only kept pace with the technological advancements of the leading US competitors but were able to engage in a process of competitive innovation which needs a more complex model of technological interaction to explain. The period of technological hegemony that existed prior to 1960, in which firms from one country set the pace for major advancements in innovation, has since given way to a period of technological convergence in which there is more balanced competition among many large firms from more than one country.

The chief proponents of the PCM of international trade and investment are the Harvard School economists. Professor Raymond Vernon, in particular, conceptualised and provided empirical support for the model in 1966, and it was further expounded and clarified in his later writings in 1971, 1974 and 1979. The model sought to formulate an effective theoretical tool to explain the composition of US trade and the existence of a persistent trade surplus in the 1950s and also to predict the likely patterns of foreign direct investment by US firms in Europe in the 1960s.

Knickerbocker (1973) employed a product cycle model both in identifying the capabilities that US manufacturing enterprises acquired, which prompted their expansion abroad, and as a critical determinant of the industrial structure in the process of international expansion. Similarly, Stobaugh (1968) and Wells (1972) applied the PCM in explaining US trade and international investment. However, empirical support for the model has extended beyond the analysis of the patterns of US direct investment. Graham (1975, 1978) and Franko (1976) have examined the significance of the PCM and other theories relevant to an understanding of the development of foreign manufacturing operations in Europe by firms from the USA compared to both the early and recent growth in the USA of some of the largest industrial firms from the western part of Continental Europe. In particular, Graham (1978), Flowers (1976) and Hymer and Rowthorn (1970) have extended the product cycle model to consider the concept of rivalry between firms from different countries.

Leontief (1954), Johnson (1958), Linder (1961), Posner (1961), Hufbauer (1965, 1970) and Douglass (1966) have contributed valuable related insights to the emergence and development of the PCM. Leontief, Johnson and Posner emphasised the important role of technological factors in explaining the pattern of US trade. In particular, Leontief signified the embodiment of higher skills in the pattern of US export products while Johnson referred to the presence of a slower rate of innovation in Europe compared to that in the USA in explaining the existence of a persistent dollar shortage in Europe. The presence of a technological gap as an important factor in explaining patterns of trade became even more apparent in the early 1960s with the work of Posner who pioneered the technology gap theory of trade. His work emphasised the different rates of innovation and learning among different firms and countries. Finally, Linder suggested similarity of income levels, factor endowments and demand patterns as the important determinants of the pattern of trade flows.

The interrelationship between product cycle and technological gap theories has also been emphasised in the work of Hufbauer (1970). The theories of both Hufbauer and Vernon stress the sequential development of production history; the major difference lies in their emphasis. While Hufbauer's technological gap theory, which drew on the work of Posner, is a supply-determined model, Vernon's product cycle model, which drew on the work of both Linder and Posner, is a demand-determined model in which entrepreneurs respond to demand and relative factor prices in the extent and form of their innovation and product development and subsequently in the location of their production activities.

Hufbauer (1965) concluded that the theories relating to technological gap and scale economies are most relevant in explaining trade in a large number of synthetic materials. In a similar vein, Douglass (1966) considered the suitability of the technological gap theory to the motion picture industry, with favourable results. Hirsch (1967) re-examined the theoretical foundations laid out by Hufbauer and assessed their significance to some sectors of the electronics industry.

The basic theory of, and empirical support for the PCM which was elaborated comprehensively in 1966, was subsequently modified in the 1970s to take account of oligopolistic market structures associated with the progressive internationalisation of industries. It is convenient for the purposes of this analysis to adopt the terminology of Buckley and Casson (1976) in differentiating between the first version of the PCM, Mark I, and that of the later version, Mark II.

The product cycle model (PCM) Mark I

The fundamental assumptions of the PCM Mark 1 conceptualised by Vernon are as follows: first, that enterprises in the advanced countries of the world have equal access to scientific knowledge and equal capacities to comprehend scientific principles; second, that entrepreneurs' consciousness of, and responsiveness to, entrepreneurial opportunities are a function of ease of communication with the market place, which in turn is a function of geographical proximity; third, that products undergo foreseeable changes in production technology and marketing methods; fourth, that production processes undergo phases through time and economies of scale are inevitable; and fifth, that tastes vary according to income and therefore products are capable of standardisation at various income levels. A sixth implicit assumption of the model is the presence of imperfections in the market for technological know-how.

The main thesis of the PCM is that the extent and form of innovation and product development are determined by demand and relative factor prices which exist in the market particular to the home country of the innovating firm. For example, the presence of a large market favours entrepreneurial opportunities in the research and development, production and marketing of new products and processes. The presence of US consumers' high income levels in the 1950s and 1960s was responsible for encouraging the generation of ownership advantages of US firms in the production of high-value consumer durable and industrial products. In particular, the presence of high labour costs in the USA relative to production creates a specific kind of entrepreneurial innovation, i.e. factor-saving innovation.

The PCM delineates three principal stages in the life cycle of a product. The first stage is that of the *innovative new product*, resulting from the awareness of unique entrepreneurial opportunities and the identification of a novel demand, or the adoption of new methods of production. The model postulates that high average income levels are instrumental in supporting new wants and that high unit labour costs greatly influence technological innovation, leading to the development of labour-saving consumer goods and industrial products.

The model further assumes that US entrepreneurs are first aware of opportunities to fulfil new wants by new products concomitant with high average income levels or high unit labour costs. These US entrepreneurs are expected to have a consistently higher rate of expenditure on product development than entrepreneurs from other countries, at least in product lines that fulfil high income wants and that substitute capital for labour. These higher rates of expenditure on product development are attributable not to some sociological drive for innovation but to more effective communication between the potential market and the potential supplier of the market.

The unstandardised nature of the new product implies that inputs cannot be determined in advance with absolute certainty and therefore producers value the degree of freedom and flexibility in improving production technology through experimentation with alternative inputs taking account of their cost. Second, the high degree of product differentiation or the existence of monopoly in the early stages means that there is a low price elasticity of demand for the output of individual firms. Third, as a result of the indeterminable nature of the new product and the market, and the likelihood of rival competition, there is a need for expeditious communication between producers on the one hand and customers, suppliers and competitors on the other. These factors emphasise the importance of a location in which there are external economies and in which the costs of communication are at a minimum. Since costs of communication increase directly with geographical distance (an explicit assumption of the model), a location which is close to the market is favoured.

The second stage in the life cycle of a product is that of the *maturing product*, the result of a certain degree of standardisation. The importance of flexibility, brought about by the integration of research, production and marketing activities at the site of innovation in the first stage, decreases. The possibility of economies of scale through mass output increases with the specification of product and process technology. Thus, in contrast to the first stage where product specifications were fundamental, production costs now become far more important. Moreover, as buyer knowledge increases, demand for the product correspondingly increases and becomes more price elastic. In time, the demand for the product in relatively advanced countries such as those in Western Europe with similar demand patterns increases, especially since the product has a high income elasticity of demand and is labour saving. These markets are first served through exports while the marginal production and transport costs of the goods exported from the home market are below the average cost of establishing a production facility in the export market where factor costs, appropriate technology and scale economies are divergent from those in the home market.

Apart from cost considerations, the threat to the large-scale export business in manufactured products in the form of local competition within the export market becomes an important stimulus to the initial import-substituting investment of a firm. Subsequent investment by other rival firms may result in a threat manifested in the form of a declining global share of the market with respect to the initial investor. The relocation of production abroad increases the possibility of exports to third-country markets and even the home market if differences in factor costs surpass transport costs.

The third and final stage in the product life cycle is that of the *standardised product*. The nature of the product at this stage means that accessibility to market information is greater and competition is largely, if not solely, on the basis of price. The search for the lowest cost source of supply therefore becomes the priority of investor firms. At this stage, the ownership advantages of the firm are based mainly on marketing and distribution, unlike the earlier stages where ownership advantages were based on the abilities of the firm to engage in technological innovation.

A major feature of the PCM is its implicit reply to the *Leontief paradox* that US firms export more labour-intensive goods than capital-intensive goods, with which the USA has a comparative advantage. The PCM characterises the research-intensive innovative stage of a product and the establishment of a pilot plant as particularly labour intensive because of the demand for research staff and marketing personnel. However, as the product reaches the stand-ardised stage, scale economies become far more important. Mass production of the standardised product necessitates greater capital intensity compared to the greater labour intensity of the innovative stage.

In addition to the substitution of capital-intensive means of production for labour-intensive means of production in the standardised product stage, there is also a substitution or displacement of higher skilled labour by less skilled or unskilled labour. A cost-determined equilibrium regulates the shift of these lower skilled and unskilled labour stages of production of standardised products to developing countries where labour costs are lowest and where incomes begin to catch up. A fourth stage in the product life cycle can therefore be envisaged in which there is a shift in the location of production to Third World countries and specifically the newly industrialising countries. However, there is a major difference in the nature of investment between US FDI in the developing countries at this stage as compared to US FDI in developed countries at an earlier stage. US FDI in the developing countries is more likely to be of an export-oriented kind which is not demand driven. By comparison, US FDI in Europe is more likely to be of an import-substituting kind which is prompted mainly by demand factors in the host country.

The product cycle model (PCM) Mark II

A modified version of the PCM Mark I was presented by Vernon in 1974 to take account of the fact that the growth of multinational enterprises has increased the possibilities of oligopolistically structured industries which operate on a global basis. Moreover, a more balanced process of technological competition between the USA, Europe and Japan has occurred in the period since the 1960s by comparison to the 1950s and 1960s when the USA fulfilled a technologically leading role in the world economy. A new three-stage product life cycle has therefore emerged in the light of these developments which takes account of different kinds of barriers to entry.

The first stage of the cycle, *innovation-based oligopoly*, is based on the PCM Mark I but recognises the impact of differences in national environment on the generation and development of innovation. While US firms tend to spe-cialise in innovations that are consistent with high incomes and high labour costs, European firms tend to specialise in innovations that are consistent with high land and material costs, while Japanese firms tend to specialise in innovations that are consistent with high material costs.

The second stage of the PCM Mark II is that of *mature oligopoly*. In contrast to the first stage, in which oligopoly is based on product innovation as a barrier to entry, in the second stage of mature oligopoly, innovation is replaced by economies of scale in production, transportation or marketing, which enable firms to maintain their ownership-specific advantages. In a second stage mature oligopoly, decisions pertaining to prices and investment by one firm are deliberately taken to disturb the existing market equilibrium. As a result, mature oligopolistic firms, especially those operating in industries characterised as having high fixed costs relative to total production costs, are inclined to adopt strategies that will maintain stability in the industry.

For example, when stability is threatened, as in the event of subsidiary price cutting in the domestic market of each large firm in the oligopoly, leader firms may engage in intra-industry production, which is explained by Vernon in terms of taking hostages, i.e. production in each other's principal markets and exchange of threats in the form of price cutting, with the objective of minimising the risks that arise from oligopolistic destabilisation. The exchange of hostages that pertains to an international oligopoly may be associated with declining rather than increasing concentration of investment at a world level.

The third and final stage of the product life cycle in PCM Mark II is that of *senescent oligopoly*. In this stage, the importance of economies of scale as an effective barrier to entry declines and despite attempts by oligopolistic firms to prolong the equilibrium by erecting new barriers to entry, such as entering into cartel arrangements and engaging in product differentiation, such barriers are not sufficiently high to maintain stability and the firm is faced with competitive pressures. Some firms are forced to exclude the senescent product from their product lines while other firms with special resources that are transferable to other products or which enjoy externalities from the sale of the senescent product may continue. However, the firms that decide to continue have to be subject to some form of genuine price competition and therefore the location of production is largely determined by competitive market forces and cost differentials.

Criticisms of the product cycle model

The product cycle model represented a renewed emphasis on classical trade theory, to the extent that innovation in the form of introduction of new products is an important determinant of the growth of production and international economic activity. However, elements of neoclassical trade theory are also seen in the importance given to comparative costs as a determinant of the location of production at each stage of the product cycle.

The PCM emphasised the impact of the following factors on the location of production: demand (market) factors associated with technological innovation and product improvement; and communication between the market and the firm and transportation costs. In fact, the changes in market demand give the dynamic nature to the model by showing how the pattern and direction of trade and investment are likely to evolve over time with the product cycle. The model effectively establishes the systematic relationship between trade and investment of US firms on one hand and product innovation on the other, as well as the response of firms to changing market conditions as domestic and foreign demand for the product increases, matures and declines.

A major advantage of the PCM is the method by which trade and investment are considered part and parcel of the process of foreign market exploitation. The relationship between these two modes of foreign economic involvement are examined in a dynamic context. Indeed, the PCM posed the question 'when' to the theory of foreign investment in addition to the 'why' and 'where' (Dunning, 1981a). However, although the PCM is explicit about changing locational advantages being important determinants of the exports versus international production modality of servicing foreign markets, the analysis of internalisation advantages in determining licensing versus international production as alternative routes for replacing exports to foreign markets is undertaken implicitly. The implicit assumption made in the model is that firms internalise because of the presence of imperfections in the market for technology. Such internalisation advantages may be analysed in terms of Dunning's (1988a) concepts of appropriability and co-ordination. The firm integrates research, production and marketing in order to appropriate a full return on its remaining ownership advantages as a technological leader (appropriability) as well as to co-ordinate the use of its complementary assets in marketing and distribution (co-ordination). The extent to which licensing as opposed to international production is used as an alternative route to exports in servicing foreign markets in the PCM Mark I may be taken to be implicitly determined by the extent to which a competitive fringe of firms in the export market are able to erode the technological leadership of the innovating firm (Dunning and Cantwell, 1982).

The PCM provides an invaluable analytical tool for examining the emergence of MNEs in the manufacturing sector. The PCM Mark I is particularly useful in explaining the early post-Second World War direct investment by US manufacturing firms in other advanced countries, such as parts of Western Europe, especially in industries characterised as having high technological intensity. The model implies that US foreign investment is a response made either to bridge the gap between product innovation and standardisation or to stimulate consumer tastes for new standardised products (Servan-Schreiber, 1967; Galbraith, 1967). The model also has some general explanatory power for highly innovative firms undertaking their initial foreign investment and for direct investments involving final products. The model shows how the demand for a product becomes more elastic over time and across countries. It is also useful in explaining the phenomenon of offshore production in countries that have low labour costs (Moxon, 1974).

However, the explanatory power of the PCM has itself undergone maturity and decline (Giddy, 1978). Recent developments in international trade and investment have necessitated more general explanations of the method by which large firms cope with barriers to entry in world markets. The range of countries, industries and firms involved in international manufacturing and marketing of products has dramatically increased in the twenty-five years since the inception of the basic model in 1966. On the country level, the USA has a diminished role as a source country of foreign investment. The statistics on foreign direct investment (FDI) presented in Chapter 2 show that the source countries of FDI have increased markedly in the post-Second World War period with the growth of European, Japanese and, more recently, developing country or Third World MNEs. The growth of Third World MNEs in particular shows that technological dynamism can take place in developing as well as in developed countries.

The impact of differences in national environment on the generation and development of innovation in the first stage of the cycle in PCM Mark II reasserts the fundamental role of home country-specific or market demand factors in the extent and form of innovation and product development. Increasing demand for the maturing product in the follower countries determines the extent to which a competitive fringe of firms is able to catch up in the process of technological development through the subsequent imitation of the innovation of the technologically leading firm. Where there are competing innovations among firms from the follower country in which the process of catching up is determined by the inherent potential of firms for innovation rather than limited to their scope for imitating the innovation of the technological leader, then the PCM is on weaker ground. Hence, the PCM is often regarded as an unsatisfactory analytical tool in examining the European or Japanese response to the American challenge characterised by technological rivalry between US, European and Japanese firms in the same industries.

The PCM is an application of Schumpeter's (1934) model in which innovation is regarded as a discontinuous process of major technological breakthroughs in scientific research. New entrepreneurs fulfil the role of implementors of such innovation by identifying profitable business opportunities in the new marketable product, the embodiment of scientific research. Therefore in a Schumpeterian view, innovation is undertaken only by new firms that produce entirely new products. Every new product is seen as a radical innovation, and successive improvements to both the product and production process are the incremental changes that bring the product to maturity. A new product is seen as a radical departure from the old product and destined to follow a similar evolution. The PCM, by following this Schumpeterian model of discontinuous technological innovation, has applications for explaining long waves in the historical pattern of technological innovation in the period prior to 1914 and in the immediate post-Second World War era when scientific discoveries led to the creation of new products and industries. The model has limited explanatory power in analysing the continuous forms of innovation undertaken by established firms, the sectoral pattern of innovation across countries or the process of technological competition across sectors. The capability of large firms in the twentieth century for endogenous innovation has been incorporated in Schumpeter's later model of innovation in 1943, while the relationship between the discontinuous long waves in innovation and the sectoral pattern of innovation across countries has been established by Pavitt and Soete (1982). In particular, a dynamic model with which to explore the relationship between the disruption of a major new wave of innovations by US firms and the subsequent path of technological competition by European firms across sectors has been formulated by Cantwell (1987e).

Even PCM Mark II drew on the framework of major discontinuities in the pattern of technological innovation rather than the gradual and continuous process of adaptation in innovation. Freeman and Perez (1988) and Perez (1988) have shown that the innovative process is such that products build upon one another and are interconnected in technology systems. In this view, successive products within a system are equivalent to successive improvements to a product as firms acquire knowledge, skills, experience and the externalities required for the various products within the system. A series of interrelated technology systems constitute techno-economic paradigms which evolve in time from an early phase through growth to maturity. Changes in the techno-economic paradigm will affect the whole range of technology systems which evolved and matured under the previous paradigm.

The continuous process of innovation in the research, production and marketing of products has decreased the possibility of product and process standardisation and increased the possibility of product and process differentiation in response to local market demand. The acceleration of the product development cycle has been commensurate with a shorter product life cycle (Cahill, 1989). As a result, a high degree of organisation in the process of product development and innovation has been achieved (Buckley and Casson, 1976).

Shifts in the location of production may result in increased innovation even at the level of the product. This means that although the PCM may be accurate in explaining the life cycle of some products from the new product stage to the standardised product stage, the model cannot explain the life cycle of all products in all industries. The growth of the motor vehicle and colour television industries in Japan provide excellent cases in point. The continuous innovation introduced by Japanese firms in the production of motor vehicles enabled Japan to become the most important producer in the 1960s and 1970s, taking over from US, German and other European firms. The production of motor vehicles did not undergo maturity and standardisation as Japanese firms introduced considerable innovative improvements, especially in the area of electronics engineering (Walker, 1979). This trend away from product standardisation towards sustained efforts in innovation to achieve product differentiation is also seen in the production of colour televisions in Japan

(Peck and Wilson, 1982; Althuser *et al.*, 1984; Baba, 1987; Ohmae, 1987). As a result of the higher technological embodiment of the motor vehicle and colour television, Japanese firms relocated the production of these products in the USA and Europe rather than in the lower-income developing countries.

The model also implicitly suggests that the dynamic process of technological accumulation is the same for all product groups although small firms are characterised as having higher rates of product innovation and large firms as having slower rates of process innovation. The empirical evidence does not offer convincing confirmation of the widespread existence of such a process except for specific product models or marks (Pavitt, 1988). Product innovation in textile machinery and many other classes of capital goods has taken place over many years while firms have remained small and in recent years there has been an acceleration in the pace of product innovation (Walker, 1979). In addition, process innovation in the production of standard bulk materials (i.e. glass, paper and steel), particularly in the exploitation of latent economies of scale, have been concentrated in large firms from the beginning. Therefore, although product life cycle curves represent the dynamics of national production in broad product groups, they may reflect changing demand elasticities over different ranges of per capita income rather than autonomous acceleration and deceleration of innovation.

The specification of the market as the prime determinant of innovation in the PCM is also subject to scrutiny. Innovative activities are not determined entirely by present or anticipated market signals as predicted by the PCM but also by evolving technologies (Mowery and Rosenberg, 1979; Atkinson and Stiglitz, 1969). Market factors relating to relative factor prices and structure of demand are likely to be fundamental in influencing both the rate and direction of technical progress within the boundaries defined by the nature of technological paradigms. These technologies determine the range within which products and processes can adjust to changing economic conditions (Dosi, 1988). These more recent theories of innovation show that the firm, rather than the product, is the prime determinant of innovation (Dosi, 1982, 1984; Pavitt, 1988). The relegated importance of the firm and industry to that of the product constitutes a major limitation in the effectiveness of PCM in explaining patterns of innovative activities in firms and countries (Pavitt, 1988) and the internationalisation of industries characterised by horizontal specialisation and the presence of both intra-industry trade and production (Cantwell, 1987e).

Through the emphasis placed on the influence of the product on firm strategy, the model has analysed only one form of interdependence between firms of various nationalities which pertains to the exchange-of-threat behaviour and not the follow-the-leader or other forms of behaviour undertaken by firms to maintain stability in an international oligopoly. Large MNEs have increasingly pursued global strategies that are inconsistent with the model. The combined effects of the high costs of R&D, the convergence in the technological capabilities between firms in the developed countries and the rapid rate of technological diffusion in these countries has meant that the firms' networks must introduce newly developed products in all sales territories simultaneously (Ohmae, 1985). At best, 'the PCM should be regarded as a strategic business concept that can be anticipated, followed or even reversed by alert international product managers' (Giddy, 1978).

Furthermore, the model disregards the increasing proportion of foreign investment that is not trade replacing, and the sources of ownership advantages that enable MNEs to compete effectively with local firms are not clearly identified. The model assumes that decisions made in the different stages of the product life cycle pertaining to investment in product innovation, modes of foreign economic involvement and methods of coping with barriers to entry in world markets are distinct and unrelated. Finally, the model is regarded as being sequential rather than dynamic as it fails accurately to specify the rate or the time lags that distinguish each sequence. The limitations of the PCM Mark I elucidated in this chapter are consistent with those of Buckley and Casson (1976), Giddy (1978), Hood and Young (1979), Buckley (1985) and Cantwell (1989a). Even Vernon in his later writings concluded that the PCM had the tendency to discard or distort empirical facts in order to conform to the specifications of the model (Vernon, 1979).

An important feature of PCM Mark II is the emphasis placed on the global strategy of the firm in response to world-wide competition. However, PCM Mark II does not completely consider the criticisms made of PCM Mark I because firm strategies are still determined by the maturity of the industry when technology creation is diffused abroad. Competition and intra-industry trade and production only occur at the mature oligopoly stage when there is a threat to international market equilibrium. But in practice, MNEs compete in industries that are often in the forefront of technological innovation and are sustained in part through intra-industry investment to gain access to complementary forms of foreign innovation that can be used to upgrade and extend indigenously generated innovation (Cantwell, 1987f). The process of technological accumulation and competition provided the strongest driving force behind the expansion of international production in the industrialised countries until the early 1970s (Cantwell, 1987e).

Since the mid-1970s, the slow down in the growth of markets, innovation and productivity has meant that intra-industry trade and production are increasingly carried out in response to the global strategy adopted by MNEs and not to threats of instability made by rival firms. The period since the mid-1970s has necessitated the rationalisation of international production by MNEs by which the firm gains additional advantages from an international division of labour established among the global network of the firm. Each affiliate in the global network has a greater degree of product and process specialisation based on its different structure of factor endowments and market characteristics. The process of technological accumulation is frequently organised within internationally integrated networks of the MNE which have become global organisers of economic systems, including systems for allied technological development in different parts of the world (Cantwell, 1991b). The gains from these investments are explained in the eclectic paradigm in terms of the MNE gaining additional ownership and internalisation advantages from the common ownership and control of separate but interrelated production facilities located in different countries. These advantages are apart from those that accrue to the firm based on the ownership of distinctive intangible assets. The sequential nature of PCM Mark II does not fully consider the significance of such forms of global competition and the interdependence between investment in product innovation, modes of foreign economic involvement and the presence of ownership advantages that enable firms to compete effectively in world markets.

Vernon (1979) acknowledged the declining explanatory power of the product cycle hypothesis owing to two major changes in the international environment: the increasing number of countries that engage in innovative activities as a result of the establishment of overseas subsidiaries; and the increasing convergence in income, market size and factor cost patterns of advanced industrial countries, leading to a greater similarity in their markets. Hence, the relevance of the PCM is increasingly becoming limited in analysing the relationship of the US economy to other industrialised countries as well as the relationship between advanced industrialised countries and developing countries. Nevertheless, Vernon indicated that the hypothesis would still be relevant in explaining the innovative activities of smaller firms that have not yet acquired a capacity for global scanning through a network of foreign manufacturing subsidiaries. The course of particular products at a microeconomic level may conform to a PCM but not the behaviour of a globally integrated firm as a whole.

Perhaps the most important criticism of the PCM is that recent forms of foreign investment, especially those undertaken by developing countries in developed countries, do not conform to the Schumpeterian analysis of the PCM that results in a chain of events which starts with innovation in the leading country and leads to a technology transfer and diffusion of such technology to a competitive fringe of firms in the host country. The latter chapters of this book present evidence that firms from developing countries, like those of the developed countries, may engage in foreign direct investment or licensing activities as a form of backward technology transfer in order to gain access to complementary foreign technology which is generated in a particular location but which these firms can combine with their indigenous technological innovation (Dunning and Cantwell, 1990). These forms of FDI to gain access to technology created in a particular location cannot be analysed within the framework of the PCM which tackles FDI geared to transfer technology created in the home market. Nevertheless, Vernon pointed to the continued relevance of the PCM in explaining trade and investment among developing countries.

The product cycle model and Third World MNEs

Some of the earliest work on Third World MNEs had been carried out within the framework of the product cycle model. Vernon (1979) has clearly indicated that although the PCM is losing its relevance in explaining trade and investment among advanced industrialised countries, the model maintains some relevance with regard to developing countries since some form of divergence still exists in their national markets. For example, there is some scope for foreign subsidiaries in LDCs to innovate and manufacture products intended for richer and larger markets. Empirical investigation has shown that some of these foreign subsidiaries, as well as domestic firms in the more industrialised of the developing countries, have a capability for innovation that responds to the unique conditions of their home markets and results in the creation of entirely new products or processes or a modification of existing products or processes.

Some of these firms have then explored the possibilities of foreign markets, first through exports and then through direct investment. Their foreign markets are typically the less industrialised developing countries, which conforms with the prediction of the PCM that countries would trade and invest in other countries that are lagging behind in the industrial pecking order. The ownership advantages of these firms need not be based on technological innovation or other oligopolistic advantages but on the development of special skills in the maintenance, repair and supply of spare parts for second-hand machinery, the use of lower-cost labour-intensive production processes and low salary payments to managers who are none the less adept at organising in developing country conditions.

Wells (1977) conveniently viewed the internationalisation of firms from developing countries as a stage in the product life cycle or technological gap model. The technological gap model proposed by Hufbauer (1965) considers countries as falling into a pecking order according to their ability to produce a particular product. The pecking order explanation is based on the different abilities for technological innovation and production costs. The model predicts that countries currently manufacturing the product export to those countries that have not established manufacturing plant, i.e. those countries further down the pecking order. The direction of foreign direct investment is also believed to follow such a pattern. Hence, the advanced countries are the first sources of exports and FDI followed by the more advanced of the developing countries. The empirical evidence tends to support the phenomenon. First, shifts in production to different countries may result in the adaptation of products in response to the needs of a somewhat lower-income market. These producing countries then develop an advantage in the manufacture of lower-income products. Second, and relatedly, these producers are willing to offer more attractive terms to lower-income countries.

As with developed country firms, the competitive advantages of developing country firms seem to derive from the peculiar nature of their home markets.

As Japanese and European firms develop a competitive advantage embodied in adapting lower-income products, developing country firms develop a competitive advantage of a different kind. They base their advantages on adapting the manufacturing technique rather than the product. These firms innovate by way of developing small-scale technology just as American firms innovate by way of developing high-income and labour-saving products and European firms innovate raw-material saving and capital-saving technologies. However, some of these firms are also capable of innovating products as well as technologies designed to suit the needs of their national markets.

Even as the pecking order explanation accounts for the trade and investment cycle of countries across time, there are some important limitations which the model has recognised although not taken account of. For example, some developing country firms have established export platform investments in other developing countries to gain access to low-cost labour in order to maintain their competitiveness in export markets. Furthermore, some developing country firms engage in resource-based investments and/or service sector investments in other countries.

Wells (1981, 1986) has expounded on the competitive advantages of firms from developing countries which derive from the peculiar nature of their home markets. First, firms from developing countries are engaged in industries characterised by low expenditure on R&D and low product differentiation. Second, firms from developing countries develop small-scale, labour-intensive processes and products and have a higher tendency to source inputs locally. Third, firms from developing countries develop technology for manufacturing in small volumes. This small-scale technology may result from genuine process innovation or may represent adaptations of large-scale technology for small-scale processes. Fourth, and as a result of adaptation, most small-volume plant designed by firms from developing countries have greater flexibility. Vernon concludes that these innovations, made in response to small markets and scarce foreign inputs which are exploitable in foreign markets, are in line with the expectations of the PCM.

Although the PCM as applied to Third World MNEs takes account of the capacity of developing countries for technological innovation, the scope of such innovation is limited to the imitation and adaptation of foreign technology in accordance with the requirements of Third World markets and production conditions. However, although Wells (1983) contends that Third World MNEs are essentially based in such standardised product sectors, he does not entirely exclude the possibility of local innovation. He suggests that Third World MNEs may have advantages that come from adapting foreign technology to the circumstances of smaller plant and smaller firm size, as well as a technological advantage in their ability efficiently to utilise locally available natural resources rather than imports. According to Wells, the descaling of foreign technology in the manufacture of traditional products and the use of local resources as novel inputs in the production process still amount to only a limited innovativeness on the part of Third World MNEs as these activities are generally confined to the tail end of the product cycle. What is more, there is presumably no reason why the developing country affiliates of US or European MNEs cannot imitate and copy the technological improvements achieved by Third World firms, since these innovative activities can be viewed simply as a different way of adapting an essentially foreign technology and product development.

The modality of foreign economic involvement of firms from developing countries was also further elaborated by Wells (1981, 1983). These firms are likely to prefer exports over other modes of serving foreign markets in the first instance, since this modality incurs fewer risks, capital and information costs. However, at some later stage, firms might find the opportunities offered by exports to be restrictive. Transportation costs as well as the presence of tariff barriers may make their exports less competitive in foreign markets. Therefore, at this stage, the previously exporting firm may choose to engage in FDI.

The reason firms from developing countries choose to internalise their competitive assets across national boundaries rather than license is very similar to the internalisation advantages experienced by firms from the industrialised countries. However, there are some reasons unique to firms from developing countries which stem from the nature of their ownership advantages. The product cycle model predicts that firms from developing countries will choose to internalise since most of their skills and technology are embodied in machinery or in the knowledge of managers which cannot be easily codified. Wells writes:

> Very similar are the problems of evaluating an asset about which the buyer knows few details and the seller cannot reveal more without losing control over the asset. In the case of equipment, the problem is further complicated. In many cases of developing country skills, much of the skill is embodied in the machinery itself. In a world of perfect markets, the machinery could be sold to a would-be user abroad for a price that reflected the competitive edge its ownership would convey. Like the buyer of such assets from a firm from the industrialised country, a potential purchaser would need some assurance that the value of the asset would not be eroded by further sales to local competitors. More important, however, buyer and seller are simply not likely to come together. Although the machinery markets are well developed for equipment from industrialised countries, marketing networks are not generally in place for equipment from the developing nations. Moreover, it would be a rare local entrepreneur to whom it would seem a sensible proposition to seek out his plant in countries outside Japan, Europe, or North America.
>
> (Wells, 1979, reprinted 1981, p.30)

However, the real underlying issue is that the skills and technology of firms in developing countries are embodied in the methods used in exploiting machinery and in the knowledge and experience of managers as a result of learning by doing and learning by using and *not* in the machinery itself. It is the nature of these forms of competitive assets that cannot be easily codified and transferred to other firms.

THE THEORY OF LOCALISED TECHNOLOGICAL CHANGE

The alternative theory of Third World MNEs supposes that Third World enterprises have a wider scope for innovation, based on a model of localised technological change advanced by Lall (1981, 1982b, 1983a, 1983c). The idea of localised technological change can be traced back to Atkinson and Stiglitz (1969) and Nelson and Winter (1982), and has been more recently developed by Stiglitz (1987).

The model provides a valuable theoretical framework within which to view the nature of proprietary or firm-specific advantages of developing country firms *vis-à-vis* developed country firms. These advantages derive from the ability of developing countries to innovate on essentially different lines from those of the more advanced countries, i.e. innovations that are based on lower levels of research, technology, size and skill. While the competitive assets of firms from developed countries are derived from 'frontier' technologies and sophisticated marketing, those from developing countries are derived from widely diffused technologies, special knowledge of marketing relatively un-differentiated products or special managerial or other skills. These assets may have resulted either from some adaptation or improvement in the product or process technology (otherwise known as 'minor' innovation) which would be costly for other firms to produce or from a cost advantage in providing standardised technology, or both.

The capability of a firm in a developing country to innovate for a unique proprietary asset stems from the nature of technical progress. Departing from abstract neoclassical theory and adopting an evolutionary theory of techno-logical change, Lall follows Atkinson and Stiglitz (1969), Nelson and Winter (1977, 1982) and more recently Arthur (1988, 1989) in concluding that al-though technical changes are largely determined by the market and scientific advance, such change is localised at the micro level and is path dependent and irreversible. Such localisation of technical change means that firms will only undertake a very limited range of techniques since any shifts on a theoretical production function require substantial costs. Such technical change affects not only the innovating firm but also the whole range of linked industries. Technical change is also described as path dependent and irreversible because older technologies cannot be efficiently reproduced or transferred once an entire industry has progressed to new technologies and become firmly estab-lished.

Small firms from developing countries that have low levels of technology and skill may have proprietary advantages that can sustain competitiveness *vis-à-vis* firms from the developed countries. These proprietary advantages are derived from three sources. First is the localisation of technical knowledge around an entirely different range of techniques owing to the different factor costs and availability in LDCs. Innovation may take the form of adaptation of imported technology or specialisation in some foreign outdated technology. Second is the suitability of the product to developing countries. Innovation in this case may result indigenously through the improvement of local products or the adaptation of imported products. Third is innovation that results in smaller-scale technique as opposed to the large-scale technique of developed countries. These proprietary advantages are strengthened by access to low-cost skilled manpower in the home market and by special assets that derive from being part of a conglomerate group. Technical innovation on the part of firms from developing countries results in ownership advantages which can be exploitable abroad through FDI.

Since technical change is largely dependent on country-specific factors such as trade and industrial strategies, the prediction is that different developing countries will generate different kinds of MNEs. The sources and manifestations of what Lall refers to as monopolistic advantages of firms from developing countries are shown in Figure 4.1. However, although Lall and

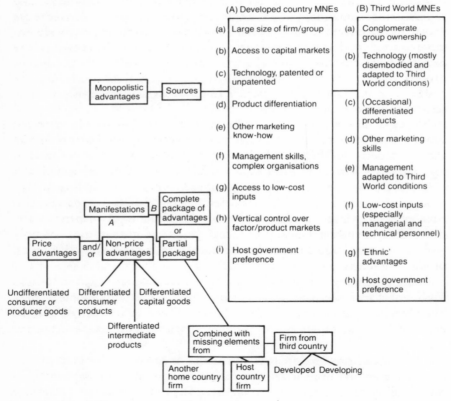

Figure 4.1 Manifestations and sources of monopolistic advantages of MNEs
Source: Lall, 1983c.

Kindleberger refer to the presence of monopolistic advantages, Chapter 3 of this book has pointed out that these advantages are really of an oligopolistic rather than a monopolistic kind. The existence of oligopolistic as opposed to monopolistic advantages is necessary for foreign investment.

Lall has also attempted to modify the static conventional Hecksher-Ohlin trade theory into a dynamic theory of trade and investment for developing countries. The model specifies that the comparative advantages of developing countries are changing dynamically and that these countries are increasingly capable of producing goods characterised as having high levels of technology, skills and scale. These technical changes are in part brought about by technological and organisational innovations introduced by inward foreign direct investment in developing countries to take advantage of low factor costs, and in part by the indigenous technological and marketing innovation undertaken by local firms. An examination of the sources of comparative advantage would lead to the prediction that different developing countries would generate different comparative advantages.

The rapid evolution in the pattern of activity of Third World MNEs has meant that theories may have to take account of the possibility that these firms can also embark on independent technological accumulation. The theories have to recognise the increasing role of innovation within firms which become more sophisticated as international production develops. As a result, a disagreement has emerged between two schools of thought on Third World MNEs.

The theoretical explanation provided by Lall seeks to emphasise the distinctive nature of proprietary advantages of Third World MNEs in industrial technology and thereby provides a more general outlook on the perceived conventional wisdom endorsed by previous analyses of Third World MNEs and specifically the Wells (1977, 1981, 1983) product cycle model. The conventional wisdom identifies the sources of ownership advantages of developing country firms as arising primarily from know-how in low cost production associated with small production runs and inexpensive labour and from technology and management adapted to Third World conditions. Furthermore, these ownership advantages are manifested more in terms of price competition than product differentiation and represent only a partial rather than a complete package.

The limitations of the prevalent product cycle approach to Third World MNEs stem from the generalisation made about the nature of these MNEs. The model fails to explain marked variations in the characteristics of MNEs between developing countries which are as great as, if not greater than, the variations in the characteristics between developed and developing countries. The model excludes other important sources of competitive advantages for developing country firms and, specifically, the genuine indigenous technological capability for competitive innovation. The product cycle model stresses that the technological advantages of Third World MNEs lie in small-scale, labour-intensive technologies compared to the large-scale, capital-intensive technologies of the developed countries.

In the theoretical framework of Lall, this small-scale labour-intensive technology may result from any of the following conditions. First, the firm has undertaken some technological changes in the nature of equipment or production process resulting in changes in factor intensity. Second, the firm has assimilated a foreign but obsolete technology through scaling-up, alteration or improvement, resulting in an adapted technology which foreign MNEs cannot reproduce or transfer without costs. Third, these small-scale labour-intensive technologies are more efficient than those offered by competitor firms. For these reasons, greater labour intensity can only be considered a unique proprietary advantage of Third World MNEs when such technology arises from genuine technological adaptation and not from the use of dated technology. Similarly, the notion that Third World MNEs have small-scale advantages must be looked at in the context of the actual process of technical change within the firm.

The various causes, *loci* and methods of technical change in a Third World manufacturing firm under an import-substituting regime are summarised and presented in a schematic diagram in Figure 4.2. The competitive advantages of Third World MNEs may stem either from some unique minor product or process innovation which is difficult for other firms to emulate, from a unique marketing strategy or from some historical accident. The specific source of these advantages varies between countries and firms in developing countries.

Third World MNEs have a greater likelihood of engaging in joint venture arrangements in their foreign investments than MNEs from developed countries since their unique proprietary advantages are more limited than those of MNEs from the developed countries and are more specific to Third World conditions. Third World firms therefore need to gain access to foreign technology and engage in complementary activities with MNEs from developed countries in developing their technology. The pressure applied by home or host governments on the division of economic rent through joint ventures is an equally significant factor.

The theory of localised technological change is an idea implicit in the more general theory of technological accumulation which suggests that firms follow a technological course that is to some extent independent and a function of their own unique learning experience. The steadily rising number of examples of Third World multinational enterprises that have become genuinely innovative necessitates a broader theoretical explanation of their indigenous technological capabilities. Similar to the growth of international production established by industrialised country MNEs in the manufacturing sector, the growth of Third World MNEs has become increasingly dependent on the existence of innovation within the firm. Although the different type of innovation they pursue requires an empirical investigation that looks at technology creation broader than the sphere of research and patenting activity, the theory of technological accumulation is still a useful means of analysing the interna-

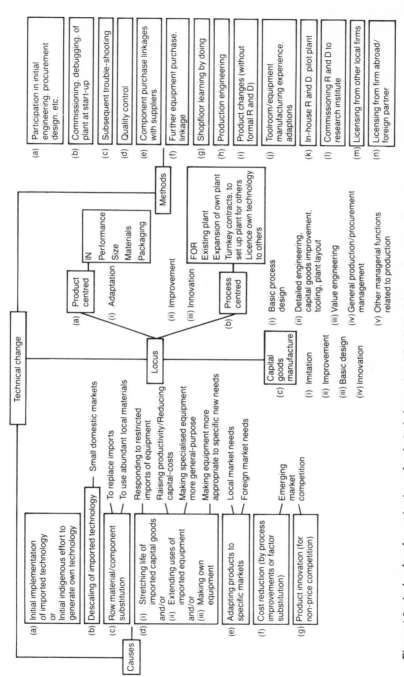

Figure 4.2 A simple schematisation of technical change in a Third World manufacturing firm
Source: Lall, 1983c.

tional growth of manufacturing firms from quite different environments, and at different stages of technological development and capacity (Cantwell and Tolentino, 1990).

THE THEORY OF TECHNOLOGICAL ACCUMULATION

The theory of technological accumulation which was originally proposed by Pavitt (1988) suggests that innovation is a firm-specific, cumulative and differentiated process. Such a theory has previously been applied to explain the historical evolution of the sectoral pattern of comparative advantage in the innovative activity of national groups of firms in industrialised countries (Cantwell, 1990) and in work on competition between rival technologies (Arthur, 1988, 1989). These innovative activities, especially in research and development, have been found to be statistically significant determinants of differences in export and productivity performance (Soete, 1981b; Fagerberg, 1988) and also of internationalisation of business among the major OECD countries (Vernon, 1974, 1979; Cantwell, 1989a). More recently, the theory was further elaborated and incorporated in a more general theory of technological competence to explain the differential growth rates of firms and the determinants of international competitiveness (Cantwell, 1991a).

The theory of technological accumulation proposes that firms follow distinctive technological trajectories in the course of innovation, which are at least to some extent independent of one another. This view of technology took root from the work of Lindbeck (1981) and Vernon (1966) who both stressed the importance of the technology factor in international competition and discussed the determinants of firm-specific and country-specific technological advantages. In particular, Lindbeck suggested that technology necessary for innovation consists of very specific knowledge of particular products, processes and markets. In Vernon's product cycle model, such innovative activities are undertaken by industrial firms with the objective of matching technological opportunity and market need.

The concept of technological accumulation argues that the high proportion of technology for innovation generated within the innovating firms reflects the existence of a highly differentiated range of techniques and related technical knowledge and therefore there exists a complex and continuing interaction between the creation of technology and its use in production. Such a view of technology as having a highly differentiated nature within the firm is not akin to a view of technology as a body of information or a blueprint that can be costlessly passed between firms or locations, providing an appropriate price for the transfer can be agreed (Arrow, 1962; Teece, 1977), and where firms can produce and use innovations through widely accessible knowledge of scientific principles (Vernon, 1966). The nature of technology as being firm specific, differentiated and often tacit means that major costs are incurred when implementing a technology outside the environment in which it was first created. Firms will as a rule only wish to buy or imitate the technology created by others when it is complementary to the path taken by their own innovation, and can therefore be integrated beneficially and adapted within their own programme of technology generation. The more complementary the technology is to their own, the lower will be the costs relative to the benefits of adoption and integration.

The highly differentiated nature of technological innovation within the firm means that improvement and diversification of technology are dependent on the existing technological base, i.e. that technological and technical changes within the firm are a cumulative process (see also Dosi, 1988). Once the cumulative and firm-specific nature of technology is recognised, the theory of technological accumulation predicts that its development ceases to be a random process but becomes constrained to zones closely related to existing technological activities (see also Nelson and Winter, 1977, and the recent work of Dosi and Orsenigo, 1988). As a result, the likely future patterns of innovative activities in firms and countries can in principle be predicted.

In the case of industrialised country multinationals the concept of building upon existing experience implies that the industrial composition of the innovative activity of the firms of any given country changes only gradually over time, and that world market shares are regulated by such reasonably stable technological strengths (Cantwell, 1989a). On the face of it, the theory of technological accumulation appears to have less relevance when speaking of firms based in Third World countries. The earlier technological stage of development of Third World MNEs has meant that their innovation has assumed a rather different form. Such firms spend far less on research and development, and are largely dependent upon foreign technology for their growth. Without a strong reliance on research and development, a greater part of their technological activity has been given to production engineering, learning by doing and using, and organisational capabilities, although – as discussed in some cases in this book – there is a steadily rising number of examples of Third World multinational enterprises that have become genuinely innovative through their progression to the early stages of research and development activities. The earlier stage of their technological development has led to innovative activities that are less scientifically refined and do not generally involve frontier technologies. With higher rates of innovation and growth, combined with investments in production engineering and in some cases research and development, the technological latecomers from the Third World are increasingly able to catch up with the leaders from more developed countries.

Evidence is also presented in this book that, although Third World firms have a larger dependence on foreign technology, they have followed a technological course that is to some extent independent and a function of their own unique learning experience. Due to the localised nature of technological change these firms have acquired an innovative capacity that is not easily copied by the affiliates of industrialised country MNEs in developing regions. Although these firms are still at a relatively early stage of development, there is every reason to believe that their abilities in technological accumulation will continue to expand. Some Third World MNEs now operate in sectors whose products are far from standardised, and indeed their activities are helping to generate a fresh pulse of innovation in such cases.

Chapters 8 and 9 present evidence on the growing significance of Third World outward investment, and on how innovative activity among Third World MNEs, although currently at an early stage, has tended to become increasingly important and more significant as international production develops and as their home countries advance through progressively higher stages of industrial development. Chapter 5 considers how such an evolutionary pattern in Third World outward investment relates to other macroeconomic developmental approaches previously developed to describe investments from the industrialised countries. The chapter describes how the general trend towards the internationalisation of business, owing in part to the rise of outward investment from other new source countries including those in the Third World, has affected the relationship between countries' investment position and their national level of development. It outlines how the sectoral and geographical structure of Third World outward investment might be expected to evolve. The trend in the industrial composition of Third World outward investment as development proceeds is analysed in Chapters 8 and 9, while Chapter 10 looks at the equivalent trend in the geographical distribution of outward investment. In each case, attention is drawn to the underlying accumulation of technological advantages as a means by which Third World MNEs have transformed the significance and the character of their international economic activity.

CONCLUSION

This chapter compared and contrasted the two distinct theories of FDI that specifically explain the emergence and growth of Third World MNEs: the product cycle model and the theory of localised technological change. The initial and subsequent versions of Vernon's product cycle model were presented along with an analysis of their inherent strengths and weaknesses including their application to Third World MNEs. The PCM essentially views the internationalisation of firms from developing countries as a stage in the product life cycle. The product life cycle proposes that countries fall into a pecking order according to their ability to produce a particular product, which reflects their different abilities in technological innovation and differences in production costs.

The competitive advantages of firms from developing countries is predicated to derive from their ability to imitate and adapt foreign technology in accordance with Third World markets and production conditions. First, these firms are engaged in industries characterised by low expenditure on R&D and low product differentiation. Second, they develop small-scale, labour-intensive processes and products and have a tendency to source inputs locally. Third, they develop technology for manufacturing in small volumes. Fourth, and as a result of adaptation, their plants have greater flexibility.

On the basis of these competitive advantages, the product cycle model predicates that firms from developing countries choose to internalise since most of their skills and technology are embodied in their machinery or in the knowledge of managers, which cannot be easily codified. Technology that is embodied in machinery is, in fact, easily codified, but these firms internalise because the methods used in exploiting machinery and their accumulated experience as a result of learning by doing and learning by using are not codifiable and therefore cannot be easily transferred to other firms.

Perhaps the major limitation of the PCM is in failing to consider the case when firms from the more advanced developing countries increasingly generate the capacity for localised technological change. The theory of localised technological change formualated by Lall shows the capability of indigenous firms in developing countries to generate genuinely unique innovation based on some occasions on licensed foreign technology. The dynamic theory of technological competence embodied in Lall's theoretical explanation determines the way ownership advantages are generated, which is distinct from the predications of the static theories of international production.

The nature of ownership advantages in the dynamic theory stresses that the advantages of firms from developing countries are not necessarily in descaling and increasing the labour intensity of sophisticated technologies. Rather, their advantages derive from their ability to innovate on essentially different lines from those of the more advanced countries, i.e. innovations that are based on lower levels of research, size, technological experience and skills, and to achieve improvements by modernising an older technique, including foreign outdated technology. The capability of a firm in a developing country to innovate for a unique proprietary advantage conforms to the evolutionary theories of technical change formulated by Atkinson and Stiglitz (1969), Nelson and Winter (1977, 1982) and Stiglitz (1987), and also to the path-dependent nature of technology advanced by Arthur (1988, 1989) which explains that technical change is localised at the firm level and is partially irreversible.

The factors that influence the internalisation advantages of firms from developing countries reflect the nature of the technology transferred by enterprises, which is largely implicit in the skills and experience of their employees and is therefore not easily codified or embodied in patents, blueprints or trademarks. However, internalisation does not preclude the marketing of particular ownership advantages through non-equity forms of foreign involvement. The locational factors that determine the exports versus FDI modality are broadly in tandem with those of firms from developed countries and result from the considerations of advantages of production in particular locations.

The next chapter sets the debate between these two competing theories of Third World multinationals within the context of the investment development cycle.

5 The concept of an investment development cycle: some econometric testing

INTRODUCTION

This chapter seeks to place the debate between the two competing theories of Third World multinationals analysed in the last chapter within the context of the concept of an investment development cycle. The idea of an investment development cycle has been advanced by Dunning (1981a, 1981b, 1986a, 1986c, 1988a) and is based on the proposition that the level of inward and outward investment of countries, and the balance between the two, is a function of their stage of economic development as measured by gross national product (GNP) per capita. These factors influence the ownership, internalisation and locational advantages of firms within countries at any stage of development.

The existence of an investment development cycle is consistent with both the theoretical explanations of Third World multinationals: the product cycle model (PCM) formulated by Vernon and applied to Third World MNEs by Wells and the model of localised technological change adopted by Lall. The product cycle model predicates that as countries increase their income per capita and demand for higher quality products, their firms are better able to generate ownership advantages through the imitation and adaptation of the technology of the technologically leading firms, resulting in greater outward direct investment. On the other hand, the model of localised technological change predicates that as countries occupy higher stages of development, their firms generate greater ownership advantages because their capacity for localised technological innovation increases, through indigenous creation, adaptation of foreign technology, or through the specialisation in techniques which were used at an earlier stage in advanced countries, leading to greater outward investment (Lall, 1983c). The concept of the investment development cycle is therefore a framework within which these competing theories can be analysed. The major aim of this chapter is to investigate in econometric terms the concept of an investment development cycle formulated by Dunning in order to throw light on these two different theoretical approaches to the growth of Third World outward investment.

Cross-sectional analysis of net outward investment and GNP per capita of thirty countries is undertaken for the periods 1960–75 and 1976–84, as well as for the entire period 1960–84. The analysis is also undertaken for five sub-periods within this time span: 1960–5, 1966–70, 1971–5, 1976–80 and 1981–4. The quadratic functional form provides a means of testing whether a J-shaped or inverted L-shaped investment development curve gives a good fit of the cross-section data. However, additional measurable and classificatory variables are introduced to improve the fit of the quadratic equation.

As a necessary supplement to the static cross-section analysis, the next stage of the econometric investigation is based on a time-series analysis of net outward investment (NOI) flows and GNP per capita of the 30 countries in the sample for the period 1960–75 and 1970–84 in order to test in dynamic terms the underlying dynamic theoretical exposition of the investment development cycle pertaining to the different levels of NOI as countries advance through higher stages of development. Such analysis is undertaken using linear and log-linear equations with the NOI flows examined both in untransformed terms as well as through the moving averages method. The results of time-series analysis enable countries to be arranged accordingly into three distinct groups depending on whether their NOI flows are negatively related, unrelated or positively related to GNP per capita.

THE INVESTMENT DEVELOPMENT CYCLE

The concept of an investment development cycle states that there is a relationship between NOI and a country's relative stage of development as measured by GNP per capita. Since worldwide inward investment must equal outward investment, by definition, the measure of economic development is considered more in relative than in absolute terms (Dunning, 1986a).

Although the concept of an investment development cycle represents a relationship between NOI and a country's relative stage of development with a balanced investment position at the early and late stages of the cycle, the term 'cycle' seems inappropriate when referring to the presence of a J-shaped investment development curve. The terms 'stage' or 'path' seem more appropriate as countries at higher levels of development have not returned but have passed through and have progressed beyond the NOI position of countries at lower levels of development.

Dunning suggested that the plotted data of the NOI and GNP of different countries, both variables normalised by the size of the population, show the presence of a J-shaped investment development curve with countries classified as belonging to four main groups corresponding to four stages of development. However, a fifth group corresponding to the fifth stage of development was later added (Dunning, 1988a).

The first group of countries belonging to the first stage of development is taken to comprise those that have little inward and no outward investment and therefore a zero or small negative NOI. The low level of inward investment is largely due to the lack of adequate location-specific advantages offered by the host country to attract foreign investment. The absence of outward investment, on the other hand, is due to the lack of ownership-specific advantages on the part of indigenous firms to make the direct investment route feasible. The few ownership advantages that these firms may possess may be better exploited through other routes, namely minority direct investment, contractual resource transfers and/or exports.

The second group of countries is taken to comprise those whose inward investment may be rising but outward investment remains negligible and therefore NOI is becoming more negative. In this stage, the host country is able to attract inward direct investment as a result of improvement in location-specific advantages with the expansion of the domestic market and the decline in the variable costs of servicing such a market. As in the first stage, outward direct investment remains small as indigenous firms have not generated adequate ownership-specific advantages to overcome the initial barriers to entry in foreign production.

The third group of countries corresponding to a third stage of development are those in which NOI is still negative but is becoming smaller as a result of either of two factors. First, inward investment may be falling with respect to outward investment or, second, outward investment may be rising faster than inward investment. NOI per capita declines in this stage because the original ownership-specific advantages of foreign firms may decline as these firms move through the industry technology cycle (Magee, 1977a) or the ownership-specific advantages of indigenous firms improve further as a result of, *inter alia*, the presence of foreign affiliates, larger markets and host government assistance. As a result of the improvement in their ownership-specific advantages, these firms may have a greater capacity to exploit foreign markets through direct investment.

The fourth group of countries are those in which NOI per capita is positive and rising, either as a result of outward investment surpassing inward investment or the growth of outward investment being faster than that of inward investment. The NOI of these countries is a manifestation of the strong ownership-specific advantages of indigenous firms as well as an increasing propensity to exploit these advantages through internalisation as a result of their growing size and geographical diversification and the increased opportunity for regional or global products and process specialisation.

A fifth group corresponding to a fifth stage has recently been identified by Dunning, which is associated with the rise of cross-investments between developed countries. As a consequence of rising intra-industry production between the major industrialised countries, there is a fall back in NOI towards zero at this stage, as a result of the levelling off in the growth of NOI in the fourth stage. This stage is characterised as one in which the ownership advantages become firm specific rather than country specific, i.e. ownership advantages are regarded as being more of the transaction cost minimising kind than the asset kind. In addition, at this stage, locational decisions by foreign and domestic MNEs become based less on the distribution of factor endow-

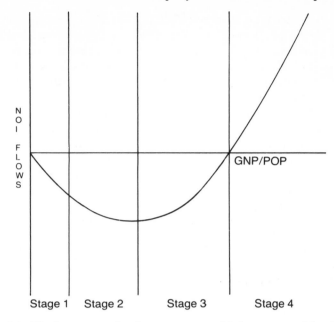

Figure 5.1 The investment development curve with four stages of development

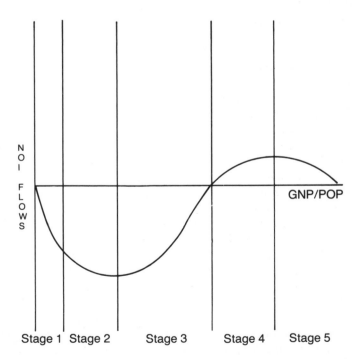

Figure 5.2 The investment development cycle with five stages of development

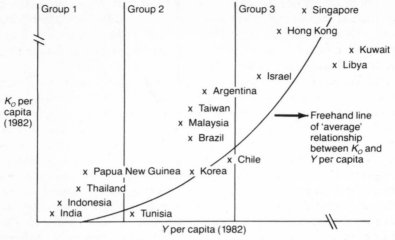

Figure 5.3 Outward direct investment stake (K₀) and income levels (Y) for selected
developing countries
Source: Dunning, 1986a.

ments and more on the differentiation of consumer tastes, economies of scale
and the gains from internalising related activities over space. In contrast to the
first four stages, the fifth stage is taken to be less dependent on the stage of
development of a country but more on its economic position *vis-à-vis* that of
other countries. In sum, this stage takes account of the increasing convergence
of countries' income levels and economic structure and their increasing propensity for cross-investment flows.

An idealised form of the investment development curve with the presence
of the original four groups of countries is illustrated in Figure 5.1. The idealised
form of such a curve with the presence of a fifth group of countries is illustrated
in Figure 5.2, which suggests that the curve turns downwards and moves back
towards zero on the right-hand side due to substantial cross investments
between the high income countries that comprise this group. The change in
the level of net outward investment that accompanies a more advanced stage
of development is in part based upon a change in the nature and composition
of outward investment as the level of national development rises. Early foreign
ventures are frequently resource based or sometimes import substituting, but
in each case have a quite specific location associated with a particular type of
activity. However, upon maturity, firms move beyond investment in a single
activity or product, and adopt a more international perspective on the location
of their different types of production activities. At this stage cross-investments
between countries become more common, and the visible hand of the direct
organisation of an international division of labour by firms increasingly
replaces the invisible hand of co-ordination by international market trans-

actions. The character of their overseas investment activity and their owner-ship advantages then becomes less determined by conditions in their home countries, although a recent study by Patel and Pavitt (1991) shows that home country-specific factors dominate over firm-specific factors in determining national technological performance.

Figure 5.3 is taken from Dunning (1986a) and suggests that most developing countries can be classified as belonging to one of the first three stages in the investment development cycle. For example, countries such as Papua New Guinea, Thailand and Indonesia are characterised as having a large negative NOI and hence belong to the first stage of the cycle, i.e. Group 1 countries. On the other hand, Taiwan, Malaysia and Brazil are characterised as having a larger negative NOI because, although inward investment is rising, outward investment remains negligible. They therefore belong to the second stage of the cycle, i.e. Group 2 countries. Finally, countries such as Hong Kong and Singapore are characterised as having a smaller negative NOI owing to either inward investment falling with respect to outward investment or outward investment rising faster than inward investment. They belong to the third stage of the cycle, i.e. Group 3 countries.

The emerging outward direct investment of firms from developing coun-tries is explained by the investment development cycle in terms of the growing ownership advantages which are, in part or in whole, country specific in origin. These ownership advantages are then internalised rather than sold in a licens-ing arrangement in order to maximise the economic rent from these assets, discounted for risks. Finally, the route of direct investment is chosen over exports because firms find that the combination of their ownership advantages with other, usually immobile, resources in a foreign market is a profitable means of exploiting that market.

The major difference between the operations of Third World MNEs and those from developed countries is explained by the investment development cycle in terms of the different nature and composition of ownership and internalisation advantages which reflect the greater maturity and sophistica-tion of MNEs from the developed countries. The speed and direction with which various countries move along different stages of the investment devel-opment cycle is then taken to depend on the structure of indigenous resource endowments, the extent of interaction with the rest of the world, the size of the local market, the economic system, government policy and, finally, the nature of the markets for the kind of transactions firms wish to engage in with foreign entities. These factors influence the extent and kind of ownership, location and internalisation advantages that different firms from different countries possess, the industrial pattern of their inward and outward invest-ments and, consequently, the deviations of countries around the average J-shaped NOI or investment development curve at a specific point in time. However, the model also postulates that a dynamic analysis of the influence of these factors on the ownership, location and internalisation advantages of firms would enable predictions to be made about countries' likely future investment position.

ECONOMETRIC TESTING OF THE INVESTMENT DEVELOPMENT CYCLE

The econometric investigation of the investment development cycle consists of both cross-section and time-series analysis of the net outward investment flows and GNP per capita for a sample of thirty countries for which data on NOI flows were available for the period 1960–84 from the various issues of the IMF *Balance of Payments Statistics*. The sample consisted of nineteen developed countries and eleven developing countries. The full list of countries included in the sample is shown in Table 5.1.

Cross-section analysis

The cross-section analysis between NOI flows and GNP per capita for the sample of thirty countries was undertaken for the following periods: 1960–75, 1976–84 and 1960–84. Five sub-periods are also analysed at the second stage:

Table 5.1 Full list of countries included in the cross-section regression to test investment development cycle

Developed countries	Developing countries
Western Europe	*Africa*
EC	Tunisia
Belgium	*Asia and Pacific*
Denmark	Korea
France	Philippines
West Germany	Thailand
Italy	*Latin America*
Netherlands	Argentina
Spain	Brazil
Portugal	Chile
UK	Colombia
Other Europe	Costa Rica
Austria	*Middle East*
Finland	Israel
Norway	Libya
Sweden	
North America	
Canada	
USA	
Other developed countries	
Australia	
Japan	
New Zealand	
South Africa	

1960–5, 1966–70, 1971–5, 1976–80 and 1981–4. The NOI flows for each of the thirty countries were calculated through the summation of annual flows and the average GNP per capita for each of the thirty countries was calculated for the period under consideration. Since the plotted data of NOI and GNP are reflected in the investment development cycle as having a J-shaped NOI or investment development curve, the regression equation used to estimate the cross-section relationship is the quadratic equation:

$$NOI_i = \alpha + \beta(GNP_i/POP_i) + \gamma(GNP_i/POP_i)^2 + \mu_i$$

where NOI is Net Outward Investment flows and GNP/POP is Gross National Product divided by the population for each country i.[1]

To test whether a J-shaped investment development curve exists with a minimum turning point at some negative level of NOI requires a test of the prediction that α is zero, β is negative and γ is positive. The results of the regression equation are as follows:

The period 1960–75

$$NOI_i = 7065 - 17.97 \, (GNP_i / POP_i) + 0.0057 \, (GNP_i / POP_i)^2$$
$$\quad\quad (1.58) \quad (-3.64) \quad\quad\quad (4.95)$$

$R^2 = 0.574$ R-bar squared (adjusted for degrees of freedom) = 0.542
F-statistic (2,27) = 18.22 Mean of Dependent Variable = 2019.41

The result is in line with the predictions of an investment development cycle. The observed F-statistic, 18.22, is greater than the theoretical F value with $\upsilon_1 = k-1 = 3-1 = 2$ and $\upsilon_2 = n-k = 30-3 = 27$ degrees of freedom of 3.36 at the 95 per cent level of significance which leads to the acceptance of the hypothesis that the regression equation is significant, i.e. that both (GNP/POP) and (GNP/POP)^2 are significant explanatory factors.

The critical value of the t-statistic for a two-sided t-test at the 1 per cent significance level with $n - k = 30-3 = 27$ degrees of freedom is 2.77. This leads to the conclusion that although the estimated coefficient on the constant term is not significantly different from zero, those of the independent variables (GNP/POP) and (GNP/POP)^2 are. The regression results also reveal the right signs on the estimated coefficient of the explanatory variables, a negative β coefficient on (GNP/POP) and a positive γ coefficient on (GNP/POP)^2. The fact that the coefficient on the constant term is not significantly different from zero simply confirms the proposition that the J-shaped investment development curve begins from the origin for countries at the lowest levels of development. The graph of the investment development curve for the period 1960–75 is illustrated in Figure 5.4.

The turning point of the predicted investment development curve can be calculated through the differentiation of the quadratic equation

$$NOI_i = \alpha + \beta(GNP_i/POP_i) + \gamma \, (GNP_i/POP_i)^2$$

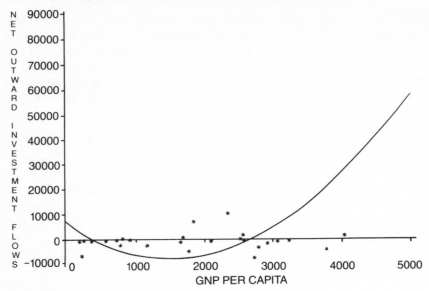

Figure 5.4 Investment development cycle, 1960–75

which yields:

$$dNOI_i/d \,(GNP_i/POP_i) = \beta + 2\gamma \,(GNP_i/POP_i)$$
$$dNOI_i/d(GNP_i/POP_i) = 0 \text{ where } (GNP_i/POP_i) = -\beta/2\gamma$$
$$d(NOI_i)^2/ \, d^2 \,(GNP_i/POP_i) = 2\gamma$$
$$d(NOI_i)^2/d^2 \,(GNP_i/POP_i) > 0 \text{ where } \gamma > 0 \text{ and the turning point is a}$$
minimum and not a maximum.

Applying the formula for a minimum turning point to the quadratic equation in the cross-sectional analysis for the period 1960–75 yields:

$$(GNP/POP) = -(-17.9673) / 2 \,(0.0057)$$
$$= 1580.5292$$

The period 1976–84

$$NOI_i = -6322 + 2.73 \,(GNP_i / POP_i) - 0.0002 \,(GNP_i / POP_i)^2$$
$$(-1.28) \,(1.25) \qquad\qquad (-0.92)$$
$$R^2 = 0.124 \text{ R-bar squared (adjusted for degrees of freedom)} = 0.059$$
F-statistic (2,27) = 1.92 Mean of Dependent Variable = 1879.57

In this case, the observed F-statistic is less than the theoretical F value with $\upsilon_1 = 2$ and $\upsilon_2 = 27$ degrees of freedom at the 95 per cent level of significance which leads to the rejection of the hypothesis that the regression equation is significant. Similarly, the observed t-statistics on the estimated coefficients on

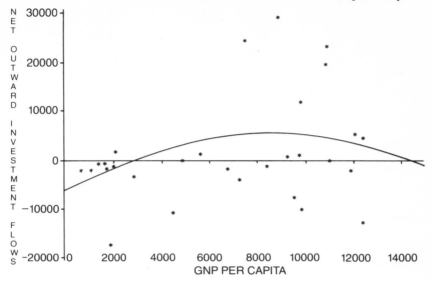

Figure 5.5 Investment development cycle, 1976–84

the constant term and the explanatory variables fall within the acceptance region of the null hypotheses as defined by the critical t-value of 2.05 at the 5 per cent level of significance which leads to the conclusion that none of the estimated coefficients of the regression equation is significantly different from zero.

The econometric testing of the investment development curve shows a structural change in the relationship between NOI and GNP per capita in the mid-1970s. The graph of the investment development curve as estimated by the above equation for the period 1976–84 is illustrated in Figure 5.5. The turning point of the curve as calculated by the formula $(GNP_i/POP_i) = -\beta/2\gamma$ is:

$$(GNP/POP) = -(2.73) / 2 (-0.0002)$$
$$= 8551.99$$

The shape of the curve obtained for the period 1976–84 is a reversal of that obtained for the earlier period, 1960–75. While the proper negative sign and positive sign are obtained on β and γ respectively for the period 1960–75 giving a J-shaped NOI curve, for the period 1976–84 the signs are reversed to yield an inverted J-shaped NOI curve.

The next stage of the investigation is to determine the impact of the structural change that occurred from the mid-1970s on the overall relationship between NOI flows and GNP per capita for the entire period 1960–84.

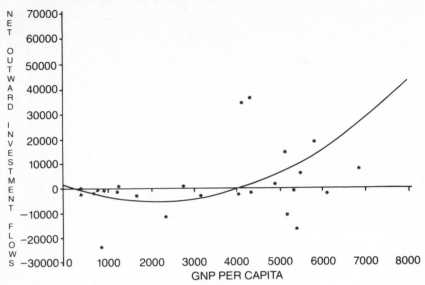

Figure 5.6　Investment development cycle, 1960–84

The period 1960–84

$$NOI_i = 1268.09 + - 6.11 (GNP_i / POP_i) + 0.001 (GNP_i / POP_i)^2$$
　　　(0.18)　(–1.15)　　　　　　　　　(1.8750)
　　$R^2 = 0.305$ R-bar squared (adjusted for degrees of freedom) $= 0.254$
　　F-statistic (2,27) $= 5.924$ Mean of Dependent Variable $= 3,992.64$

The observed F-statistic, 5.9236, is greater than the theoretical F value with $\upsilon_1 = 2$ and $\upsilon_2 = 27$ degrees of freedom of 3.36 at the 95 per cent level of significance although such an observed F-statistic in this period is substantially less than that calculated for the period 1960–75 at 18.22. Nevertheless, the regression equation is still significant, with both the explanatory variables accounting for only 30.5 per cent of the variation of the dependent variable, as compared to 57.4 per cent for the period 1960–75. Even a comparison of the R-bar squared statistics reveals that the regression equation accounts for 54.2 per cent of the variation in the dependent variable during the period 1960–75 compared to only 25.4 per cent for the period 1960–84.

However, although the regression equation may be significant, the observed t-statistics on the estimated coefficients on the constant term and the explanatory variables fall within the acceptance region of the null hypotheses as defined by the critical t-value of 2.052 at the 5 per cent level of significance. This leads to the conclusion that none of the estimated coefficients of the regression equation is significantly different from zero even though the coefficients on β and γ show the right signs.

Such findings show that the structural change that occurred during the period from the mid-1970s was sufficiently large to nullify the previously significant statistical relationship between NOI flows and GNP per capita. The graph of the investment development curve in the period 1960–84 is illustrated in Figure 5.6. Applying the formula to determine the turning point of the curve yields:

$$(GNP/POP) = -\beta/2\gamma$$
$$= -(-6.1075)/2(.0014)$$
$$= 2147.3027$$

The changing shape of the investment development curve is an indication of a structural change in the relationship between NOI and GNP per capita. In order to investigate this phenomenon further, a cross-sectional analysis of five distinct sub-periods is undertaken, namely: 1960–5, 1966–70, 1971–5, 1976–80 and 1981–4.

The period 1960–5

$$NOI_i = 1816.46 + -7.14\,(GNP_i/POP_i) + 0.0035\,(GNP_i/POP_i)^2$$
$$(2.41)\quad(-5.52)\qquad\qquad(7.51)$$
$R^2 = 0.744$ R-bar squared (adjusted for degrees of freedom) = 0.725
F-statistic (2,27) = 39.21 Mean of Dependent Variable = 362.24

The observed F-statistic ratio is greater than the theoretical F value with υ_1 = 2 and υ_2= 27 of 3.36 at the 95 per cent level of significance, which leads to the conclusion that the regression equation is significant, with both explanatory variables accounting for 74.4 per cent of the variation in NOI flows. The observed t-statistic on the estimated coefficient of the constant term falls within the rejection region of the null hypothesis as defined by the critical value of 2.771 at the 1 per cent level of significance, which leads to the conclusion that the estimated coefficient on the constant term is not significantly different from zero. On the other hand, the coefficients on the explanatory variables, β and γ, are significant at the 1 per cent level and exhibit the right signs.

The period 1966–70

$$NOI_i = 1745.99 + -4.74\,(GNP_i/POP_i) + 0.0016\,(GNP_i/POP_i)^2$$
$$(2.03)\quad(-5.31)\qquad\qquad(8.24)$$
$R^2 = 0.798$ R-bar squared (adjusted for degrees of freedom) = 0.783
F-statistic (2,27) = 53.375 Mean of Dependent Variable = 523.177

The observed F-statistic, 53.375, is greater than the theoretical F value with υ_1 = 2 and υ_2 = 27 degrees of freedom of 3.36 at the 95 per cent significance level, which leads to the conclusion that the regression equation is significant. The independent variables account for 79.8 per cent of the variation in the dependent variable. The t-statistics on the estimated coefficients of the explanatory variables, (GNP/POP) and (GNP/POP)^2, fall within the rejection region of the null hypotheses as defined by the critical t value of 2.77 at 27 degrees of freedom and 1 per cent level of significance. This leads to the conclusion that the coefficients on the independent variables, β and γ, are significantly different from zero. Furthermore, an examination of these coefficients reveals the right signs. On the other hand, the t-statistic on the constant term falls within the acceptance region of the null hypothesis which means that the estimated coefficient is not significantly different from zero and the curve originates from zero.

The period 1971–5

$$NOI_i = 1677.46 + - 3.48 \, (GNP_i \, / \, POP_i) + 0.0008 \, (GNP_i \, / \, POP_i)^2$$
$$(0.549) \quad (-1.410) \qquad\qquad (1.957)$$
$R^2 = 0.219$ R-bar squared (adjusted for degrees of freedom) = 0.161
F-statistic (2,27) = 3.785 Mean of Dependent Variable = 1130.79

The observed F-statistic, 3.785, is greater than the theoretical F value of 3.36 with $\upsilon_1 = 2$ and $\upsilon_2 = 27$ at the 95 per cent level of significance, which leads to the conclusion that the regression equation is significant, with the independent variables accounting for 21.9 per cent of the variation in the dependent variable. However, the t-statistics on the estimated coefficients of the constant term and β fall within the acceptance region of the null hypotheses as defined by the critical t value of 2.052 at 27 degrees of freedom at the 5 per cent level of significance. However, the t-statistics on γ, 1.9568, are close to the rejection region of the null hypothesis at the 5 per cent level.[2]

The period 1976–80

$$NOI_i = - 1210.01 + - 0.077 \, (GNP_i \, / \, POP_i) + 0.00008 \, (GNP_i \, / \, POP_i)^2$$
$$(-0.321) \quad (-0.045) \qquad\qquad (0.568)$$
$R^2 = 0.164$ R-bar squared (adjusted for degrees of freedom) = 0.102
F-statistic (2,27) = 2.649 Mean of Dependent Variable = 1977.03

The observed F-statistic, 2.649, is less than the theoretical F value of 3.36 with $\upsilon_1 = 2$ and $\upsilon_2 = 27$ degrees at the 95 per cent level of significance, which leads to the conclusion that the regression equation is *insignificant*, i.e. that (GNP/POP) and (GNP/POP)^2 are not significant explanatory factors of the variation in NOI flows. An examination of the t-statistics confirms the breakdown of the regression equation in this period. The t-statistics on the estimated coefficient of the constant term and both the explanatory variables fall within the acceptance region of the null hypotheses, which means that none of the estimated coefficients in the regression equation is significantly different from zero at the 5 per cent level of significance.

The period 1981–4

$$NOI_i = -4007.38 + 1.94 (GNP_i / POP_i) + 1.9370 (GNP_i / POP_i)^2$$
$$(-0.321) \ (-0.045) \qquad\qquad (0.568)$$

$R^2 = 0.0784$ R-bar squared (adjusted for degrees of freedom) = 0.010
F-statistic (2,27) = 1.149 Mean of Dependent Variable = -109.463

The observed F-statistic for the regression equation in this period deteriorated further to 1.15 from 2.65 in the previous period. This low value of the F-statistic is therefore much less than the theoretical F value with $\upsilon_1 = 2$ and $\upsilon_2 = 27$ degrees of freedom of 3.36 at the 95 per cent level of significance and leads to the conclusion that the regression is *insignificant*. The examination of the t-statistics on the estimated coefficients in the regression equation confirms this finding. The value of the t-statistics on the constant term and the two explanatory variables falls within the acceptance region of the null hypotheses, which means that none of the estimated coefficients in the regression equation is significantly different from zero at the 5 per cent level of significance.

The cross-section regression analysis of the five different sub-periods shows that the presence of a J-shaped NOI or investment development curve is confirmed only for the three earlier periods, 1960–5, 1966–70 and 1971–5 as shown by the results of the F-test for the significance of the regression equation at the 95 per cent level. However, the t-statistic on the estimated coefficients of the regression equation is statistically insignificant from zero at the 5 per cent level from the period 1971–5 which means that additional variables ought to be introduced into the regression equation to improve the goodness of fit and to increase the value of the t-statistics on the estimated coefficients of the regression equation. This exercise is undertaken in the next section.

Improvement of fit of the quadratic cross-section equation

In the previous section, the quadratic equation $NOI_i = \alpha + \beta(GNP_i/POP_i) + \gamma (GNP_i/POP_i)^2$ was used to estimate the relationship between NOI flows and GNP per capita for the thirty countries in the sample for the period 1960–75, 1976–84 and 1960–84. The *a priori* expectation is that α must be zero, β must have a negative sign and γ must have a positive sign in order to yield the J-shaped investment development curve proposed by Dunning (1981a, 1981b, 1986a, 1986c, 1988a).

The quadratic equation for the period 1960–75 yielded a statistically significant negative estimated coefficient on (GNP/POP), a statistically significant positive coefficient on (GNP/POP)^2 and a statistically insignificant estimated coefficient on the constant term in broad conformity with the expectation that the investment development curve originates from zero. The equation had an R^2 value of 0.574. This section aims to improve the goodness of fit of the quadratic equation used in estimating the cross-section equations in the previous section.

The first stage consists of introducing additional measurable variables which were adopted by Dunning (1986a, c), namely:

1 Percentage of educated population defined as the number enrolled in secondary school as a percentage of age group plus the number enrolled in higher education as a percentage of population aged 20–4, divided by 2.

2 Percent of labour employed in industry and services (i.e. non-agricultural services).

3 Value of exports plus imports of goods, divided by 2, as a percentage of GNP.

4 Urban population as a percentage of total population.

Data on these four measurable variables for the thirty countries in the sample were obtained for the median year 1968 from various issues of the *World Development Report* by the World Bank and from the *International Financial Statistics Yearbook* of the IMF. These variables were then introduced to the quadratic equation individually and in combination. A total of fifteen regression equations were tested.

The regression equations showed an improvement in the R-bar squared value (R^2 adjusted for degrees of freedom) in three out of the fifteen cases and an actual reduction in the R-bar squared value in twelve out of the fifteen cases. More importantly, although the t-statistics on the estimated coefficients improved in some of the equations tested, not one of fifteen regression equations tested exhibited a significant F-statistic on the additional explanatory variables tested as a whole. In addition, on the basis of t-tests, not one of the fifteen regression equations had a significant estimated coefficient on the additional measurable variable introduced.

On the basis of these findings, new additional explanatory variables had to be introduced. An inspection of the size of the residuals obtained in the 1960–75 quadratic cross-section equation revealed that countries such as Canada, Japan, Sweden, the UK, the USA and Brazil had the largest residuals among the countries in the sample. Japan, the UK and the USA were way above the investment development curve while countries such as Canada, Sweden and Brazil were way below it.

Three hypotheses were then tested, of which two pertain to countries such as Japan, the UK and the USA which were above the curve and whose NOI flows are relatively high, given GNP per capita. The first hypothesis is suggested by Japan and pertains to the impact on NOI flows of policy constraints on inward investment. The presence of such constraints on inward investment has a negative impact on inward investment flows which would lead to higher NOI flows given GNP per capita. The second hypothesis is suggested by the UK and the USA and pertains to the positive impact on NOI flows of being traditional outward investors. Traditional outward investors which have a large stock of outward investment are expected to have large flows of outward investment if only on the basis of reinvested profits. As a result, traditional outward investors are expected to have higher NOI flows.

The third hypothesis to be tested pertains to countries such as Canada and Brazil whose NOI flows were relatively low given average GNP per capita. This hypothesis is the impact on NOI flows of being traditional host countries to FDI. Traditional host countries which have a large stock of inward investment are expected to have large flows of inward investment through reinvested profits. As a result, traditional host countries are expected to have lower NOI flows.

The information sources for the values of the dummy variables are *Policy constraints on inward investment*, Dunning and Cantwell (1987); *Traditional Host and Home Countries of FDI*, US Department of Commerce (1960) which gives data on US outward investment in 1957; and the UK Department of Industry (1974) which gives data on UK outward investment in 1962. These sources reveal that countries such as Canada, Brazil and Chile are traditional host countries to US FDI while Australia is a traditional host country to UK FDI. Dunning (1983a) shows that the USA, the UK and the Netherlands are traditional source countries for FDI.

Each of these three hypotheses was then tested on the basis of three classificatory dummy variables respectively. For example, the presence of policy constraints on inward investment would give a value of 1 to the first dummy variable and a value of zero otherwise. The UK, the USA and the Netherlands take the value of 1 in the second dummy variable because these countries are traditional outward investors while other countries assume a value of zero. Finally, Canada, Brazil, Australia and Chile have a value of 1 in the third dummy variable since these countries are considered traditional host countries to FDI. The *a priori* expectation is that the coefficient on dummy variables 1 and 2 is positive and the coefficient on dummy variable 3 is negative.

The regression results with the introduction of each of these dummy variables to the quadratic equation obtained for the period 1960–75 are as follows:

The period 1960–75

a) The introduction of dummy variable 1 (DV1)

$$NOI_i = 5809.83 + - 17.65(GNP_i/POP_i) + 0.006(GNP_i/POP_i)^2 + 4876.18 \text{ (DV1)}$$
$$(1.224) \quad (-3.543) \qquad\qquad (4.922) \qquad\qquad (0.839)$$
$R^2 = 0.586$ R-bar squared (adjusted for degrees of freedom) = 0.538
F-statistic (2,27) = 12.249 Mean of Dependent Variable = 2019.41

The F-test undertaken to check the significance of the additional explanatory variable, DV1, reveals that the coefficient on the DV1 is not statistically different from zero. In addition, the t-test on the coefficient of DV1 yielded insignificant results. The t-statistics on the constant term and explanatory variables were actually lower when compared to the original quadratic equation. Finally, a comparison of the R-bar squared values actually shows that the extended equation has a lower value. These results show that the presence of policy constraints on inward investment has no significant impact on the NOI flows.

b) The introduction of dummy variable 2 (DV2)

$$NOI_i = 6349.12 + 16.66\,(GNP_i/POP_i) + 0.005\,(GNP_i/POP_i)^2 + 21836.2\,(DV2)$$
$$\quad (1.742)\,(-4.134) \qquad\qquad (5.114) \qquad\qquad (3.852)$$
$$R^2 = 0.729 \text{ R-bar squared (adjusted for degrees of freedom)} = 0.698$$
$$\text{F-statistic } (2,27) = 23.32 \text{ Mean of Dependent Variable} = 2019.41$$

The F-test on the significance of the additional explanatory variable, DV2, shows that the coefficient on the DV2, is statistically different from zero. This finding is confirmed by the t-test on the estimated coefficient on the DV2, which is positive and significantly different from zero at the 1 per cent level. The t-statistics on the constant term and the explanatory variables are higher than in the original quadratic equation and show that at the 1 per cent level the constant term is not significantly different from zero although β and γ are significantly different from zero and show the right signs. A comparison of the R-bar squared values of the two equations shows a much higher R-bar squared value of 0.698 for the extended quadratic equation with DV2 as an additional explanatory variable compared to 0.543 for the unextended quadratic equation.

All these results show that being a traditional outward investor has a significantly positive impact on NOI flows.

c) The introduction of dummy variable 3, DV3

$$NOI_i = 8942.54 + -18.78\,(GNP_i/POP_i) + 0.006\,(GNP_i/POP_i)2 + -10421.8\,(DV3)$$
$$\quad (2.046)\,(-3.979) \qquad\qquad (5.356) \qquad\qquad (-1.95)$$
$$R^2 = 0.629 \text{ R-bar squared (adjusted for degrees of freedom)} = 0.586$$
$$\text{F-statistic } (2,27) = 14.675 \text{ Mean of Dependent Variable} = 2019.41$$

The F-test on the significance of the additional explanatory variable, DV3, shows that the coefficient of the additional explanatory variable is not statistically different from zero at the 95 per cent significance level. The F-statistic on the additional explanatory variable is 3.80 which is slightly less compared to the theoretical F value of 4.20 with $\upsilon_1 = 1$ and $\upsilon_2 = 26$ degrees of freedom at the 95 per cent significance level. The t-test on the estimated coefficient of DV3 shows that although such a coefficient is not statistically different from zero at the 5 per cent level, it is statistically different from zero at the 10 per cent level and exhibits the right negative sign. The t-statistics on the constant term, (GNP/POP) and (GNP/POP)^2, have increased compared to the unextended quadratic equation. Such statistics show that at the 1 per cent level, the constant term is not significantly different from zero while β and γ are significantly different from zero. Finally, a comparison of the R-bar squared values between this equation and the quadratic equation yields a higher value of 0.586 for this equation compared to 0.543 for the unextended quadratic equation. All these results show that countries that are traditional hosts to FDI have statistically lower NOI flows at the 10 per cent level of significance.

Since DV2 and DV3 are the significant additional explanatory variables, the last part of the econometric exercise is to introduce these two additional explanatory variables together into the quadratic equation.

d) The introduction of DV2 and DV3

$$NOI_i = 7786.55 + - 17.35 \ (GNP_i/POP_i) + 0.005 \ (GNP_i/POP_i)^2 + 20234.2 \ (DV2)$$
$$\quad (2.15) \quad (-4.44) \qquad\qquad (5.48) \qquad\qquad\qquad (3.65)$$
$$+ - 7685.58 \ (DV3)$$
$$+ (-1.72)$$

$R^2 = 0.757$ R-bar squared (adjusted for degrees of freedom) = 0.719
F-statistic (2,27) = 19.549 Mean of Dependent Variable = 2019.41

The F-test shows that the regression is significant, i.e. that the right-hand variables are significant explanatory factors of the variation in NOI flows. The t-tests on the estimated coefficient of the additional classificatory variables DV2 and DV3 show a statistically significant coefficient at the 1 per cent and 10 per cent level of significance respectively. The t-statistics on the estimated coefficients of the constant term and explanatory variables have increased and show that at the 1 per cent level, the constant is not significantly different from zero while β and γ are significantly different from zero. All the estimated coefficients exhibit the right signs, negative on β and DV3 and positive on γ and DV2. More importantly, an examination of the R-bar squared values shows the value of 0.719 which is considerably higher than 0.543 for the unextended quadratic equation.

We now turn to the examination of the period 1976–84 using our new extended quadratic equation with two dummy variables.

The period 1976–84

$$NOI_i = - 5731.07 + 2.58 \ (GNP_i/POP_i) + - 0.0002 \ (GNP_i/POP_i)^2 + 3795.71 \ (DV2)$$
$$\quad (-1.107) (1.150) \qquad\qquad (-0.862) \qquad\qquad\qquad (0.580)$$
$$+ - 2543.14 \ (DV3)$$
$$(-0.458)$$

$R^2 = 0.146$ R-bar squared (adjusted for degrees of freedom) = 0.009
F-statistic (2,27) = 1.065 Mean of Dependent Variable = 1879.57

The value of the F-statistic is less than the theoretical F value of 2.69 with $\upsilon_1 = 4$ and $\upsilon_2 = 25$ degrees of freedom at the 95 per cent level of significance, which leads to the conclusion that the regression equation is insignificant, i.e. that the right-hand variables are insignificant explanatory factors of the variation of NOI flows. In particular, the F-test undertaken to assess the significance of the estimated coefficients on the additional explanatory variables, DV2 and DV3, reveals that such coefficients are not statistically different from zero.

The results of the F-test are reinforced by the low values of the t-statistics which show that the estimated coefficients on all the right-hand variables are not significantly different from zero. In fact, the t-statistics on the constant term, (GNP/POP) and (GNP/POP)^2, have actually decreased from those in the unextended equation without the dummy variables. Nevertheless, the signs of the estimated coefficients on the explanatory variables still represent a reversal of the signs obtained for the period 1960–75. These findings confirm and emphasise the presence of a structural change in the relationship between NOI flows and GNP per capita from 1976, which coincided with the surge of new sources of outward direct investment from both the developed and developing countries.

We now turn to an examination of the period 1960–84 using our new extended quadratic equation with two dummy variables.

The period 1960–84

$$NOI_i = 2174.83 + -5.76(GNP_i/POP_i) + 0.0012 (GNP_i/POP_i)^2 + 30850.6 (DV2)$$
$$(0.383) \quad (-1.396) \qquad\qquad (1.984) \qquad\qquad\qquad (4.06)$$
$$+ -8490.28 (DV3)$$
$$(-1.338)$$

$R^2 = 0.617$ R-bar squared (adjusted for degrees of freedom) = 0.555
F-statistic (2,27) = 10.049 Mean of Dependent Variable = 3992.64

The value of the F-statistic is greater than the theoretical F value at the 95 per cent level of significance, leading to the finding that the regression is significant. However, although the t-statistics on the constant term and all the explanatory variables have increased from those of the unextended quadratic equation, these t-statistics are still not high enough to reject the null hypotheses that the estimated coefficients are equal to zero at the 95 per cent level of significance.

The analysis of the regression results in the periods 1960–75, 1976–84 and 1960–84, with the introduction of additional classificatory explanatory variables to the quadratic equation, emphasises the structural change that occurred in the relationship between NOI flows and GNP per capita in the period from 1975 onwards. An examination of the individual five-year regression equations from 1960–5, 1966–70, 1971–5, 1976–80 and 1981–4 would enable us to verify the occurrence of such structural change.

The period 1960–5

$$NOI_i = 1930.84 + -5.76(GNP_i/POP_i) + 0.0012 (GNP_i/POP_i)^2 + 30850.6 (DV2)$$
$$(0.383) (-6.856) \qquad (8.487) \qquad (3.263)$$
$$+ -8490.28 (DV3)$$
$$(-2.33)$$

$R^2 = 0.857$ R-bar squared (adjusted for degrees of freedom) = 0.834
F-statistic (2,27) = 37.393 Mean of Dependent Variable = 362.240

The high value of the F-statistic enables the conclusion to be drawn that the regression equation is significant, i.e. that the right-hand variables are significant explanatory factors of the variation in NOI flows. The t-tests undertaken show that the estimated coefficients on all the explanatory variables apart from DV3 are statistically different from zero at the 1 per cent level of significance. The explanatory variables which show the right signs explain 85.7 per cent of the variation in NOI flows.

The period 1966–70

$$NOI_i = 1931.07 + -5.76(GNP_i/POP_i) + 0.0015 (GNP_i/POP_i)^2 + 3556.08 (DV2)$$
$$(2.789) \quad (-6.461) \qquad (9.204) \qquad (2.973)$$
$$+ -2329.07 (DV3)$$
$$(-2.537)$$

$R^2 = 0.884$ R-bar squared (adjusted for degrees of freedom) = 0.865
F-statistic (2,27) = 47.498 Mean of Dependent Variable = 523.177

The high value of the F-statistic leads to the conclusion that the regression equation is significant at the 95 per cent level of significance. The t-tests on the estimated coefficients of the explanatory variables apart from DV3 show that these coefficients are significantly different from zero at the 1 per cent level of significance. Hence, as in the period 1960–5, the explanatory variables in the equation explain a large proportion of the variation in NOI flows.

The period 1971–5

$$NOI_i = 2565.44 + -4.11(GNP_i/POP_i) + 0.0008 (GNP_i/POP_i)^2 + 13981.3 (DV2)$$
$$(2.789) (-6.461) \qquad (9.204) \qquad (2.973)$$
$$+ -3329.79 (DV3)$$
$$(-2.537)$$

$R^2 = 0.572$ R-bar squared (adjusted for degrees of freedom) = 0.503
F-statistic (2,27) = 8.346 Mean of Dependent Variable = 1130.79

The value of the F-statistic, which is greater than the theoretical F value of 2.69 with $\upsilon_1 = 4$ and $\upsilon_2 = 25$ degrees of freedom at the 95 per cent level of

significance, shows a significant regression equation. The t-tests on the estimated coefficients show that the coefficients on the measurable variables are
statistically different from zero at the 5 per cent level of significance and show
the right signs. The coefficient on DV2 is significantly different from zero at
the 1 per cent level. However, the estimated coefficient on the constant term
and DV3 are not statistically different from zero at the 5 per cent level.
Notwithstanding these results, the explanatory variables in the equation explain 57.2 per cent of the variation in the dependent variable, NOI flows.

The period 1976–80

$$NOI_i = 48.92 + -0.68\ (GNP_i/POP_i) + 0.0001\ (GNP_i/POP_i)^2 + 15888.6\ (DV2)$$
$$(0.016)\ (-0.503) \qquad (0.88) \qquad\qquad (4.117)$$
$$+ -2481.01\ (DV3)$$
$$(-0.761)$$

$R^2 = 0.521$ R-bar squared (adjusted for degrees of freedom) $= 0.444$
F-statistic $(2,27) = 6.79$ Mean of Dependent Variable $= 1977.03$

The F-statistic shows that the regression equation is significant, i.e. that the
right-hand variables are significant explanatory factors of the variation in NOI
flows. However, the t-tests on the estimated coefficients show that apart from
the coefficient on DV2, the other coefficients on the equation are not statistically different from zero at the 5 per cent level even though the explanatory
variables explain 52.1 per cent of the variation in NOI flows. Hence, during
this period the relationship between NOI flows and the explanatory variables
specified, which was significant in the periods 1960–5, 1966–70 and 1971–5,
has broken down and undergone some structural change.

The period 1981–4

$$NOI_i = -4734.92 + 2.38\ (GNP_i/POP_i) + -0.0002\ (GNP_i/POP_i)^2 +$$
$$(-1.033)\ (1.711) \qquad\qquad (-1.716)$$
$$-12515.6\ (DV2) + -2481.01\ (DV3)$$
$$(-1.840) \qquad\qquad (0.043)$$

$R^2 = 0.191$ R-bar squared (adjusted for degrees of freedom) $= 0.061$
F-statistic $(2,27) = 1.473$ Mean of Dependent Variable $= -109.463$

The F-statistic for this equation is below the theoretical F value of 2.76 with
$v_1 = 4$ and $v_2 = 25$ at the 95 per cent level of significance, which leads to the
acceptance of the null hypothesis that the regression equation is *insignificant*,
i.e. that the right-hand variables are not significant explanatory factors of the
variation in NOI flows. The t-tests on the estimated coefficients show that
none of the coefficients is statistically different from zero at the 5 per cent level
of significance.

The analysis of the cross-section data for the different time-periods from
1960 until 1984 show that the structural change that led to the breakdown of

the relationship between NOI flows and the specified explanatory variables started to occur in the mid-1970s and that such change has been sustained ever since. A comparison of the results obtained on the basis of the unextended quadratic equation and that of the extended quadratic equation with the introduction of two classificatory dummy variables, DV2 and DV3, to account for traditional home and host countries, shows that the structural change that was predicted to occur in the mid-1970s did occur in the regression equation for the period 1976 onwards only for the extended quadratic equation. In the unextended quadratic equation, the structural change seemed to begin much earlier, in the period 1971–5, even though the F-statistic during this period still revealed a significant regression. However, the t-tests on the estimated coefficients on the explanatory variables, (GNP/POP) and $(GNP/POP)^2$, show that these coefficients are not statistically different from zero at the 5 per cent level.

The data on foreign direct investment in the second chapter has shown that the surge of outward investment from the newer source countries such as Japan, Germany, Italy, Spain and France, as well as from a number of developing countries, occurred in the period from the mid-1970s onwards and therefore the extended quadratic equation seems to conform more to geographical trends in foreign direct investment. The existence of a structural change in the relationship from 1976 seems to conform to the dynamic *theory of technological accumulation* (Cantwell, 1989a) and the *theory of localised technological change* applied by Lall (1981, 1982b, 1983a, 1983c) to MNEs from developing countries. These theories suggest that the ownership advantages of firms and their capacity for internationalisation are related more to absolute than to relative levels of development. On the other hand, the investment development cycle, which suggests a relationship between NOI and a country's relative level of development, is only certain to be sustained over time if the determining factor is the country's relative stage of development, i.e. where firms at lower levels of development never really gain a major capacity for the internationalisation of their operations. Wells' product cycle model (1977, 1981, 1983, 1986) comes closer to suggesting that such relative level of development of countries and firms affects their competitive position and their capacity for sustained internationalisation.

The existence of an investment development cycle therefore throws light on the debate on Third World multinationals which suggests that in analysing the growth of outward investment over time, theories predicated on relative levels of development, which is a strong feature of both the investment development cycle and the product cycle model, provide inadequate explanations of the fact that firms at some threshold absolute lower level of development may also have the capacity for internationalisation. However, although the dynamic theories of technological competence and localised technological change predicate that the capacity for technological development and internationalisation increases with the absolute level of development, such theories do not necessarily imply that such internationalisation will occur. For example, although firms from Germany, Japan and Canada and other newer source countries of outward investment had developed the capacity for internationalisation in the earlier part of the 1960–75 period, such internationalisation has only been undertaken in a major way from the 1970s onwards.

A theory which relates the absolute capacity of countries and their firms to engage in outward investment is therefore necessary although not sufficient to explain the generalised rise in internationalisation of firms which caused the structural change that occurred in the relationship between NOI and GNP per capita in the mid-1970s. Cantwell (1987c) postulates reasons for the general trend towards internationalisation which depend fundamentally, although not exclusively, on the changing characteristics of the new technology paradigm as referred to by Freeman and Perez (1986), the ability to weaken the bargaining power of trade unions and political circumstances. The statistical investigation of the investment development cycle undertaken in this chapter serves to reinforce the argument that there has been a general trend towards internationalisation, with the newer multinationals increasingly generating the capacity to catch up with the more mature and historically established ones.

The predication of the product cycle model is that the competitive advantages of present Third World outward investors are based on an imitation or adaptation of the technology of advanced countries in accordance with the requirements of Third World markets and production conditions. These advantages are therefore described as resulting from small-scale, labour-intensive manufacture, substitution of local materials, and other such skills. Because these advantages have low or no barriers to entry, new MNEs in other developing countries may gain the capacity to copy or adapt these advantages and replace those firms whose advantages have eroded. Such a predication has limited explanatory power in the light of the recent developments in international investment and in particular the increasing sources of outward investment from other developing countries.

Indeed, as a result of their greater experience in international production, the earlier sources of outward investment have developed more complex and more significant forms of competitive ownership advantages, helped by the increasing importance of their innovative activities. By contrast, the competitive ownership advantages of the newer sources of outward investment are much simpler because their innovative activities are at an early stage. However, these advantages may be expected to become increasingly complex and sophisticated as their experience in international production is broadened.

Hence, the major difference between MNEs from the major industrialised countries and those from the Third World, and indeed between the newly industrialising countries and the lower-income developing countries, is in the form and extent of their innovation. While the form of innovation of firms from more industrialised countries is increasingly geared towards frontier product and process innovation, and advanced marketing skills obtained through the continuous process of R&D, the form of innovation of firms from less developed countries is characterised more in terms of simpler and less research-intensive technologies embodied in production engineering, and organisational capabilities gained through learning by doing and using. Evidence of such simpler forms of innovation and technology creation, which are geared to solving the problems specific to producing in Third World conditions, is provided in Chapters 8 and 9.

The key element is therefore the different forms of innovation undertaken by firms from different countries which ensure that newer sources of outward investment can be expected to emerge in the future *in addition to* the present sources. This results from the accumulation of ownership advantages of indigenous firms from new source countries which enables them to grow faster *vis-à-vis* other firms in their industry, not only in the domestic market but also in international markets. An examination of the time-series data between NOI flows and GNP per capita of each of the thirty countries in the sample is essential to enable us to understand the concept of the investment development cycle further and to perceive the debate regarding the nature of Third World multinationals within an overall perspective.

Time-series analysis

The time-series analysis of the relationship between net outward investment flows and GNP per capita for each of the thirty countries in the sample is undertaken for the period 1960–75 and 1970–84. There is a slight overlap between the time frames adopted for the time series because the period 1976–84 would yield results obtained on the basis of only nine observations on NOI flows and GNP per capita for each of the thirty countries.

The relationship is estimated by the linear regression equation:

$$NOI_t = \theta + \upsilon(GNP_t/POP_t) + \mu_t$$

where NOI, the dependent variable, is Net Outward Investment flows and GNP/POP is Gross National Product divided by the population for each country.

A log-linear regression equation is also fitted for comparison:

$$\log NOI_t = \omega + \varepsilon \log (GNP_t/POP_t) + \mu_t$$

However, since a majority of the net outward investment flows are in negative terms, a constant term was added in order to make these flows positive. Although such data transformation alters the value of the estimated coefficient on the constant term, the significance of the estimated coefficient on the independent variable remains unaltered. The results of the regression equation for the two time periods specified are as follows. For purposes of the analysis, only the estimated coefficient and the t-statistic on the independent variable are given, since by definition there is a strict relationship between the R^2 and F-statistic of the overall regression equation and the t-statistic on the estimated coefficient of the sole independent variable in a simple bivariate regression. Only the results of the linear regression are given since the results of the log-linear regression made no substantive difference to the results of the linear equation.

The period 1960–75

The t-statistics are given in brackets. The critical t-value with $n - k = 16 - 2 = 14$ degrees of freedom is 2.145 at the 5 per cent level of significance.

Countries	v, the estimated coefficient on the independent variable		T-test at the 5% level
Australia	− 0.0532	(−.8951)	insignificant
Austria	− 0.0117	(−1.6433)	insignificant
Belgium	− 0.1257	(−10.1041)	significant
Canada	0.0718	(1.5189)	insignificant
Denmark	− 0.0189	(−2.8084)	significant
Finland	− 0.0056	(−1.8469)	insignificant
France	− 0.0833	(−1.7216)	insignificant
W. Germany	0.1177	(1.8647)	insignificant
Italy	− 0.1568	(−1.4107)	insignificant
Japan	0.4519	(9.3328)	significant
Netherlands	0.1109	(4.6374)	significant
New Zealand	− 0.0391	(−6.5849)	significant
Norway	− 0.0243	(−3.6488)	significant
Portugal	− 0.0572	(−6.2563)	significant
South Africa	− 0.2509	(−2.2109)	significant
Spain	− 0.0847	(−5.3346)	significant
Sweden	0.0602	(10.0024)	significant
UK	0.6094	(5.8209)	significant
USA	1.1349	(4.1577)	significant
Tunisia	− 0.0754	(−5.2561)	significant
Korea	− 0.1965	(−4.774)	significant
Philippines	− 0.3141	(−2.6165)	significant
Thailand	− 0.5081	(−6.0946)	significant
Argentina	0.0234	(1.466)	insignificant
Brazil	− 1.4854	(−8.069)	significant
Chile	0.3165	(1.4841)	insignificant
Colombia	− 0.0347	(−0.636)	insignificant
Costa Rica	− 0.0864	(−10.172)	significant
Israel	− 0.2661	(−2.063)	insignificant
Libya	0.0851	(3.66)	significant

On the basis of the econometric results obtained in the individual time-series analysis of the thirty countries in the sample, a three-group classification is formulated. *Group 1 countries* are those whose NOI flows are negatively related to GNP per capita. *Group 2 countries* are either those whose NOI flows are not related to GNP per capita, i.e. there is a structural change in the relationship between NOI flows and GNP per capita, or they are close to the turning point of the investment development curve. Finally, *Group 3 countries* are those countries whose NOI flows are positively related to GNP per capita.

Two essential hypotheses are then tested. The first hypothesis is that the average GNP per capita of Group 1 countries is less than that of Group 2 countries, which in turn is less than that of Group 3 countries. The second hypothesis is that the turning point of the quadratic equation in the cross-section analysis, i.e. the point at which NOI flows rise much faster than GNP per capita, is approximately equal to the average GNP per capita of Group 2 countries.

The full list of countries belonging to each of the three groups as estimated by the linear relationship is as follows:

Group 1	*Group 2*	*Group 3*
Belgium	Australia	Japan
Denmark	Austria	Netherlands
New Zealand	Canada	Sweden
Norway	Finland	UK
Portugal	France	USA
S. Africa	W. Germany	Libya
Spain	Italy	
Tunisia	Argentina	
Korea	Chile	
Philippines	Colombia	
Thailand	Israel	
Brazil		
Costa Rica		

The average GNP per capita of each of the countries was calculated in the previous cross-section analysis for the period 1960–75. The overall average GNP per capita of the different countries that belong to a distinct group can then be calculated. The results are as follows:

Group	*Average GNP per capita*
Group 1	US $1,196.41
Group 2	US $2,008.35
Group 3	US $2,833.40

This finding confirms our first hypothesis, that Group 1 countries as a whole have the lowest average GNP per capita among the three groups, followed by Group 2 countries. Group 3 countries, taken as a whole, have the largest average GNP per capita among the three groups. The second hypothesis consists of verifying whether the turning point of the quadratic equation in the cross-sectional analysis undertaken for the period 1960–75 is approximately equal to the average GNP per capita of Group 2 countries. The turning point of the quadratic equation in the cross-sectional analysis was calculated in the previous section to be US $1,580.53. Hence, the critical GNP per capita above which NOI flows increase much faster than GNP per capita, US $1,580.53, is approximately equal to the average GNP per capita for Group 2 countries as a whole, which is US $2,008.35.

Recalculating the NOI flows on the basis of a three-year moving average yields the following grouping of countries:

Group 1	*Group 2*	*Group 3*
Austria	Australia	Japan
Italy	Belgium	Netherlands
New Zealand	Canada	Sweden
Norway	Denmark	UK
Portugal	Finland	USA
S. Africa	France	Libya
Spain	W. Germany	
Tunisia	Korea	
Thailand	Philippines	
Brazil	Argentina	
Costa Rica	Chile	
	Colombia	
	Israel	

The major difference in this grouping of countries, obtained on the basis of three-year moving average NOI flow data, is the composition of countries belonging to Groups 1 and 2. Countries such as Belgium, Denmark, Korea and the Philippines, which belong to the first group on the basis of untransformed NOI flow data, occupy the second group on the basis of NOI flow data recalculated on the basis of a three-year moving average. On the other hand, Austria and Italy, which belong to the second group on the basis of untransformed NOI flow data, occupy the first group of countries on the basis of NOI flow data recalculated on the basis of a three-year moving average. As a result, the number of countries occupying the first group has declined from thirteen to eleven and the number of countries occupying the second group has correspondingly increased from eleven to thirteen. The number and composition of countries belonging to Group 3 remain unchanged. The average GNP per capita of these groups of countries obtained on the basis of NOI flows recalculated on the basis of a three-year moving average are as follows:

Group	*Average GNP per capita*
Group 1	US $1,177.37
Group 2	US $1,899.54
Group 3	US $2,833.40

The above figures again confirm our hypothesis that Group 1 countries would have the lowest average GNP per capita compared to Group 2 countries and that Group 3 countries would have the highest GNP per capita. The average GNP per capita of Group 2 countries, US $1,899.54, obtained on the basis of the NOI flows recalculated on the basis of a three-year moving average method is closer to the turning point of GNP per capita at which NOI increases much faster than GNP per capita at US \$1,580.53 than the average GNP per capita of Group 2 countries obtained on the basis of untransformed NOI flow data. However, an examination of the composition of each group obtained on the basis of NOI flows recalculated on the basis of the moving average method is less satisfactory than that obtained on the basis of untransformed flow data. It seems highly inappropriate to classify countries such as Austria and Italy, which are significant sources of outward investment, in Group 1 when the Philippines and Korea, which are less significant sources of outward investment than Austria and Italy, are classified as belonging to Group 2.

On the basis of these findings, the classification of countries obtained on the basis of untransformed NOI flow data is adopted.

The period 1970–84

The analysis of the linear and log-linear relationship between NOI flows and GNP per capita for each of the thirty countries in the sample for the period 1970–84 constituted the second stage of the time-series analysis. The results obtained in this time period would enable us to determine whether the original grouping of countries obtained for the period 1960–75 had undergone some changes.

The critical t-statistic at $n - k = 15 - 2 = 13$ degrees of freedom is 2.160 at 5 per cent level of significance. The individual t-statistics for the v, the estimated coefficient on the independent variable, GNP per capita, is given in brackets.

Countries	v, the estimated coefficient on the independent variable		T-test at the 5% level
Australia	0.0354	(0.4996)	insignificant
Austria	− 0.0064	(−1.6520)	insignificant
Belgium	− 0.0793	(−1.9517)	insignificant
Canada	0.5516	(3.5091)	significant
Denmark	0.0107	(1.082)	insignificant
Finland	0.0269	(3.6158)	significant
France	0.0563	(0.7633)	insignificant
W. Germany	0.3538	(6.8922)	significant
Italy	0.2495	(3.8092)	significant
Japan	0.4584	(6.3247)	significant
Netherlands	0.1753	(4.3799)	significant
New Zealand	− 0.0054	(−0.3339)	insignificant
Norway	0.0109	(0.5090)	insignificant
Portugal	− 0.0673	(−2.6911)	significant
S. Africa	0.3017	(2.3391)	significant
Spain	− 0.3251	(−6.3689)	significant
Sweden	0.0533	(3.0681)	significant

UK	0.3924	(2.2019)	significant
USA	− 2.1050	(−4.3018)	significant
Tunisia	− 0.2130	(−4.3371)	significant
Korea	0.0468	(2.6342)	significant
Philippines	0.2233	(1.4160)	insignificant
Thailand	− 0.3664	(−4.6918)	significant
Argentina	− 0.1583	(−4.3381)	significant
Brazil	− 0.7295	(−5.8649)	significant
Chile	− 0.2353	(−2.9782)	significant
Colombia	− 0.2917	(−4.3933)	significant
Costa Rica	− 0.0150	(−1.7325)	insignificant
Israel	0.0499	(2.7557)	significant
Libya	0.0302	(1.7203)	insignificant

On the basis of these econometric results, the individual countries are then classified into three distinct groups in the same manner as for the previous time period.

Group 1	Group 2	Group 3
Portugal	Australia	Canada
Spain	Austria	Finland
USA	Belgium	W. Germany
Tunisia	Denmark	Italy
Thailand	France	Japan
Argentina	New Zealand	Netherlands
Brazil	Norway	S. Africa
Chile	Philippines	Sweden
Colombia	Costa Rica	UK
	Libya	Korea
		Israel

The major results of the time-series analysis of the period 1970–84 compared to the earlier period 1960–75 show that the number and composition of each of the three distinct groups of countries have changed. The number of countries in Group 1 has declined from thirteen to nine and those in Group 2 from eleven to ten. However, the number of countries in Group 3 has increased from six to eleven from the earlier to the later period.

The average GNP per capita of each of the groups of countries in the period 1970–84 is as follows:

Groups	Average GNP per capita
Group 1	US $2,407.81
Group 2	US $5,536.34
Group 3	US $6,328.28

The results still confirm the earlier hypothesis that countries belonging to Group 1 have the lowest average GNP per capita, followed by Group 2 countries. Group 3 countries have the highest GNP per capita of the three. However, compared to the period 1960–75, the average GNP per capita of Groups 1, 2 and 3 has increased from a level of US $1,196.41, US $2,008.35 and US $2,833.40 respectively.

The recalculation of NOI flows on the basis of a three-year moving average yields the following grouping of countries:

Group 1	*Group 2*	*Group 3*
Belgium	Australia	Canada
New Zealand	Austria	Finland
Portugal	Denmark	W. Germany
Spain	France	Italy
USA	Norway	Japan
Tunisia	Argentina	Netherlands
Thailand		S. Africa
Brazil		Sweden
Chile		UK
Colombia		Korea
Costa Rica		Philippines

The above classification of countries obtained on the basis of NOI flow data recalculated on a three-year moving average method differs from that obtained on the basis of untransformed NOI flow data in several respects. The number, as well as composition, of each group of countries differs markedly. The number of countries in Groups 1 and 3 has increased from nine to eleven and from ten to thirteen respectively. As a result, the number of countries belonging to Group 2 has declined from eleven to six when the dependent variable, NOI flows, is re-estimated on the basis of a three-year moving average.

The major difference in the composition of countries is that Belgium, New Zealand and Costa Rica, which belong to Group 2 on the basis of untransformed NOI flow data, belong to Group 1 on the basis of NOI flow data recalculated on the basis of a three-year moving average. In addition, Argentina, which belongs to Group 1 on the basis of untransformed flow data, belongs to Group 2 on the basis of NOI flow data recalculated on the basis of a three-year moving average. Finally, the Philippines and Libya, which belong to Group 2 on the basis of untransformed NOI flow data, belong to Group 3 on the basis of NOI flow data recalculated on the basis of a three-year moving average.

The average GNP per capita of the new grouping of countries obtained on the basis of a three-year moving average is as follows:

Groups	*Average GNP per capita*
Group 1	US $3,024.48
Group 2	US $6,730.72
Group 3	US $5,553.78

The above results show that although Group 1 countries have the lowest GNP per capita of the three, the average GNP per capita of Group 2 countries surpasses that of Group 3, which does not conform to our previous hypothesis that Group 3 countries have the highest GNP per capita of the three. The econometric results based on NOI flow data recalculated on the basis of a three-year moving average do not therefore yield a satisfactory classification of countries in broad correspondence to our findings in the previous period.

The next stage of the analysis would therefore focus on changes in the composition of each group, obtained on the basis of untransformed NOI flow data for the periods 1960–75 and 1970–84. The results of such time-series analysis support the results of the cross-section analysis that a structural change occurred in the relationship between NOI and GNP per capita in the mid-1970s. In order to test the presence of a structural change in the time-series analysis, the next stage is to introduce a time trend dummy variable to the regression equation.

The estimation of the coefficient of the time trend dummy variable for the period 1960–84

Given that countries have undergone changes in their groupings between the periods 1960–75 and 1976–84, the introduction of a time trend dummy variable with value 1, 2, 3, 4, etc. from 1976 onwards but with the value of zero from the period 1960–75 would enable the testing of whether countries that have moved up from a lower to a higher group would have a positive coefficient on the time trend dummy variable. On the other hand, countries that have moved down from a higher to a lower group would have a negative coefficient on the time trend dummy variable. Finally, countries that have remained within the same group during the two periods would have an insignificant coefficient on the time trend dummy variable.

The linear and log-linear equations to be tested for time-series data on NOI flows and GNP per capita for each of the thirty countries in the sample for the period 1960–84 are as follows:

$$NOI_t = \theta + \upsilon(GNP_t/POP_t) + \lambda T_t$$
$$\log NOI_t = \omega + \varepsilon \log (GNP_t/POP_t) + \lambda T_t$$

where NOI is NOI flows, GNP/POP is Gross National Product divided by the population and T is the time trend dummy variable with value 1 for the period 1976, 2 for 1977, 3 for 1978, etc. This variable would have a value of zero for the period 1960–75. The analysis of the grouping of countries obtained for the periods 1960–75 and 1970–84, and the actual econometric results with the introduction of a time trend dummy variable, are as follows:

Countries expected to have a positive coefficient on the time trend dummy variable	*Actual econometric result*
Belgium (from Group 1 to 2)	insignificant
Denmark (from Group 1 to 2)	insignificant
Norway (from Group 1 to 2)	$+\lambda$
S. Africa (from Group 1 to 3)	insignificant
Korea (from Group 1 to 3)	$+\lambda$
Philippines (from Group 1 to 2)	$+\lambda$ (log-linear)
Costa Rica (from Group 1 to 2)	insignificant
Israel (from Group 2 to 3)	$+\lambda$ (log-linear)
W. Germany (from Group 2 to 3)	insignificant
Italy (from Group 2 to 3)	$+\lambda$
Canada (from Group 2 to 3)	insignificant
Finland (from Group 2 to 3)	$+\lambda$
New Zealand (from Group 1 to 2)	$+\lambda$ (linear)

Countries expected to have an insignificant coefficient on the time trend dummy variable	*Actual econometric result*
Portugal (from Group 1 to 1)	$-\lambda$
Spain (from Group 1 to 1)	$-\lambda$
Tunisia (from Group 1 to 1)	insignificant (linear) $-\lambda$ (log-linear)
Thailand (from Group 1 to 1)	insignificant (linear) $-\lambda$ (log-linear)
Brazil (from Group 1 to 1)	insignificant
Australia (from Group 2 to 2)	insignificant
Austria (from Group 2 to 2)	insignificant
France (from Group 2 to 2)	insignificant
Japan (from Group 3 to 3)	insignificant
Sweden (from Group 3 to 3)	insignificant (log-linear) $+\lambda$ (linear)
UK (from Group 3 to 3)	insignificant (linear) $-\lambda$ (log-linear)
Netherlands (from Group 3 to 3)	$+\lambda$

Countries expected to have a negative coefficient on the time trend dummy variable	*Actual econometric result*
Argentina (from Group 2 to 1)	$-\lambda$ (linear) insignificant (log-linear)

Chile (from Group 2 to 1)   insignificant
Colombia (from Group 2 to 1) $-\lambda$
USA (from Group 3 to 1) $-\lambda$
Libya (from Group 3 to 2)   insignificant

The above results show that of the thirteen countries that moved from a lower to a higher group during the periods 1960–75 and 1970–84, the time trend dummy variable registered a positive coefficient for seven of them. An insignificant result was registered for the other countries. In addition, of the twelve countries that remained in the same grouping during the two periods under review, the time trend dummy variable registered an insignificant coefficient for eight of them. Two countries had a positive coefficient while another two had a negative coefficient on the time trend dummy variable. Finally, of the five countries that moved from a higher to a lower group during the two periods under review, a negative coefficient on the time trend dummy variable was registered for three of them. An insignificant result was registered for the other countries.

In any case, the grouping of countries becomes less relevant when the equation provides a maximum rather than a minimum, especially for the later period, 1970–84, when the relationship between NOI flows and GNP per capita has undergone a structural change.

CONCLUSION

The statistical investigation of the existence of the investment development path has shown that a structural change has occurred in the relationship between NOI and the country's relative stage of development as from the mid-1970s, as a result of the general rise in the internationalisation of firms from countries above Group 1. The existence of such structural change renders the theories of technological accumulation and localised technological change that relate more to absolute levels of development as having greater relevance over time *vis-à-vis* the investment development cycle and product cycle model that relate more to relative levels of development. These relative theories of development lead to the predication that the concept of an investment development curve is maintained over time even though the curve may undergo an overall shift to the right with countries' changing relative positions. However, the econometric tests undertaken in this chapter have suggested that these relative theories are inadequate, owing to the general trend towards internationalisation as the newer multinationals from countries at intermediate stages of development, including the richer developing countries, have acquired the capacity and the incentive to follow the earlier outward multinational expansion of the traditional source countries for outward investment, the USA and the UK. This phenomenon may be better explained by absolute theories that relate more to development as a cumulative process where the form of localised technological change or accumulation, rather than its existence, varies with the stage of development.

However, although the theory of technological accumulation suggests that the ownership advantages of firms and their capacity for internationalisation relate more to the absolute than the relative level of development, such a theory also has relative elements in suggesting that the form of innovation and technological accumulation of First World MNEs is different from that of the Third World. The technological accumulation of Third World MNEs, which are at an earlier stage of development, pertains more to simpler and less research-intensive forms of technology creation and organisational capabilities as a result of learning by doing and using rather than deliberate efforts made in research and development. By contrast, the technological accumulation of First World MNEs is fundamentally research intensive and consists of frontier product and process innovation and advanced marketing skills. The findings of this chapter suggest that these relative elements cannot adequately explain the structural change that has occurred as a result of the general trend towards internationalisation which in recent years has been common to firms of all countries. The rising number of countries belonging to Group 3, i.e. countries whose NOI flows are positively related to GNP per capita, confirms the increased significance of outward investments from newer source countries from both developed and developing countries and provides first-hand evidence of the general trend towards internationalisation.

6 The role of foreign technology in the development of Third World multinationals

INTRODUCTION

This chapter seeks to examine the role of foreign technology in strengthening indigenous technological development and growth of international production of indigenous firms in developing countries.[1] The theoretical framework adopted is based on the Schumpeterian premise that economic growth and performance are dependent on the creation of new technology, diffusion of technology and efforts related to the economic exploitation of innovation and diffusion (see, for example, Fagerberg, 1988; Badulescu, 1991; and applications for developing countries in Kim, 1980). The idea is implicit in the theory of technological competence as an important determinant of international competitiveness and the differential growth rates of firms (for further elucidation of the theory, see Cantwell, 1991a).

The process of technological change in developing countries is fundamentally derived from the diffusion of new and existing technologies originating within developing countries or abroad (Soete, 1986), and this chapter argues that apart from Asian NICs that have developed some national innovative capabilities, by and large the main source of growth of developing countries in the catch-up phase of technological development is found in their ability to access technology from abroad through diffusion, and in their indigenous capacity to exploit the benefit of technology created abroad. As a result, apart from a few exceptional countries based in the NICs, research expenditure by developing countries geared to the achievement and improvement of international competitiveness is related more to technology importation than to autonomous research spending on domestic R&D. The analysis of the industrialisation and growth of developing countries in the late twentieth century on the basis of international diffusion of foreign technology has historical antecedents in the industrialisation of Europe and the USA in the nineteenth century and Japan in the twentieth century.

This chapter attempts to show that firms from some developing countries with high levels of indigenous technological capabilities have demonstrated their ability to absorb rapidly the more advanced technology generated in the West and to catch up in the dynamic process of international investment. Their direct investment activities and other technology export activities demonstrate the attainment of a certain level of technological competence (see Lall, 1984b). Other studies on the important role of the transfer and effective implementation of new technology in the development of indigenous technological innovation and the growth of technology exports of developing countries are evident *inter alia* in the works of Kojima and Ozawa reviewed in Chapter 3, Cooper and Sercovich (1970), Helleiner (1975), Ranis (1976), Germidis (1977), Stewart (1979), Kim (1980), Lall (1982c), Jenkins (1984), Shrivastava (1984), Dahlman and Sercovich (1984), Teitel and Sercovich (1984), Rodrigues (1985), Juilland (1986), Das (1987) and Katrak (1989). The gradual process of technological and capital accumulation that underlies the process of outward investment of the more traditional source countries is seen to apply equally to the growth of outward investment from the developing countries, with suitable qualifications.

Apart from inward foreign direct investment which contributes to economic growth through the augmentation of resources for capital formation, the improvement in the efficiency of investments, the growth of internationally competitive industries (Rana, 1987; Hiemenz, 1987) and their potential positive interaction on human capital and technology diffusion (Wang, 1990), this chapter will show some examples of strategic alliances or forms of cooperative arrangements with foreign firms. These take the form of joint ventures, licensing agreements, turnkey plant, technical assistance, subcontracting arrangements and other forms of non-equity investments, as well as forms of foreign technology transfer which have also become very important complementary modalities for the transfer of more advanced and frequently non-proprietary technology in developing countries (Lenae, 1985; Perlmutter and Heenan, 1986). The typologies of these various forms of technology transfer to developing countries by multinational enterprises has also been discussed in UNCTC (1987).

The main thesis is designed to show that more advanced foreign technology transfer has acted as a trigger mechanism for modern economic growth in some developing countries which are on a lower level of economic and social development, in the way predicated by Kuznets (1968). Since developing countries have limited technological and capital abilities, the transfer and implementation of packaged technology from abroad provide the initial basis for their technological development. Foreign technology helps to develop local technical and entrepreneurial capabilities which would provide the major sources of innovation at a more intermediate and advanced stage of technological development, where research and development within the firm, in addition to engineering efforts obtained in the early stages, provide the basis for technological change and growth of production in new technologies (Kim, 1980; Kwon and Ryans, 1987; Lee, Bae and Choi, 1988).

The focus of this chapter is to show how in certain cases foreign technology has been an agent in supporting dynamic changes in the comparative advantage of firms and industries in developing countries towards higher value-added and more technology intensive industries, in accordance with policies implemented by governments in these countries. In particular, the growth of more complex sectors of industry associated with more advanced stages of development is reflected in the growth of international production of local firms. The technology transferred from abroad has helped some local firms in developing countries to strengthen their technological capabilities and to generate competitive forms of ownership advantages necessary for carrying out their international production activities.

This chapter has several parts and draws from several theories that shed light on the role of foreign technology in the accumulation of technological competence of firms based in developing countries. The literature on the economics of technological change and the empirical studies on technological innovation of developing countries have been consulted and integrated to provide a better understanding of the conceptual and empirical process of technological change of firms from developing countries as opposed to firms from the more advanced, industrialised countries. Second, and stemming from this, a conceptual explanatory framework had to be adopted to explain the different impact of foreign technology on technological development and growth of Third World firms, in particular, the different relative rates of success of Third World MNEs in sectors characterised by rapidly growing technological opportunities. The theory of technological competence proved to be a useful explanatory framework. It supposes that technological competence plays a central role in firms' rate of growth, so that those with a greater degree of technological competence are predicated to have a faster rate of growth. Such a theory was developed to explain the growth of firms or MNEs based in the industrialised countries and the process by which these firms sustain competition in international industries.

In order to make the theory operational in explaining the growth of firms in developing countries characterised by less sophisticated levels of technological competence, it had to be modified in the context of theories of development and underdevelopment to throw light on the role of foreign technology in the achievement of technological competence of firms from developing countries. The concept of cumulative causation implicit in the theory of technological competence allows for a variable impact of foreign technology on local technological competence and growth by assuming an intermediate stance between the theories of development that allow for a positive developmental role for foreign technology and the theories of underdevelopment that allow for a more negative impact of foreign technology on local development. Such a concept is applied to explain the different levels of technological competence and rates of growth of Third World MNEs at different stages of technological development, in the latter part of the chapter. The main conclusions and policy implications are drawn in the final section.

THE ROLE OF FOREIGN TECHNOLOGY IN THE DEVELOPMENT OF TECHNOLOGICAL COMPETENCE IN DEVELOPING COUNTRIES

Studies have shown that Third World MNEs have based their initial techno-logical development on foreign technologies mainly, but not exclusively, from the more industrialised countries where the vast majority of new and more advanced technologies originate (see, for example, Lall, 1984c and Unger, 1988). In particular, multinational enterprises from the more industrialised countries constitute an essential source of supply of new technologies in sectors of growing importance to developing countries such as microelectro-nics, biotechnology, material sciences and renewable energy (UNCTC, 1990a). However, firms from other developing countries are important in terms of providing basic industry and infrastructural needs in lower techno-logically intensive, capital-intensive and raw material-intensive industries, where these firms have accumulated considerable experience in production in their home developing countries (see, for example, Kaplinsky, 1984). The technology provided by developing countries differs from that of developed countries in that it is stable and mature, competitive and essentially similar, but of smaller scale, less automated and localised to conditions in developing countries, leading to lower costs or to concentration in sectors in which developed countries either have no experience or no longer provide technol-ogy (Lall, 1984b; Singh, 1986).

Effective technology transfer from abroad can harness the development of indigenous capabilities, namely those of technology search and selection, negotiation and acquisition, assimilation and adaptation, replication and, ulti-mately, innovation (UNCTC, 1987). These foreign technologies represent an important means of achieving rapid technological progress in developing countries through the improvement of the productivity or efficiency of exist-ing industries as well as through the introduction of new areas of industrial production. As Slaybaugh (1981) has stated:

> Much technology can be learned over a period of time without outside technical assistance; the basic purpose, therefore, of procuring technology is usually to save time and expense that would be required to develop it alone.
>
> (p.302)

The import of foreign technology and the international diffusion of technol-ogy exert crucial importance for autonomous national growth, industrial development and the process of technological leapfrogging in follower nations and firms (Freeman, Clark and Soete, 1982; Soete, 1985). This occurs in part because the adoption of technological decisions of earlier innovators or a widely diffused technology encourages the development of complementary technologies resulting in lower costs (Arthur, 1989).

In particular, if foreign direct investment is the modality of foreign technological transfer, then the MNE becomes an agent of dynamic comparative advantage for new technology generation as well as for mature technology diffusion (Lall, 1979). However, the stability of the sectoral pattern of comparative advantage held by each national group of firms in the medium term dictates the path of technological development and diffusion (Dosi and Orsenigo, 1988). This means that firms and countries are locked in to a specific course of technological development which is conditioned by the prevailing techno-economic paradigm and only changes when shifts occur in the paradigm as a result of a radical restructuring of the fields of technological opportunities that offer the fastest growth (Freeman and Perez, 1988; Cantwell, 1990).[2] Periods of change in the techno-economic paradigm represent excellent windows of opportunity for developing countries to break out from the vicious circle of technological and economic dependency and specialisation in technologically mature products and industries, and effectively to catch up in the process of technological advancement. Apart from the innovative efforts in R&D sourced through the network of the firm, which may be limited for Third World firms, this may additionally require licensing and/or other collaborative arrangements with organisations responsible for pioneering the new paradigm (Teece, 1988; Chesnais, 1988).

The possibility for developing countries to catch up in the production of new products using the new technological paradigm may result in part because firms from these countries have not been locked in to old technology systems and the concomitant infrastructure, which are difficult to change. The rapid international diffusion of British technologies to foreign countries such as the USA and Germany and the decline of its absolute technological leadership around the turn of the twentieth century can be analysed in relation to the hampered diffusion of new technologies in the UK market due to previous investment outlays in the existing technology, the commitment of management and the skilled labour force, and development research geared towards the improvement of existing technology.

The specific features of the present new microelectronics paradigm and in particular the potential for small-scale production, their impact on capital productivity and their more limited requirements for highly specialised technical, mechanical and electrical skills, provide strong incentives for their rapid diffusion not only in advanced countries, but also in semi-industrialised and newly industrialising countries where growth has been hampered by the scarcity of capital and labour that possesses specialised skills (Soete, 1982). The extent of leapfrogging and catching up is greater where there are initial capabilities in closely related types of activity (Cantwell, 1991a) which in turn is determined by the national system of innovation such as, *inter alia*, the supply of entrepreneurial skills and scientific and technical knowledge involved in the transfer of technology related to know-why as opposed to know-how capabilities; a reasonable level of knowledge necessary for the assimilation and eventual commercialisation of the imported technology; and favourable locational advantages related to the general infrastructure and other economic and institutional conditions exogenous to the firm. Some of these pertain to the development of the local capital goods sector which functions as source and locus of technology capacity, the social and political changes in the international community such as increased protectionism in the developed countries which determines the future development of comparative advantage, and countries' specific technological and industrial strategies (Unger, 1988; Freeman, 1988b; Perez, 1988; Perez and Soete, 1988).

THE COSTS OF FOREIGN TECHNOLOGY TRANSFER

The relative importance of the various components of the national system of innovation increases or decreases over the life cycle of the technology, with the phases of introduction and maturity requiring the lowest thresholds of entry (Perez and Soete, 1988). Immediate adoption of the new technology not only implies a significant contribution to the monopoly profits of the supplying innovator but also a commitment to a technology which might well undergo significant improvements in the not too distant future (Soete, 1982). In any event, early investment in new technologies is important *inter alia* to gain *first mover advantages* and to prevent a lock-out in technological paths which may be useful in the development of areas of comparative advantage and specialisation (Freeman, 1988b; Cohen and Levinthal, 1990).

Apart from the costs incurred in the importation of foreign technology, firms in developing countries, like those that experienced earlier technological growth in the more industrialised countries, have made considerable efforts in facilitating indigenous technological change, and have incurred significant costs in assimilation.[3] Assimilation results in the process of technological change through adaptation and accumulation of technical expertise (for related studies, see among others Quinn, 1969; Stewart 1981; Dahlman and Sercovich, 1984; Monkiewicz and Sosnowski, 1986; Kwon and Ryans, 1987; and Lee, Bae and Choi, 1988). As a result, the national system of innovation is enhanced through a cumulative process of learning by producing, using and changing, as well as learning by the training, hiring, searching and interaction of producers and users (Bell, 1984; Andersen and Lundvall, 1988).[4] Through the combination of newly acquired skills and existing levels of indigenous skills, a synergistic combination of expertise is achieved (Wilmot, 1977).

Developing countries differ greatly in their technological absorption capacity, technological needs and goals in seeking technology transfer (Prasad, 1981a). In essence, fundamentally the success of a foreign technological transfer depends upon both supply and demand conditions (Ranis, 1984).[5] On the one hand, the ability of the recipient firm or country to select, adapt, absorb, combine and develop the transferred technology with its existing level of demand is dictated by its indigenous technological capabilities. The selection environment may also be influenced by growth in demand that exerts an important influence on the diffusion of competence across firms in an industry (Teece, Pisano and Shuen, 1990). The size of a developing country, the extent of labour surplus and other demand-side typological characteristics such as the nature of the industrial structure, the levels of overall competitiveness and policies also crucially affect the intensity of the search for indigenous technological activity (Ranis, 1984). Equally important determinants are supply-side factors such as a suitable cultural and socio-economic environment, an adequate infrastructure which includes scientific institutions, research and development facilities, vocational, technical and management training institutes, and skilled personnel of different specialisations within the recipient country who support technological negotiation, modification and capacity development (Mansour, 1981; Dore, 1984; King, 1984; Freeman and Perez, 1988). The interrelation between science, technology, educational system and organisation of work in the development of indigenous technological capacity has been central to the work of King (1984) and Caillods (1984).

These findings are in consonance with the view that since technology is a differentiated undertaking between firms and locations (Pavitt, 1988), the process of technology transfer between firms in different countries is necessarily costly not only in terms of adaptation to local levels of technological capabilities but also in terms of market structures and policies on the role of foreign technology in host countries, resistance to foreign technology, large differences in infrastructure, and the geographical distance which entails travel and communication costs (Teece, 1977; Findlay, 1978; Contractor, 1979). Clearly, the institutional and social context of innovation has also to be understood as exerting an important influence on the process of technological change. The fundamental matching between the existing and required socio-institutional framework for a particular technological revolution influences the rate of diffusion and growth potential of the new technologies (Freeman and Perez, 1988; Perez, 1988; Johnson, 1988). These constitute important elements of the national system of innovation that supports industrial learning by doing and learning by interacting in the key sectors of the economy (Andersen and Lundvall, 1988).

The process of technology transfer is even more costly between firms in countries at different stages of development (Rosenberg, 1976). In developing countries, in particular, a number of social, organisational and economic characteristics either preclude or make highly uneconomic any attempt at straightforwardly replicating 'from-the-shelf' technology previously used in developed countries. Among these distinctive characteristics of Third World technology in comparison to that of the more developed, industrialised countries are smaller domestic markets, higher rates of tariff protection, weaker competitive atmosphere, stronger distortions in technical information and market imperfections, shortages of skills and more dramatic levels of uncertainty, etc. (Katz, 1984b). The costly process of technology transfer between developed and developing countries is evident *inter alia* in terms of difficulties in communication and training owing to lower levels of entrepreneurship, education or absorptive capacities, and the need for more specialised and updated technology (Slaybaugh, 1981; Lall, 1984a).

TRANSFER OF KNOW-HOW VS KNOW-WHY: IMPLICATIONS FOR LOCAL TECHNOLOGICAL DEVELOPMENT

Developing countries have relied on foreign direct investment and imports of disembodied and embodied or patented technology to varying degrees to promote their indigenous technological development and economic growth (for related findings see Unger, 1988). Only a small number of firms, mostly based in the NICs, have the inventive capacity to develop competitive technology in-house. Disembodied technology is often referred to as *know-how* or *production engineering* and represents the most elementary stage of technological development which consists of minor adaptation to assimilated technology to conform to local sales, product mix and raw materials. The engineering efforts may consist of substitution of raw materials and modification of equipment and tools, quality control, improved plant layout and production practices, trouble shooting and the like (Lall, 1985a) which form the basis of a significant amount of innovation and improvement (Rosenberg, 1976). Such non-patentable and non-codifiable technology, which consists of general manufacturing experience that leads to practical improvements as a result of learning by doing and using, becomes specific or localised to the firm and is of limited value to other firms. Such a gradual process of learning and the need for critical adaptation and the development of complementary technology are consistent with a view of technology as a cumulative process (Pavitt, 1988).

On the other hand, the acquisition of embodied technology such as technical information, drawings, tools, machinery, process information, specifications and patents is derived from inventions that emerge from basic research. Such technology represents a more advanced stage of technology development that leads to the acquisition of *know-why* capabilities which reflect a deeper understanding of the nature of the underlying process and product technologies and lead to substantial adaptation, improvement and even replacement by new products or processes (Ozawa, 1981; Lall 1985b). Know-why capabilities embody skills in basic design, research and development and are referred to as the foundations of technological paradigms which enable the production and development of new industries and products along the trajectory anticipated for the new technology.

The role of imported technologies in the acquisition of know-how is seen as particularly crucial as a source of technological progress in developing countries (see, for example, Katz, 1978). The high level of technological competence and faster rate of growth of MNEs in international industries that offer favourable technological opportunities, and the integration of their unique technological characteristics with local systems in their host countries, may exert a positive impact on the know-how of related local firms through linkages and competition (Findlay, 1978).

Latin American countries such as Argentina, Brazil and Mexico have demonstrated a greater reliance on disembodied forms of foreign technology than Asian countries such as India or Korea. Korea, for example, has relied heavily on the imports of embodied technology in the form of capital goods in the development of its industrial sector (Dahlman and Sercovich, 1984). These imports of embodied foreign technology have promoted know-why capabilities, especially in the NICs with advanced technological skills and industrial experience. However, such know-why technological capabilities require continuous injections of fresh foreign technologies to keep pace with rapid developments abroad (Lall, 1983d). This is especially important in sectors characterised by faster growth of technological change which allows leasers to increase their technological superiority relative to weaker firms trying to catch up (Cantwell, 1991a). These technological capabilities in R&D are increasingly becoming a more important determinant of technology exports of developing countries than know-how capabilities. However, their efforts in R&D are generally distinguished from those of the more advanced countries by their concentration on minor innovation, design imitation and assimilation, raw material adaptation, process down-scaling, equipment modification, slight product change, upgrading of components, product diversification and the like. These technical efforts reflect more advanced capabilities than production engineering (Lall, 1984a) but are nevertheless a reflection of the nature of innovation as an incremental process in their gradual progression to related and more advanced technological activities (Pavitt, 1988).

MODALITIES OF FOREIGN TECHNOLOGY TRANSFER

The transfer of proprietary know-how in developing countries where the level of technological sophistication of transferrer and transferee is significantly large may be expected to be undertaken more efficiently through MNEs as opposed to alternative modes (Teece, 1981c). Foreign direct investment is the major modality of technology transfer in most developing countries (Findlay, 1978; Lall, 1983d), especially in sectors characterised by product differentiation resulting from R&D, the use of patents, brand names, trade marks, and intensive promotion and marketing, or where absorptive capacity of the recipient country is low (UNCTC, 1987). However, for developing countries that have assumed rather more restrictive policies towards MNEs and/or higher forms of local technological capabilities, contractual resource flows or non-equity forms of technology transfer which represent arm's length international exchange of the intermediate product, such as technology, have become much more important modalities for the transfer of technology. Generally, the main determinants of the form of technology transfer are the technological content of the operations in the industry, the extent of barriers to entry, the degree of competition, and the bargaining power and policies of host countries (UNCTC, 1987).

Teece (1981c) has identified four linkage mechanisms with which developing countries can gain access to the world stock of technical and organisational knowledge, namely multinational firms, equipment vendors, foreign aid agencies and an indigenous science community. Since technology is highly concentrated within multinational corporations, their licensing, technical assistance and other activities are the major modalities of direct foreign technology transfer in developing countries. In addition, their direct investment and international subcontracting activities have become important indirect sources of technology in these countries (Prasad, 1981a). The process of adaptation and modification of foreign technology to local operating conditions by foreign MNEs in developing countries where there are few indigenous firms capable of innovation and imitation is a significant channel in the international diffusion of technology to developing countries (Pavitt, 1988). Multinational enterprises fulfil an especially important role in developing countries as transferrers of advanced technology that is subject to rapid technological change (Lall, 1984a).

By contrast, licensing agreements and other forms of non-equity investments have been more common means of technology acquisition in lower technologically intensive industries or in sectors where technology is standardised, widely diffused and easily accessible (Mansfield and Romeo, 1979; UNIDO, 1981). Licensing has been viewed as an attractive alternative to direct investment because of its lower costs and less damaging impact on technological self-reliance (Prasad, 1981b). A large part of the industrial success of the NICs such as India, South Korea, Taiwan and Hong Kong is attributable to local firms licensing or copying foreign technologies rather than to the modality of foreign direct investment.[6]

The relative importance of licensing and FDI as modalities of foreign technological transfer in the post-war period is not the same in developing countries as it is in Japan, owing to their different levels of technological development which exert an important influence on their level of technological self-reliance. Perhaps most fundamental is that the primary agent in local technological development in Japan in the post-war period has been the private firm.[7] Active domestic competition among local firms generated strong pressures for technology import, diffusion and R&D. These competitive forces dictated the choice for technology importation and development (including commercialisation), while the main role of government has been to expand the scope of technological opportunities for local firms (Nagaoka, 1989). By contrast, in a number of the Asian NICs that are in an earlier stage of development, the government has assumed a more active role in targeting particular commercial applications of technologies on behalf of local enterprises. The absorption of foreign technology in Korea, for example, has been aided by targeted intervention by the government in negotiating terms with foreign suppliers which were output and employment oriented rather than profit or publicity oriented (Enos and Park, 1988).

Owing to the expanded capacity of Japan to adopt and adapt advanced foreign technologies, which nurtured the development of indigenous production technologies through the Second World War, the growth of modern growth industries in Japan in the post-war period could be sustained through the inflow of proprietary technology acquired not through foreign direct investments but through licensing agreements. Such imports of disembodied technology through licensing agreements, especially from the USA, were an important means of acquiring unpatentable technical information and know-how which are extremely valuable for product development. These licensing agreements were supplemented by the importation of modern plant machinery and equipment, the provision of basic training by licensors for the acquisition of embodied technology and the foreign visits of Japanese businessmen and students for the purpose of acquiring new, non-proprietary knowledge (Ozawa, 1974, 1981).

Still another important difference between foreign technology transferred to Japan and to the developing countries, which is indicative of their different levels of development, lies in the nature of the technology transferred. Using Magee's concept of an industry technology cycle of invention, innovation and standardisation in describing changes in competitive market structures and the level of standardisation of trade in intermediate products as a result of diffusion and changes in product technology, Ozawa (1981) clearly illustrates that the technological imports of Japan in the post-war period, as in most industrialised countries, were undertaken during the innovative, pre-commercialisation stage of technology development, to sustain the growth of new industries. The transfer of innovative technology has enabled Japanese industry rapidly to expand its technological capacity in product development and the ability to commercialise the production of new products and processes through adaptive R&D. By contrast, imports of technology by developing countries were similar to Japan's earlier technological imports to modernise pre-war origin-ated industries. These technologies are frequently already in the standardisation or diffusion stage of technological development within the economy where they originated.

Apart from licensing, other forms of strategic alliances with foreign firms in the form of joint ventures, turnkey plant, technical assistance, subcontract-ing arrangements and other types of non-equity investments have also become very important complementary modalities for the transfer of more advanced and frequently non-proprietary, mature and standardised technology in de-veloping countries (Lenae, 1985; Perlmutter and Heenan, 1986; UNCTC, 1987).[8] For example, process industries such as chemicals and petroleum have been successfully developed with turnkey contracts and engineering contracts. The basic semiconductor technology, where patents have expired, has been acquired through non-equity alternatives (Prasad, 1981b). Technical assistance has frequently provided not only embodied technology in the form of draw-ings, specifications and services but, more importantly, confidential know-how accumulated through learning by doing (Slaybaugh, 1981). In addition, the indigenous production of more advanced, intermediate-sized, 32-bit super mini-computers in Brazil and South Korea has shown the import-ant role of alliances with foreign proprietors of super mini-computer technology and the active collaboration between the state and the local private sector (see Woog, 1986; Evans and Tigre, 1989).

In addition, as will be shown in the next section, other means of foreign technology transfer are also important. These include imitation and copying or reverse engineering (purchase of samples of machinery and their dismantling and copying), the import of modern plant machinery and equipment, the practice of sending students abroad, feedback from foreign customers, the employment of foreign experts, consultancy agreements and exchanges of information and staff under international programmes for technical cooperation. Also important are the research-intensive outward investments of developing countries to take advantage of local technological competence in a particular location. These investments in centres of innovation enable firms to widen the scope of their technological specialisation through the development of related skills and technologies. This subject is further dealt with in Chapter 9.

THE THEORY OF TECHNOLOGICAL COMPETENCE AND THE THEORIES OF DEVELOPMENT AND UNDERDEVELOPMENT

The examination of the role of foreign technology in strengthening the technological capabilities of MNEs from the Third World is undertaken within the framework of the theory of technological competence. The model of cumulative causation implicit in the theory has previously been used to explain the variable impact of inward foreign investment in achieving a developmental upgrading role in the internationalisation of industries (Cantwell, 1987d; Cantwell, 1991a). With suitable modifications, the chapter attempts to show the relevance of the theory to determine the more general role of foreign technology in the development of Third World multinationals.

The theory of technological competence suggests that the impact of foreign technology on local development is dependent upon the level of domestic technological competence (Cantwell, 1991a). As a result, the relationship between domestic and imported technology may contain elements of both substitution and complementarity (Blumenthal, 1979). Three possible consequences are envisaged in the theory, which apply to the impact of inward investment on local technological development of firms in the industrialised countries but which are still generally useful for the analysis of firms based in the less industrialised developing countries if suitable modifications are made to conform to lower levels of technological development. The first is where local industry has a high technological capability, as in the NICs, and foreign MNEs provide a helpful competitive stimulus to the innovation activities of local firms. Such a beneficial impact is enhanced when some form of applied research facilities are established by foreign-based MNEs in the host developing country. The second is where local firms are technologically weak and the entry of foreign MNEs helps to upgrade their production capabilities *inter alia* through joint ventures and strategic alliances established with local firms.

The third consequence results from the existence of local firms which are technologically weak and experience difficulties in catching up, and whose technological capabilities are undermined by the local expansion of the most internationally competitive foreign firms which rely on research carried out in their home country or elsewhere in their international network.

The concept of cumulative causation implicit in this theory predicates that through the process of technological competition, foreign technology introduced through inward investment in the first two cases shown above has the effect of perpetuating a virtuous circle of technological development by increasing the slope of the indigenous technical progress function in countries where innovative domestic industries exist. In these cases, the competitive stimulus offered by the introduction of foreign technology can foster rather than constrain the technological initiatives of local firms in developing countries that are attaining some level of technological competence, and increases their chances of adapting to the requirements of international competition (for related ideas see Jenkins, 1984). This concept is in line with the view of technological progress in developing countries as dependent on their ability to narrow the gap between their indigenous level of technology and that of the more technologically advanced developed countries. In particular, the idea of contagion in empirical studies has shown that improvements in indigenous technology are achieved when developing countries are exposed to firms that have a higher level of technological efficiency (see, for example, Mansfield, 1968, and Findlay, 1978).

At a more advanced stage of technological development, foreign technology may upgrade the research intensity of production in technologically related sectors that build upon their existing capacities. As Chapters 9 and 10 of this book will show, the dynamic process of technological accumulation has enabled some local firms in developing countries to establish international production not only in small-scale or traditional industrial sectors which are not the main stamping grounds of the larger MNEs from developed countries, but also in the more international industries characterised by advanced technology and sophisticated product marketing and managerial skills.

On the other hand, inward investment also has the effect of perpetuating a vicious circle of cumulative decline by decreasing the slope of the indigenous technical progress function in countries where less innovative sectors exist, as shown by the third case above. In particular, the MNE is seen as erecting barriers against the generation of indigenous knowledge wherever its affiliates dominate the local market while choosing to centralise technology creating activities in the home country or transferring R&D to the home country (Servan-Schreiber, 1967).

The theory of technological competence may therefore be a useful framework in analysing the variable impact of foreign technology effected through the various modalities of technology transfer on the growth of outward investment by indigenous firms in developing countries. The theory allows for the possibility of a variable impact, in contrast with the extreme views presented on the one hand by the proponents of theories of development, who see a developmentalist role for FDI, and, on the other, the proponents of the market-power, dependent development, world-system, neo-Marxist and other theories of underdevelopment.

The theories of development

The developmental role of FDI is associated with the diffusionist theories of development that emphasise a linear path towards modernisation. The idea is essentially found in nineteenth-century theories of evolution and in the belief that the western world would bring about the civilisation of developing countries through the propagation of western values, capital and technology. The conventional theorists of modernisation and economic development who emphasise the developmental role of FDI in developing countries argue that inward foreign investment at an aggregate level promotes industrialisation, local capitalist development and economic growth. This view therefore sees FDI as perpetuating a virtuous cycle of growth and development in host countries.

Such a view is evident in the works of Kindleberger (1969), Drucker (1974), Knickerbocker (1976), Johnson (1977), Warren (1980) and writers entrenched in the theory of internationalisation of capital who emphasise the progressive aspects of capitalism which increasingly exert an important influence on current economic thought. The theory of the internationalisation of capital is originally attributed to Palloix (1975, 1977), and recent applications of the theory are evident in the works of Warren (1980), Barkin (1981), Howe (1981), Marcussen and Torp (1982), Emmanuel (1972) and Jenkins (1984, 1987).

In particular, Marcussen and Torp applied the theory of capital internationalisation to the changing patterns of growth of foreign direct investment and other forms of capital such as private and public export credits, bank loans and exports of commodities, as well as large producer plant and technology in developing countries. According to these writers, the changing patterns of the internationalisation of capital are responsible for shifts in the international division of labour away from exports of raw materials towards exports of manufactured products. Such changing conditions in the periphery have led these authors to argue that autonomous development might occur in these countries if the state assumed the role of the national bourgeoisie by promoting capitalism through the extraction of rents from the agricultural sector and the international financial system established the material conditions for capitalist production. These views of growth of international production as an important part of the wider process of the internationalisation of capital leading to the greater integration of the world economy is also evident in the work of Jenkins (1984, 1987).

The sharp distinction between the operations of local and foreign firms that is inherent in the developmentalist, nationalist and dependency approaches is rendered irrelevant in this view in as much as local firms are forced to adapt to the production norms and standards set by the foreign MNEs in their midst. In some cases, even when there is little or no direct competition from foreign MNEs, large local firms may adopt similar products, techniques and competitive strategies to those employed by foreign MNEs elsewhere.

The view of MNEs as part of the wider process of internationalisation of capital therefore renders the view of the nationalist and dependency schools, with regard to the restriction of foreign capital in order to give preference to local firms, irrelevant in solving the problem that derives from the presence of foreign MNEs *per se*. Similar although slower development spill-over effects, as a result of internationalisation of industry and the requirements of adaptation to international competition, would still have been conditioned as a result of the other forms of international capitalism.

Through his emphasis on the growth of oligopolistic competition between MNEs of different national origins, Warren (1980) similarly argues for the progressive aspects of capitalism in promoting industrialisation and economic growth. In contrast to neo-Marxist or revisionist views, he argues that the prospects for non-dependent capitalist development in many underdeveloped countries are favourable especially in the more recent era of declining imperialism and advancing capitalism. In particular, the role of foreign private investment is seen as fostering the growth of indigenous capitalism and creating the conditions for the declining importance of imperialism as a system of economic inequality between nations of the world capitalist system (Warren, 1973). He attributes the obstacles to development not to imperialism but to the internal contradictions of the Third World itself. Similar ideas were put forward by Emmanuel (1972) who rejected the idea of inappropriate technology introduced by MNEs in developing countries as a contributory factor to the economic backwardness of these countries.

These diffusionist theories of development are often criticised as ethnocentric in their bias in favour of a particular economic path, ideological in their western assumption of pluralistic politics, and dogmatic in their insistence upon a continuous progression through historical stages of development (Chilcote, 1984).

The theories of underdevelopment

The other extreme view that emphasises the regressive aspects of capitalism is evident in the work of Latin American economists entrenched in the neoclassical and Marxist schools of thought, who doubted the capacity of developing countries or peripheral countries for imitative capitalism. This has led to the growth of the dependent capitalist development school of thought advanced by a group of nationalist and reformist economists associated with the Economic Commission for Latin America (ECLA) such as Sunkel (1972), Cardoso (1973), Prebisch (1978), and Cardoso and Falleto (1979). In addition to the reformist traditions that evolved around the works of the ECLA economists, there were also parallel currents of thought advanced by writers in the revolutionary or socialist tradition such as Prado Júnior (1955), Frondizi (1957), Dos Santos (1970) and Quijano (1974), and also those of the neo-Marxist or revisionist schools of thought such as Baran (1960, 1969), Baran and Sweezy (1966), Frank (1966), Emmanuel (1972), Amin (1974) and Wallerstein (1979).

These theories of underdevelopment may be traced to the ideas of capitalist accumulation and disaccumulation of Marx, Lenin and Trotsky (Chilcote, 1984). Marx anticipated current thinking on development and underdevelopment by envisaging the worldwide expansion of British industrial capital in India to be both destructive and progressive. Such expansion may be destructive where merchant capital undermines the pre-capitalist social formations without necessarily transforming them and generating new divisions of labour that are representative of industrial capitalism in the present era. On the other hand, such expansion is progressive or regenerating in the sense of devastating non-capitalist or pre-capitalist societies (social formations) and establishing material conditions or foundations for the advancement of capitalism or capitalist industrialisation outside Europe, or the transformation of the non-capitalist societies into the world market system (Marx, 1943). However, such transformation of society would determine a course of development and a structure of productive powers that is completely dependent on the UK (Mohri, 1979).

The idea of dependent capitalist development, although inherent in Marx, was elaborated by Lenin (1967) who advanced the concept of an international division of labour determined by the presence of dominant industrial and dependent agricultural parts of the world. Such a concept of dependency led Trotsky to argue that backward nations must not necessarily follow the path of advanced capitalist countries but instead follow a path of uneven and combined development en route to socialism. Trotsky argued that socialism depends largely on a worldwide permanent revolution that results in the complete liquidation of the class society (Trotsky 1959, 1964).

The common element in these writers who concentrate on the negative impacts of capitalism is their analysis of the development of the international capitalist system in relation to the processes of development and underdevelopment within a society. In particular, most reformist and all revolutionary writers characterise imperialist capital expansion and domination of developing countries as exploitative and leading to a dependent capitalist development and stagnation. In particular, Sunkel (1972) argued that the expansion of MNEs in the peripheral countries, encouraged by dependent state capitalism in these countries, has led to the introduction of highly capital-intensive technology and the promotion of consumption capitalism. This in turn has led to the destruction or co-optation of national capital and entrepreneurs, thereby depriving the society of autonomous development.

Similarly, Cardoso and Falleto (1979) consider multinational firms and other instruments of foreign imperialism, such as foreign technology, international financial systems and others, as important elements of external forces that relate to structural dependency in developing countries. The ability of these firms to reorganise the international division of labour with some parts of dependent economies included in their plans of productive investment has led to internal structural fragmentation and shift of political power in these countries. While the more advanced sectors of these dependent developing countries are connected to, and dependent upon, the international capitalist system through the activity of MNEs, the backward sectors act as 'internal colonies', which produces a new type of dualism. Foreign capital therefore stimulates development of only some segments of the economy of a dependent country by the manufacture and sale of products to be consumed by the domestic bourgeoisie. However, the absence of internal elements to sustain the accumulation, expansion and self-realisation of local capital accumulation, and in particular the lack of a fully developed capital goods sector, has meant that these peripheral countries can only tie themselves to dependent capitalism (Cardoso, 1973).

Writers in the socialist and revolutionary tradition attribute the deformation of the periphery to its dependency on monopoly capital and the negative impact of foreign firms on domestic economies. In particular, Frondizi (1957) argues that worldwide capitalist integration since the Second World War has resulted in a modification of the colonial situation rather than decolonisation. These semi-colonial and colonial situations failed to benefit from the bourgeois–democratic revolution that characterised the growth of the capitalist system in the UK, Germany, the USA and Japan because of their economic and political dependence.

In addition, Quijano (1974) has argued that the massive expansion of imperialist capital that evolved through a period of urban- and industrial-based accumulation in Mexico, Brazil and Argentina following the Second World War has resulted in the denationalisation of control over capital. The growth of imperialist capital through the assimilation of nationalist capitalists is characterised by a relationship based on exploitation. Similar ideas were forwarded by Prado Júnior (1955) who attributed the dependent and subordinate situation of peripheral countries in the capitalist system, like Brazil, to the large financial and capitalist centres. The appearance in Brazil of international trusts and monopolies undermined any effort to establish domestic industry by a progressive national bourgeoisie. In addition, Dos Santos (1970) put forward the concept of a 'new dependency' brought about by international commodity and capital markets, foreign exchange and multinational corporations since the Second World War which has resulted in the integration of the structure of industry and technology in the dependent countries accoding to the interests of the international capitalist system rather than to the needs of the population and national capital interests. The consequences of such relations, based on a monopolistic control of capital, are high concentration of income, exploitation of labour power and adoption of a technology of intensive capital which lead to the backwardness of these economies.

Writings on underdevelopment that closely relate to and parallel the socialist and revolutionary tradition originated with Baran and Sweezy, and were adopted by Frank, Wallerstein, Emmanuel and Amin, who analysed the patterns of trade in the international political economy, the circulation of merchant capital and the impact of the dominant world capitalist system on the peripheral areas. These writers are collectively referred to as revisionists as their efforts were directed to the modification and extension of Marx's classical thought. In particular, through the examination of the impact of capitalism on India, Baran agreed with Marx, Lenin, Hilferding, Luxemburg and others, who described the tendency of capitalism to evolve into stagnation, imperialism and political crisis (Baran, 1960). Through their focus on multinational and big corporate capital, which became characteristic of the form of capital accumulation and external expansion after the Second World War, Baran and Sweezy (1960) attempted to incorporate monopoly into Marxian economic theory by studying the generation and absorption of economic surplus. These writers emphasised monopoly capital as the cause of stagnation and underdevelopment in both advanced and developing countries. The differing abilities of advanced and developing countries to overcome the obstacles of monopoly capitalism and imperialism led Frank (1966, 1975) and Rodney (1972) to conceptualise capitalist underdevelopment in the context of a world capitalist system which consists of an exploitative relationship between the more developed countries, the metropoles and the undeveloped satellites. Although Frank seemed to have influenced the world systems theory of Immanuel Wallerstein, his later writings demonstrated that he was following the latter's view in analysing the declining importance of the dependence theory of underdevelopment (see, for example, Emmanuel, 1872; Frank, 1978a, b and 1979; Amin, 1974).

Although these revolutionary and socialist writers shared ideas of a dependent capitalist development with those of the nationalist and reformist schools of thought, they differed in terms of the solutions to underdevelopment brought about by dependency. For those who shared a nationalist and reformist sentiment, the solution was found in central planning to promote autonomous, national and capitalist development. However, Perroux (1968) and followers like Correia de Andrade (1967) advocated a scheme that would allow for both capitalist autonomy and multinational investment. But for writers in the socialist tradition, the prospects for autonomous capital development and a progressive national bourgeoisie are considered impossible in the face of the dominant nations and the failure of this bourgeoisie. Socialist writers such as Frondizi, Bagú and Prado therefore questioned the progressive role of a national bourgeoisie in the face of international capital and instead sought socialism as a solution to the underdevelopment and dependency that had shaped peripheral countries since the colonial era. Finally, for those writers who shared a more radical and revolutionary view, the solution was found in a popular revolutionary war or confrontation between military regimes and fascism on the one hand and popular revolutionary governments and socialism on the other.

The works of Hymer (1960, 1972), Newfarmer (1980, 1985), Bornschier and Chase-Dunn (1985), Paul and Barbato (1985), and Cowling and Sugden (1987) represent more recent applications of the theories of underdevelopment. Hymer argues that inward foreign investment at an aggregate level perpetuates a vicious cycle of technological decline by stifling domestic competition as a result of the motive of firms to seek new means of collusion and market power in order to increase profits. The increased industrial concentration resulting from the market power of foreign firms in the developing countries has led Newfarmer (1980, 1985) to conclude that FDI reinforces the underdevelopment of developing countries. Similar ideas are forwarded by Bornschier and Chase-Dunn (1985) who adopt a dependency theory and world system perspective in analysing the impact of foreign MNEs on economic development. These authors, as well as those who propose a North–South model of development, such as Paul and Barbato (1985), focus on the presence of an hierarchical world division of labour as the prime determinant of the continuing dependence and underdevelopment of peripheral countries. The process of locating high-grade production activities in the core (north) countries and low-grade production activities in the peripheral (south) countries is sustained by the operations of MNEs from the developed core countries. Arguing along these lines led such authors to conclude that although inward foreign investment exerts a short-run positive effect on the economic growth of the peripheral developing countries, a sustained level of inward foreign investment exerts a long-run negative effect on economic growth and income equality through the reinforcement of dependent development. Similar conclusions have been drawn by more recent neo-Marxist economists such as Cowling and Sugden who argue that transnational monopoly capitalism inhibits rapid economic growth in peripheral capitalist countries.

The theories of underdevelopment, like those of development, are criticised for their lack of unity, and their various and contradictory tendencies are not always relevant to the historical experience of backward nations.

The theory of technological competence

Although theories of uneven development are still relevant in explaining the changing international division of labour in the world economy, these theories are inadequate to explain the variable impact of foreign MNEs on economic growth in developing countries. Because of the presence of inherent technological capabilities in some firms and countries, inward investment or other modalities of technology transfer have fulfilled a beneficial role in their development. On the other hand, there are also examples of developing countries where, because of the absence or lack of inherent technological capabilities, foreign technology has set in motion a vicious cycle of cumulative decline. The theory of technological competence adopted in this book therefore assumes an intermediate stance between the two extreme views of technology as promoting an aggregate virtuous cycle and an aggregate vicious cycle by allowing for a variable developmental impact of foreign technology introduced by MNEs and other modalities of technology transfer on economic growth and structural change in developing countries, and in particular on the growth of international production through technological development by indigenous firms. As Desai (1987) has pointed out, there has been rapid economic growth and industrialisation throughout the developed and developing countries since the 1960s which cannot be adequately explained by the stagnation model of neo-Marxist economists. Indeed, the theory of technology competence argues that such growth accompanied by significant structural change cannot be explained solely by any theory that regards the presence of foreign investment and other modalities of technology transfer as fostering a system of uneven development in developing countries, or indeed by any theory that solely regards the presence of foreign investment as fostering a developmental spur to indigenous firms in less developed countries.

The theory of technological competence is therefore a useful framework within which to analyse the variable impact of foreign technology on achieving a developmental upgrading role in the internationalisation of firms from developing countries. The most important source of productive efficiency of foreign investment in developing countries lies in the competitive pressure generated by foreign firms on domestic firms (Blomstrom, 1986). The expectation is that since developing countries in general are at an earlier stage of development and have lower technological capabilities than firms from developed countries, the range of sectors in which indigenous firms have the capacity for an independent response to the presence of foreign MNEs is much narrower. Similar arguments on the differing ability of firms and countries to benefit from technical knowledge because of their different levels of indigenous technological capabilities have also been proposed by Lall (1982c) and Katz (1984a).

Even within developing countries, the model allows for the fact that the higher stage of development of the NICs and the higher capabilities of their firms for localised technological change mean that the range of sectors in which indigenous firms in these countries have the capacity for an independent response to the presence of foreign MNEs is much broader than in firms from lower-income developing countries. Foreign technology therefore has a comparatively higher probability of perpetuating a virtuous cycle of development in higher-income than in lower-income developing countries, where firms have lower technological capabilities.

Turning to the impact of foreign technology specifically introduced through the modality of inward foreign direct investment, the comparatively higher level of indigenous technological innovation in the newly industrialised developing countries in Asia and Latin America has enabled them to attract foreign investment, and in those particular sectors where dynamic innovative firms exist, such investment has strengthened innovation and growth of production by indigenous firms. Conversely, the comparatively lower level of production experience and capacity for localised technological innovation in developing countries in Africa, combined with their inward-looking indus-trialisation strategy, sustained by measures that discriminate against the private sector and especially against foreign firms, has resulted in less signifi-cant inflows of inward foreign investment compared to other developing countries (Lall, 1987). The foreign investments that have been attracted have perpetuated a vicious circle of cumulative decline as a result of the threat imposed by these foreign MNEs to the development of production experience and technological innovation by African firms. The inability of African firms to respond positively to the presence of foreign MNEs or to accumulate even elementary forms of technological advantages explains their less significant role in international trade and investment compared to the NICs and other developing countries.

Perhaps the only exception to this generalisation is indigenous firms from Kenya and to a lesser extent Zimbabwe and the Ivory Coast, which have achieved a sufficiently high level of industrialisation, not through the presence of foreign MNEs but through their access to Indian, British and French skills and technology, respectively. As a result of their access to Indian skills and technology, in some cases achieved through joint ventures established between a number of Indian firms and the Kenyan government, Kenyan firms have received a relatively cheap source of technological skills which embody Third World needs. A direct consequence of this technology transfer was the estab-lishment of overseas operations by a number of large local business houses in Kenya (Lall, 1987).

The next section applies the theory of technological competence in analys-ing the role of foreign technology in the technological development and growth of production of firms based in the developing countries.

THE ROLE OF FOREIGN TECHNOLOGY IN TECHNOLOGICAL CHANGE IN THIRD WORLD MULTINATIONALS

Tables 6.1 to 6.14 present the industrial distribution of inward and outward direct investment in a number of developing countries. These tables constitute one indicator of the way in which foreign technology, through the modality of inward foreign direct investment, may have fulfilled a developmental role in the accumulation of technological skills and growth of production of indigenous firms in particular sectors in developing countries.

An analysis of the industrial distribution of both inward and outward investment of the more advanced of the developing countries or the NICs (defined in this book as including Singapore, Hong Kong, Taiwan, South Korea, Argentina, Brazil and Mexico) in Tables 6.1 to 6.7 confirms the wide range of sectors in which indigenous firms have had the capacity to respond independently to the presence of foreign MNEs and to catch up in the process of exports and outward investment. A significant number of these sectors is emerging in the highly technology intensive fields whose growth is in part supported and sustained by aggressive forms of research-intensive outward investment in centres of innovation in the developed countries. The broader range of technological specialisation of firms from the NICs has enabled these countries to attract foreign investment which has the capability of fulfilling developmental roles by perpetuating a virtuous cycle of strengthened indigenous innovation and growth of production in a wide range of sectors.

Singapore

The data contained in Table 6.1 show that petroleum products, electronic equipment, mechanical engineering and chemical products were the most important manufacturing sectors for inward FDI in Singapore in 1981. By 1983, nearly three-fifths of the total cumulative stock of foreign manufacturing, valued at S $10.1 billion, was directed to petroleum refining and electrical and electronic equipment (Pang, 1984). Apart from being an international financial centre, Singapore is the location of the world's largest oil-refining centres, a major supplier of electronic components to the world market and a major centre in the Asian region for marine construction and ship repairing (Aggarwal, 1987).

Although the data do not enable an exact quantification of the particular sectors in which indigenous Singapore firms have established outward direct investment, information gathered from other studies shows a broad sectoral similarity between inward and outward direct investment, owing to the high degree of localised technological capability of local firms in a diverse range of sectors which have profited from linkages with foreign MNEs (see, for example, Lim and Pang, 1982 and Hill and Pang, 1988).[9] Singapore's sectors of multinational investment activity are: food and drink, chemical products, metal products, mechanical engineering, electrical equipment, textiles, leather and clothing, paper products, wood and wood products, rubber and rubber products, non-metallic mineral products, exploration and production of oil, construction, shipbuilding, engineering and real estate (Pang and Komaran, 1984, 1985).

The Singapore government had always considered foreign investment as a main engine of industrial growth. However, the extent of linkages established by MNEs with the domestic economy has been limited, partly because of the small size of Singapore's economy and also because a majority of the foreign investments consist of assembly activities (Wong, 1989). In the period from the mid-1960s to the mid-1970s, foreign investment in the manufacture and assembly of labour-intensive electronic products and components did not help local firms in Singapore to compete because capital and technology presented significant barriers to entry to these firms (Negandhi and Palia, 1989). Local firms had largely obtained initial technology for designing components for electrical appliances through licensing agreements with firms from developed countries which formed the basis for indigenous R&D (Lim, 1984; Lecraw, 1985). However, one local firm developed design capabilities in robot production through a joint venture with a US company and the hiring of a team of

Table 6.1   The industrial distribution of foreign direct capital stock in Singapore, 1981, Singapore $ million[1]

Industry	Inward investment	Outward investment
Primary	100.9	
Agriculture	82.2	
Mining and quarrying	18.7	
Oil	*NA*	
Secondary	8,203.1	
Food, drink and tobacco	282.5	
Textiles, leather and clothing	156.0	
Paper products	74.5	
Chemical products	842.9	*Not*
Coal and petroleum products	2,620.0	*available*
Metal products	354.6	
Mechanical engineering	1,024.6	
Electrical equipment	1,635.3	
Motor vehicles Other transportation equipment }	560.9	
Other manufacturing	651.8[2]	
Tertiary	8,481.2	
Construction	224.3	
Distributive trade	2,340.9	
Transport, storage and communication	564.4	
Finance and insurance	3,677.9	
Property	986.9	
Other services	686.8	
Total	16,785.2	

Source: Ministry of Trade and Industry, *Economic Survey of Singapore, 1984.*
Notes: 1 Represents total foreign equity investment, bothdirect and portfolio. Portfolio investment accounted for 16.1 per cent of total foreign equity investment in 1970.
2 Includes wood products, S$105.7 million.

Taiwan engineers in the development of the product (Pang and Hill, 1991). As a result, more recent policy objectives of the Singapore government since 1979, to engage in industrial restructuring, include incentives for foreign firms to assist in technological upgrading of local firms through investments in higher value-added technologically intensive industries. An analysis of the pattern of technology development, transfer and diffusion of twenty firms in Singapore's microelectronics industry shows the increasingly important role of foreign direct investment in the development of indigenous technological capability, in part through on-the-job training (Hakam and Chang, 1988; Hock, 1988b). In addition, the growth in offshore assembly plant of the US, Japanese and European semiconductor manufacturers in Singapore has spurred the emergence and probable future development of the domestic semiconductor industry (Davis and Hatano, 1985).[10] The local firm, Singatronics Singapore

Pte Limited, is one of the most significant locally based MNEs that produces electronic toys and assembles printed circuit boards in Malaysia. Wearne Brothers is another Singapore MNE which assembles printed circuit boards and electronic toys in Malaysia (ESCAP/UNCTC, 1987).

As with the other Asian NICs, Singapore has sought to dynamically change its comparative advantage towards the local design, production and marketing of products that embody more advanced technology. In particular, Singapore has chosen to be involved in biotechnology, microelectronics, robotics, information technology, communications technology, laser technology and electro-optics and communications technology. The government and private sectors of Singapore have established science towns to promote technical research. In addition, the government is also offering tax incentives, venture capital funds and low interest loans to local firms to upgrade their technological capacities (Berney, 1985; Kotkin, 1987; Karlin, 1988; Hock, 1988b).

Hong Kong

The data in Table 6.2 suggest that electronic equipment, textiles, leather and clothing, chemicals and food, beverages and tobacco are the most important manufacturing sectors for both inward and outward foreign direct investment in Hong Kong. The origins of export-oriented outward investments by Hong Kong firms in the textiles sector may be traced historically to the immigration of Shanghai entrepreneurs with considerable experience in the production and export of textiles. These local firms have grown through exports and direct investments with the use of new but outdated machinery from foreign firms (Wells, 1978). More recently, the linkages that Hong Kong firms have with foreign MNEs, especially with respect to sourcing of components, parts and semi-manufactured products, have helped to build their technological competitiveness. Over 50 per cent of the 541 foreign establishments surveyed in 1987 had subcontracting arrangements with local firms for approximately 25 per cent of their work (Hong Kong Industry Department, 1987). A number of these locally owned subcontracting companies in Hong Kong were often established on the basis of contracts concluded with firms based in the smaller developed countries that provided equipment as well as technical and marketing assistance (Davis and Hatano, 1985; Oman, 1988; Hock, 1988a). By and large, although foreign-based MNEs operate largely in the category of standardised or early-stage maturing products, these firms have employed greater capital intensity than local firms whose general hesitation over long-term investments for industrial restructuring has kept them relatively labour intensive (Chen, 1988).

As with Singapore, the growth of offshore assembly plant of US, Japanese and European semiconductor manufacturers in Hong Kong has enhanced the emergence and probable future development of a domestic semiconductor industry. The technical expertise of a number of electronics firms based in Hong Kong was enhanced by the presence of foreign-based offshore assembly

Table 6.2   The industrial distribution of foreign direct capital stock in Hong Kong, 1988, US $ million

Industry	Inward investment[1]	Outward investment[2]
Primary		268.6
Agriculture	*Not*	249.2[3]
Mining and quarrying	*separately*	19.4[4]
Oil	*available*	NA
Secondary	2,602.4	756.0
Food, drink and tobacco	205.6	24.5[5]
Textiles, leather and clothing	279.1	115.4[6]
Paper products	158.8	22.5[5]
Chemical products	217.2	65.3[6]
Rubber products	58.1	NA
Non-metallic mineral products	35.2	11.0[8]
Metal products	130.0	16.0[7]
Mechanical engineering	34.0	NA
Electrical equipment	1,308.8	28.7[8]
Transportation equipment	50.9	9.3[7]
Other manufacturing	124.7	2.2[9]
Tertiary		404.1
Construction		96.8[10]
Distributive trade	*Not*	95.1[10]
Transport, storage and communication	*separately*	NA
Finance and insurance	*available*	6.8[11]
Property		130.1[4]
Other services		75.3[12]
Total	2,602.4	1,428.7

Sources: Industry Department, Government of Territory of Hong Kong, 'Report on the Survey of Overseas Investment in Hong Kong's Manufacturing Industries, 1988'. Mah Hui Lim, 'Survey of Activities of Transnational Corporations from Asian Developing Countries', paper presented to the ESCAP/UNCTC Joint Unit on Transnational Corporations, Bangkok, March 1984.
Various national sources regarding inward foreign investment in host countries, namely *Indonesian Financial Statistics*, December 1983; Malaysian Industrial Development Authority; Board of Investments, Thailand, December 1978 and 1981; Ministry of Finance and Planning, Sri Lanka; OECD, *Investing in Developing Countries*, 5th edn, Paris, 1983.

Notes: 1 The stock of foreign investment consists of fixed assets plus working capital. Data are available for the manufacturing sector only, although the bulk of the remaining investment is thought to be concentrated in the services sector.
2 Includes only outward investments in Indonesia, Malaysia (1982), Thailand, Taiwan and Sri Lanka (1982).
3 Includes Indonesia, Thailand and Taiwan only.
4 Includes Indonesia and Sri Lanka only.
5 Includes Malaysia and Thailand only.
6 Includes Malaysia, Thailand, Taiwan and Sri Lanka only.
7 Includes Malaysia, Taiwan and Sri Lanka only.
8 Includes Malaysia, Taiwan and Thailand only.
9 Includes Malaysia only.
10 Includes Indonesia and Taiwan only.
11 Includes Taiwan only.
12 Includes Indonesia, Taiwan, Thailand and Sri Lanka only.

plant. Among the increasing number of competitive local firms that have emerged as MNEs are Atlas Electronics, Conic Electronics, Printed Circuits International, Roxy Electric Industries and Elec & Eltek, which invest in the production of various types of electrical and electronic equipment in the ASEAN region to take advantage of low labour costs (ESCAP/UNCTC, 1987). Other Hong Kong semiconductor firms that have equity participation from the People's Republic of China, such as Hua Ko, Elcap and RCL, have in turn acquired equity participation in design houses in the USA in order to gain access to sophisticated technology as well as produce semiconductor masks (Henderson, 1989).

Still another interesting point in the case of Hong Kong is the overseas investments undertaken in joint ventures with firms from the industrialised countries. The establishment of these joint ventures abroad has helped Hong Kong firms to gain access to complementary skills such as an established trade name and advanced consumer marketing skills.[11] For example, one paint manufacturer established a joint venture with an American firm in a Southeast Asian country in order to gain access to the trade name of the US firm. In addition, a Hong Kong textile firm provided 45 per cent interest to a Japanese firm in order to obtain synthetic fibres for its spinning and weaving operations (Wells, 1978).

Taiwan

The case of Taiwan similarly displays a broad sectoral correspondence between inward and outward direct investment. The data in Table 6.3 show that electrical equipment, chemical products, non-metallic mineral products, textiles, leather and clothing and rubber products are important for foreign MNEs in Taiwan as well as Taiwan MNEs. Taiwanese firms have exhibited a high degree of production experience and localised technological innovation, enhanced by the presence of foreign technology, the growth in local university training in the scientific and engineering fields and the repatriation of technical talent.

As with Singapore and Hong Kong, the emergence and probable future development of the domestic semiconductor, electronics, telecommunications and computer industry in Taiwan was spurred by the successful integration of experience by indigenous firms, partly obtained from the presence of offshore assembly plant of the US, Japanese and European semiconductor manufacturers in Taiwan (Saghafi and Davidson, 1989). A number of these indigenous firms have become MNEs. For example, Acer (formerly Multitech), the largest domestic manufacturer of computers, has planned joint ventures in Malaysia to produce personal computers and computer peripherals (ESCAP/UNCTC, 1987). The firm produced 2 per cent of the world's output of personal computers in 1988 (Moore, 1988). The relatively high technological capability of these firms in the production of computers and related products, which in 1987 comprised 30 per cent of electronics output, is reflected in their product

Table 6.3   The industrial distribution of foreign direct capital stock in Taiwan, 1986, New Taiwan $ million

Industry	Inward investment[1]	Outward investment[2]
Primary		194.0
Agriculture	*Not*	194.0
Mining and quarrying	*available*	NA
Oil		NA
Secondary	124,579.4	8,278.3
Food, drink and tobacco	2,501.8	361.8
Textiles, leather and clothing	11,648.8	543.8
Paper products	3,172.9	782.7
Chemical products	25,296.5	1,561.4
Coal and petroleum products	NA	NA
Rubber products	6,465.5	469.0
Non-metallic mineral products	14,400.2	503.3
Metal products	5,807.6	273.7
Mechanical engineering	15,349.4	61.8
Electrical equipment	39,464.7	3,576.4
Motor vehicles	NA	NA
Other transportation equipment	NA	NA
Other manufacturing	472.0[3]	144.4
Tertiary	16,486.5	1,815.1
Construction	160.0	72.0
Distributive trade	485.9	555.9
Transport, storage and communication	6,638.1	7.4
Finance and insurance	NA	616.4
Property	NA	NA
Other services	9,202.5	563.4
Total	141,065.9	10,287.4

Sources: Taiwan Ministry of Economic Affairs, Committee for Evaluation of Foreign Direct Investment, *Foreign Direct Investment Contribution to the National Economy: Survey and Analysis Report, 1986*; Taiwan Ministry of Economic Affairs Investment Commission, *Statistics on Overseas Chinese and Foreign Investment, Technical Co-operation, Outward Investment and Outward Technical Co-operation in the Republic of China*, February 1989.

Notes: 1   Represents cumulative realised foreign direct investment (paid-in capital) by overseas Chinese and private foreign investors.
2   Represents cumulative approved outward investment as from 1959.
3   Represents wood and bamboo products.

and technology exports in this area. By 1988, Taiwan had an 18 per cent share in the US market for unsophisticated low-end computers, while South Korean firms had 14 per cent (Mody, 1990). In addition, Mitac International announced in September 1989 the transfer of plant technology in the manufacture of two of their existing models of computers through a licensing agreement with Multipolar, an Indonesian firm (Liu, 1988; Moore, 1988).

Mitac and American Computer Products are former US firms which Taiwan acquired to gain access to US technology and to take advantage of the large US market as an outlet for its semiconductor products.

Apart from foreign technology accessed through inward foreign direct investment, Taiwan has also employed foreign technology through supplier contracts and technology agreements with foreign-based multinationals. A leading Taiwanese MNE in the electrical and electronics sector has grown through the acquisition of advanced technology in the production of household appliances and consumer electronics through technical assistance from firms based in the USA and Japan that had limited capital participation in the firm (Ting and Schive, 1981). In some cases, formal linkages had been established with foreign-based high-technology firms and professional and management personnel had been sought from the USA in order to be able to diversify their industries towards more research-intensive sectors (Crawford, 1987). For example, Tuntex Fiber Company, another Taiwan MNE, initially acquired the technology in synthetic fibres production through the purchase of equipment from an MNE based in the developed countries (Ting and Schive, 1981). Tatung, another Taiwan MNE which dominates production of computer monitors in Taiwan, and the main supplier of monitors for IBM personal computers, has gained from the technical assistance provided by IBM. The design of chips to clone, and in some cases improve upon, various IBM personal computers is Acer's overall aim (Mody, 1990). The relative success of the firm in the production of Dynamic Random Access Memory (DRAM) semiconductor memory chips has been enhanced by the establishment of a $250 million joint venture with Texas Instruments of the USA (Maitland, 1989).

Taiwan, like the other NICs, seeks dynamically to change its comparative advantage towards the local design, production and marketing of products that embody more advanced technology and in particular in the machine manufacturing, computer and communications industries, electronics and precision instruments. In addition to providing start-up capital for small firms, as well as exemption from tariff payments on certain equipment (Berney, 1985; Davis and Hatano, 1985; Brody, 1986; Saghafi and Davidson, 1989), the government of Taiwan is actively providing technological support to assist in the technological development of local firms.

The Industrial Technology Research Institute (ITRI) was established by the Taiwan government in 1973 to develop new ideas, or license advanced technology from abroad. The Electronics Research and Science Organisation (ERSO) emerged as an offshoot of ITRI and has been tremendously successful in the establishment of domestic semiconductors and computer memory chip-making ability to support Taiwan's personal computer industry. The overall objective of ERSO is to develop generic technologies and, through direct connections with local academic research departments, both ITRI and ERSO lend technical support to the private sector through the transfer of technologies, for modest fees, to local companies – primarily small- and medium-sized businesses since Taiwan has few conglomerate firms with the financial capacity to spend heavily on R&D (Henderson, 1989; *The Economist*, March 1989). ERSO was instrumental in the introduction of the BIPOLAR and CMOS process technologies and VLSI production facilities in 1976 and 1979 which enabled Taiwan to engage in the full-line production of semiconductors. This objective has been achieved in part through the early collaboration of ERSO with the US firm RCA in semiconductor production (Oman, 1988). In addition, ERSO acquired the E-Bean technology from the US semiconductor firm Perkins Elmer Company and with the assistance of Vitelic, a local semiconductor firm, co-developed the 64K and 265K CMOS DRAM.

Some of the three or four world class chip plant, such as that of United Microelectronics Corporation (UMC) and Taiwan Semiconductor Manufacturing Corporation (TSMC), plus half a dozen smaller ventures, either in operation or under construction in the Hsin-chu Science-based Industrial Park in Taiwan, are either spin-offs from ERSO or staffed by its graduates and have associations with foreign firms. For example, through the purchase of complementary metal oxide semiconductor integrated circuit technology by UMC from RCA in 1976, ERSO obtained a 10 per cent share in RCA. In addition, UMC has cooperation arrangements in the development of some products with the US-based research firms Mosel, Qusel and Vitelic which are overseas Chinese operated companies.[12] Philips International NV of The Netherlands currently also has a 27.5 per cent equity participation in TSMC which assists in the distribution of semiconductor products and in the supply of 2.0-micron technology for producing very large-scale integrated circuits (VLSI).

Apart from the emerging cooperation of some Taiwan firms with foreign firms and the employment of foreign experts, Taiwan is spending more than $1.2 billion in the construction of additional chip-making plant in order to fulfil the aim of supplying 4 per cent of the world demand for chips, which would enable Taiwan to become the fourth largest supplier by 1992 after Japan, the USA and Korea. Since the chip and PC industries are now fairly developed, future efforts of ERSO will focus on the development of technology to support sub-micron chip production and systems technology (Saghafi and Davidson, 1989; Yang, 1989; Johnstone, 1990a; Mody, 1990).

BY 1993, the number of Taiwanese semiconductor companies with chip production fabrication plant, which has grown from zero to 7 in the past 10 years, is expected to more than double to 16. However, despite the look of prosperity, many industry observers predict a reduction in the number of Taiwan's chip-making firms before 1993, owing to the need to survive on its own – independent of government incentive schemes and finances provided by the stock market boom. Moreover, Taiwanese firms' ability to compete globally is compromised by intellectual inbreeding, underutilised capacity, product duplication, patent battles, lack of marketing expertise and poor timing.

South Korea

Table 6.4 shows that electrical equipment, chemical products, motor vehicles, food, drink and tobacco and motor vehicles are common sectoral interests of foreign MNEs in South Korea and South Korean MNEs. The evidence indicates that foreign direct investment has been an important source of technology in only a few sectors, primarily in chemicals, electronics and petroleum refining (Westphal, Rhee and Pursell, 1984). As with Singapore and Taiwan, the emergence and probable future development of the domestic semiconductor and electronics industries in South Korea were spurred by the successful integration of experience by indigenous firms, partly obtained from the presence of offshore assembly plant of the US, Japanese and European semiconductor manufacturers in South Korea (Kim, 1980). The growth of indigenously owned electronics systems houses such as Samsung, Goldstar, Hyundai and Daewoo as divisions of the Korean conglomerate MNEs in the late 1970s has been stimulated in part by foreign investment in the semiconductor industry since the mid-1960s. However, no less important is their strategy of encouraging joint ventures and liberal licensing agreements in their negotiations with foreign-based multinational corporations (UNCTC, 1986). Joint ventures played an important innovative role in the establishment of facilities to produce basic petrochemicals and derivative synthetic fibres and resins, an important example being polyester fibre and yarn (Westphal, Rhee and Pursell, 1984). Other factors behind the growth of competitive domestic producers of semiconductors are the growth in local university training in the scientific and engineering fields and the repatriation of technical talent (Davis and Hatano, 1985; Henderson, 1989; Saghafi and Davidson, 1989). Training and education have been crucial in the technological development of South Korea (Bell, Ross-Larson and Westphal, 1984).

Apart from the important role of foreign MNEs in fostering the growth of indigenous industries associated with the development of competitive technological capabilities in indigenous firms, the growth of Korean exports and overseas investment may be viewed partly in the context of the dynamic process of the export-oriented growth phase of the Korean economy since 1965, in which foreign direct investment has played an important role. The export-oriented foreign MNEs in Korea may have played an important tutorial role for export-oriented Korean firms which have in turn begun to establish on-site service facilities in export markets and processing facilities in developing countries.

However, foreign technological licensing agreements have played a more important role in technology transfer to South Korea, partly due to earlier government policy which regulated the inflow of foreign direct investment (Enos and Park, 1988). Data from the Korean Ministry of Finance show that between 1962 and 1986, the number of technological licensing cases in South Korea was 4,055 while that of foreign direct investment was only 2,095. The USA accounted for 54 per cent and 66 per cent of the total cases of technologi-

Table 6.4 The industrial distribution of foreign direct capital stock in South Korea, 1987, US $ million

Industry	Inward investment	Outward investment[1]
Primary	20.82	463.35
Agriculture[2]	14.89	66.46
Mining and quarrying	5.93	396.89
Oil	NA	NA
Secondary	1,759.44	314.37
Food, drink and tobacco	120.03	5.8
Textiles, leather and clothing	36.27	29.89
Paper products	19.78	10.6
Chemical products	498.31	47.39
Coal and petroleum products	−36.60	0.2
Rubber products	NA	10.2
Non-metallic mineral products	34.86	31.5
Metal products	91.58	99.79
Mechanical engineering	128.20	14.8
Electrical equipment	505.89	35.2
Motor vehicles	321.50	23.9
Other transportation equipment	NSA	NSA
Other manufacturing products	39.62	5.1
Tertiary	1,061.79	188.34
Construction	49.92	31.30
Distributive trade	18.51	94.94
Transport, storage and communication	19.60	2.68
Finance and insurance	202.27	NA
Property	NA	26.59
Other services	771.49	32.83
Total	2,842.05	966.06

Sources: Korea Institute for Foreign Investment, *A Study on Korean Overseas Investment*, March 1989; Korea Ministry of Finance, International Finance Bureau, Overseas Investment Division, *Annual Report and Fiscal Financial Statistics*, 1987.
Notes: 1 Calculations based on net cumulative FDI flows as from 1968.
2 Including fishery and forestry.

cal licensing and foreign direct investment respectively. The corresponding shares for Japan are 24 per cent and 20 per cent. But even with the liberalisation of government policy on FDI after 1982, technological licensing cases in the period 1983 to 1986 numbered 1,770, which is more than three times the number of cases of foreign direct investment, at 520. By comparison, the ratio of the number of cases of technological licensing *vis-à-vis* FDI in the period of greater restriction over FDI between 1978 and 1982 was 5.4:1.

Apart from licensing, the central diversification strategy adopted by South Korean firms is similar to Taiwan's in that it is based on the utilisation of foreign technology, especially in managing supplier contracts, technology agreements, outright purchase and formal linkages through joint ventures with foreign MNEs. These technological alliances with foreign firms have helped to dynamically change comparative advantage through the acquisition of additional capabilities and new technologies, especially in research-intensive industries, and have also assisted in the acquisition of marketing and management know-how (Jo, 1981; Westphal, Kim and Dahlman, 1984; Euh and Min, 1986; Crawford, 1987). A number of these alliances is listed in Table 6.15.

The four semiconductor houses that are divisions of the largest Korean conglomerate MNEs, namely Samsung, Goldstar, Hyundai and Daewoo, all have the capabilities to process wafers for very large-scale integrated circuits (VLSI), the most complex and capital-intensive activity in the manufacture of semiconductors.[13] Although the growth of these firms may have been stimulated by foreign investment in the semiconductor industry in South Korea, their present growth has been sustained more through strategic alliances and cooperative relationships with foreign firms, some examples of which are shown in Table 6.5. For example, although the basic technology pattern for the production of DRAM chips is attributed to the US firm Texas Instruments, to which all makers of DRAM chips have to make regular royalty payments, the agreement that the Korean MNE Goldstar has concluded with Hitachi would enable the firm to keep abreast of other major Korean semiconductor firms such as Samsung and Hyundai (Clifford, 1989).

In 1989, Samsung emerged as the thirteenth largest firm in the world in terms of semiconductor sales volume and it is planning to be one of the top ten by 1995. The firm's high level of indigenous technological innovation is seen in its ability to catch up with four companies from Japan and the USA in the development of the 16-megabyte dynamic random access memory (DRAM) chip in August 1990 (*Korea Business World*, September 1990). Apart from its foreign collaborative arrangements, the ability of Samsung to develop the most research-intensive random access memory chip is also attributed to its in-house R&D which currently stands at above 10 per cent of annual sales with a workforce of 2,000 representing 20 per cent of the total workforce (Kang, 1990).

As a result of its excellent technological capabilities in semiconductor production, South Korea has become the third largest producer of DRAMs in the world (Mody, 1990).[14] Such rapid development towards the technology frontier in semiconductor production has been sustained through considerable dependence on foreign technology, unlike Taiwan firms whose semiconductor production is, by comparison, much more home grown (Mody, 1990).

In the field of computers, as a result of the provision of the manufacturing technology for the production of super-mini computers to the four largest Korean *chaebols*, Samsung, Hyundai, Daewoo and Goldstar, by the US firm Tolerant, these firms appear to have succeeded not only in assembling components supplied by Tolerant but also in producing whole computers. However, problems still arise in commercialisation of an indigenously made computer (Evans and Tigre, 1989). In motor vehicle production, the growth of the technological capacity of Hyundai has been sustained through advanced technology by acquisition, assimilation and adaptation of indirectly imported and directly transferred technological innovations, apart from the international finance and marketing agreements concluded by the firm with Mitsubishi (Kwon and Ryans, 1987). Its automobiles, which are marketed under the brand names Pony or Excel, have been developed with various foreign suppliers of design, technology and finance.

Although their scientific and technical resources are still modest, an important factor in the growth of South Korea's local electronics industry in general is expenditure on R&D, encouraged by Korean industrial policy, implemented in 1973, which promoted the growth of the industry in part through infant industry protection, the assurance of a guaranteed market and the formation of specialised research institutions. These research institutions, which are funded in part or whole by the government and industry, are the Korean Advanced Institute of Science and Technology, the Korean Electronics and Telecommunications Research Institute (ETRI), the Korean Automotive Systems Research Institute and the Korean Biogenetics Research Institute, which supplement university research but whose efforts are almost exclusively directed to industry (Porter, 1990).[15]

South Korea's major research effort in electronics, telecommunications and automobiles is a positive indication of its attempts to boost the competitiveness of some of its most important industries. Its particularly heavy investments in electronics R&D are evident. For example, in addition to ETRI projects which focus on telecommunications, semiconductors and computer architecture, DACOM was formed by a number of privately owned electronics firms and controlled by the Ministry of Communications to handle initiatives in data communications, such as value-added networks and electronic mail. The Korea Telecommunication Authority plans to invest 3 trillion *won* (US$4.5 billion) in R&D, representing 6 per cent of total expected revenues by 2001. In addition, in late 1989 the Ministry of Trade established the Korea Academy of Industrial Technology (Kaitech), which aims to build indigenous technology in every field of engineering and to serve as a focal point for projects aimed at developing high definition television (HDTV) and high-speed digital facsimile machines. As in Taiwan and Singapore, the government and private sectors of South Korea have established science towns to promote technical research.

Apart from the imposition of a policy of infant industry protection through an import ban on foreign-made electronics from 1973 to 1982, the South Korean government, like those of Taiwan and Singapore, provided substantial amounts of low interest capital for companies. The R&D allotment for 1990 was 133 billion *won* or 3 per cent of total expected revenues for 1990, which is expected to increase gradually to 5 per cent in 1996. The overall objective is to decrease import dependence on Japan, especially for precision components such as mini-bearings, video tape recorder heads, drums and other electronic parts. More importantly, the current efforts of South Korea to develop indigenous technology are intended to increase its bargaining power in response to the emerging reluctance of its main source of technology, the United States, to transfer proprietary technology. Japan has always been reluctant to transfer technology because of fears of developing potential competitors (Berney, 1985; Clifford, 1988; Evans and Tigre, 1989; Matsuura, 1989; Johnstone, 1990b; *Korea Business World*, April 1990).

South Korea provides an excellent example of a developing country that has effectively assimilated foreign technology owing to a relatively high level of domestic technology capacity as shown by high expenditure on indigenous R&D (see also Dahlman and Sercovich, 1984, and Kim and Lee, 1987). In the field of consumer electronics, Korea had become, by the late 1980s, the second largest producer, after Japan, of such capital- and skill-intensive products as video recorders and microwave ovens (Mody, 1990). Such growth has been enhanced by the interaction of state policy and private entrepreneurship which has led to the emergence of a few conglomerate groups (*chaebol*) similar to Japanese conglomerate groups (*zaibatsu* or *keiretsu*), which enjoy economies of scale comparable to the major MNEs based in the developed countries. These conglomerate groups allow major investment in high technology industries *inter alia* through their ability to raise large amounts of capital more efficiently, their greater capacity to bear risks through cross-subsidisation of production lines, their long-term incentives to upgrade product quality and their economies of scale and scope in manufacturing and R&D. These groups represent part of the institutional factors that have led to the faster technological development of the electronics industry of South Korea compared to that of Taiwan (Mody, 1990).

The major *chaebols* that have accumulated experience in the local manufacture of consumer electronics products have been able to engage in exports even in the USA where these firms have been able to sustain competition through the sales of personal computer clones that replicate standard computer architecture but sell on the basis of lower costs (Kim, 1986; Yu, 1988). Although South Korea had a 4 per cent share of the US market for unsophisticated low-end computers compared to Taiwan's share of 13 per cent in 1984, this share had considerably increased, to 14 per cent compared to Taiwan's 18 per cent by 1988, mainly because of the higher unit value of South Korean computers (Mody, 1990). The resultant trade pressures from the USA and the attainment of competitive abilities by Korean firms resulted in the reversal of the previous infant industry protection policy for consumer electronics in 1987 (Evans and Tigre, 1989).

The assimilation of foreign technology, especially in higher technologically intensive industries, is seen as part of the overall technological efforts of Korean firms to achieve productive efficiency and to compete effectively through increased market shares in overseas markets (Nishimizu and Robinson, 1984). Its aggressive pursuit of the acquisition of foreign technology through licences, outright purchase, joint ventures and other agreements, reverse engineering and feedback from its foreign buyers, which contributed to its efforts in product innovation, makes South Korea unique among the

NICs in its commitment to the development of indigenous product models and state-of-the-art process technology (Euh and Min, 1986; Porter, 1990). The continued use of foreign resources has undoubtedly been necessary for the achievement of rapid production growth and has resulted in the acquisition of technological capacity by Korean industry in basic production processes beyond product and plant design. Such accumulation of local know-how, including the effective assimilation, adaptation and diffusion within Korean industry, has proceeded at such a rate that much of what is considered technology from domestic sources consists of technology originally developed overseas but subsequently transferred or brought to Korea (Westphal, Rhee and Pursell, 1984).

Latin America

Indigenous firms from the NICs in Latin America, in particular Argentina and Brazil, also exhibit a high degree of localised technological sophistication in a wide range of sectors which has enabled them to benefit from foreign technology. Latin American countries in general have always emphasised scientific and technological development as part of their social and economic development. In the post-Second World War period, foreign private investment, imports of capital goods and technical assistance were deemed important in the promotion of indigenous technological development. But in more recent years, owing to concerns that foreign technology may be inappropriate for these countries, special policies designed to promote indigenous technological development have become more characteristic of development strategies in the region, in addition to the role of foreign private investment and trade in technology. Strengthening indigenous technology increases the capacity of these countries and firms to assimilate new knowledge and is an effective instrument in achieving technological change and rapid industrialisation in the region (Inter-American Development Bank, 1988 Report).

A very important distinction between NICs from Asia and Latin America lies in the importance assigned to the development of the electronics industry. Argentina, Brazil and Mexico have chosen to achieve economies of scale in the manufacture of components as opposed to the Asian NICs whose major objective has been the development of the electronics industry as a whole. The technological and commercial know-how of firms in the Asian NICs has been enhanced by national efforts dynamically to change their comparative advantage as regards the local design, production and marketing of electronic products that embody more advanced technology. The relatively high level of technological and commercial know-how has enabled firms to exploit profitable niches of market opportunity in the rapidly evolving field of electronics technology and to catch up rapidly with the technologically leading firms based in the developed countries.

Perhaps the different relative importance of the electronics industry and the role of foreign technology in indigenous technological development in regions of Asia and Latin America may be attributed to the fact that the early comparative advantages of Latin America were based more on natural resources than on the supply of plentiful, low-cost labour (Inter-American Development Bank, 1988 Report). The different patterns of comparative advantage in the two developing regions, as well as protectionist trade and investment policies in Latin America, may have been among the more crucial reasons for the more rapid growth of labour-intensive production and sub-contracting arrangements by foreign electronics firms in Asia in response to global competition in the production of microcomputers in the 1970s and 1980s. These investments have enabled local firms in Asia to strengthen their technological innovation *inter alia* through access to technical information in open computer architecture which enabled reverse engineering and imitation in hardware and software design to be done. The production of copies, clones and compatible IBM computers throughout the world may be attributed to efforts in this direction by firms from Asia. Taiwan in particular accounted for 10 per cent of the world's personal computer market in 1988 (Liu, 1988).

Strengthened technological innovation in Asian countries has given their firms the capacity to undertake technology exports and outward investment. In more recent years, the rising labour costs in the home country, associated with more advanced stages of industrial development, have forced Asian firms to engage in a further international division of labour to maintain their competitiveness. Chapter 8 will show some examples of firms from South Korea and Taiwan that have re-located the more labour-intensive stages of their production processes in other developing countries, including the Latin American and Caribbean region where the level of real wages is now comparatively low and where there are tax regimes favourable to profits earned by foreign MNEs.

Argentina

Table 6.5 shows that both inward and outward direct investment in Argentina have a similar industrial concentration on food, drink and tobacco, chemical products, mechanical engineering and motor vehicles. Argentine MNEs have also been significant in metal products and paper products. In addition to their direct investment activities, Argentine firms are also significant exporters of technology in nuclear energy, pharmaceuticals and agricultural machinery. These international activities are attributed to the important role played by foreign-based multinational corporations, sustained through their relatively high levels of human capital and long-established industrial infrastructure. However, Argentina's industrial policy of narrowly focusing on a few selected industries, and the emigration of qualified manpower during the period of political turmoil in the late 1970s, has not helped local firms to develop technological capabilities in a wide range of sectors (Dahlman and Sercovich, 1984).

The development of indigenously generated engineering and technology in

Table 6.5   The industrial distribution of foreign direct capital stock in Argentina, 1981–3, US $ million[1]

Industry	Inward investment, 1983	Outward investment, 1981
Primary	1,156.8	29.37
Agriculture	454.48[2]	3.34
Mining and quarrying	NA	NA
Oil	702.32	26.03
Secondary	4,268.79	45.44
Food, drink and tobacco	414.59	10.39
Textiles, leather and clothing	NA	NA
Paper products	NA	3.69
Chemical products	1,403.62	4.15[3]
Rubber products	NA	NA
Non-metallic mineral products	172.99[4]	0.11
Metal products	NA	12.24
Mechanical engineering	265.86	6.63
Electrical equipment	NA	0.42
Motor vehicles	495.44	7.81
Other transportation equipment	NSA	NA
Other manufacturing	1,516.29	NA
Tertiary	2,012.23	12.33
Construction	49.41	6.98
Distributive trade	1,462.03	3.05
Transport, storage and communication	NA	1.07
Finance and insurance	473.57	NSA
Other services	27.22	1.23
Total	7,437.82	87.14

Sources: Banco Central de Argentina y El Registro de la Inversión Extranjera Directa, Ministerio de Economía de Argentina; Dirección de Inversiones en el Exterior, Ministerio de Economía de Argentina.

Notes: 1   Estimates based on cumulative investment flows reported by the Banco Central de Argentina and El Registro de la Inversión Extranjera Directa, Ministerio de Economía de Argentina as from 1955 in the case of inward investment and 1965 in the case of outward investment.
   2   Including mining and quarrying.
   3   Including chemicals (US $2.59 million), toiletries (US $0.25 million) and plastics (US $1.31 million).
   4   Including other transport equipment.

Argentina in the area of consumer electronics, industrial electronics and information technology is an excellent case in point. In the mid-1970s, the policies adopted to achieve import liberalisation – the overvaluation of the Argentine currency combined with high domestic interest rates – discouraged local investment in improving competitiveness against lower cost and better designed imports. This led to the abandonment of local production of consumer electronics and components on the part of a number of Argentine manufacturers in favour of importing. However, efforts are currently being made by the Argentine government to establish an electronics industry under domestic control, with the formation of the National Commission on Information Technology (Comisión Nacional de Informática) in 1984. The main tenet of this policy is the generation of autonomous technological development, which includes the acquisition of necessary foreign technologies. Licensing contracts covering both joint and domestic ventures have proved to be the general means of rapid and sure access to standard information technology and to supplies for the assembly and possible manufacture of products. The strategy of foreign participation in the acquisition of new technology has proved to be worth while in that it ensures rapid entry which allows the generation of localised technology to be quickly accomplished (Inter-American Development Bank, 1988 Report).

Brazil

Although Table 6.6 does not enable an evaluation to be made of the sectoral similarity or dissimilarity of foreign MNEs in Brazil and Brazilian MNEs in the manufacturing sector, secondary sources of information show that Brazilian firms have operated successful multinational investment operations in the production of metal products, electrical equipment and textiles and clothing (see, for example, Diaz-Alejandro, 1977; White 1981; Villela, 1983; Wells, 1988). Of all the Latin American countries, Brazil has the greatest reliance on foreign direct investment and foreign licensing agreements to promote the development of its localised technological advantages and production experience, especially in the capital goods sector (Dahlman and Sercovich, 1984). Such imports of foreign technology incorporated in capital goods during the early stages of Brazilian industrialisation have been replaced in the later stages by more complex technological import requirements which are largely met through licensing and technical agreements with foreign producers (Guimaraes, 1986).[16] For example, the technology for overseas projects in petrochemical complexes, offshore drilling rigs and feroalloy plant has been acquired through foreign licensing agreements (Sercovich, 1984). The risk contracts that Petrobrás, Brazil's national oil company, has established with foreign oil companies for oil research since 1976 have been a source of valuable technological interchange. In addition, Tenenge, a Brazilian engineering firm with subsidiaries in Chile and Paraguay, has been able to deliver state-of-the art refineries through the successful combination of foreign licensing agreements and its indigenously generated technology. Engevix, another leading Brazilian consulting engineering firm, has successfully designed part of Baghdad's subway through the adaptation of technology previously acquired from Sofrerail, a French company.

The joint ventures that Brazilian MNEs have established with foreign firms in the developed countries have enabled them to expand their activity in

Table 6.6 The industrial distribution of foreign direct capital stock in Brazil, 1982, US $ million

Industry	Inward investment	Outward investment[1]
Primary	754.56	238.0
Agriculture	131.64	NA
Mining and quarrying	622.92	NA
Oil	NA	238.0
Secondary	15,493.77	170.2
Food, drink and tobacco	1,218.22	
Textiles, leather and clothing	529.89	
Paper products	459.39	Not
Chemical products	3,322.85	separately
Coal and petroleum products	453.78	available
Rubber products	523.27	
Non-metallic mineral products	379.85	
Metal products	1,515.64	
Mechanical engineering	2,029.74	31.9
Electrical equipment	1,538.40	Not
Motor vehicles	2,179.64	available
Other transportation equipment	671.28	
Other manufacturing	671.82	138.3
Tertiary	4,452.82	454.3
Distributive trade	812.57	NA
Transport, storage and communication	43.50	30.2
Finance and insurance	852.43	390.2
Property	119.77	NA
Other services	2,624.55	33.9
Total	21,176.38[2]	862.5

Source: Banco Central do Brasil, *Boletim Mensual* (21, 2) July 1985.
Notes: 1 Represents cumulative investment flows, 1977–82. The total actual investment flow for the period is US $1,416.0 million.
2 Total includes unspecified figure.

high-technology sectors. For example, Embraer has linked up with Aeritalia and Aermacchi of Italy to manufacture the sophisticated AMX-fighter bomber in Brazil and Italy. The competitive price quoted for this aircraft places Embraer in direct competition with British Aerospace's Hawk, the similar Alpha jet built by the Franco–German consortium Dassault/Dornier and the Soviet Sukhoi–25 'Frogfoot' model (Wells, 1988).

Another interesting aspect of Brazilian technological development lies in the area of information technology in which early signs of the internationalisation of domestic firms are increasingly becoming apparent. For example, Brazilian firms have concluded contracts to automate supermarkets in Portugal in addition to their efforts at technology transfer in neighbouring countries such as Argentina. The emergence of such outward-looking Brazilian firms may be attributed to their acquisition of technological capacities, achieved largely through a coherent national policy embodied in the National Informatics Law of 1984 which imposes government limitations on inward foreign direct investment and effectively imposes a policy of market reservation of the computer industry for locally owned firms. As a result, efforts to promote the development of indigenous production capacities in computers especially in mini and microcomputers and peripherals such as printers, screens and disks whose local production has been sustained through licensing with smaller computer firms such as Sycor (USA), Logabax (France), Nixdorf (Germany), Fujitsu (Japan) and Ferrati (UK) that have been prepared to supply technology through arms-length transactions (Oman, 1988). Although the law offers Brazilian firms major incentives to develop technological self-reliance, the protected market of indigenous computer producers has led to technological capacities limited to imitation, and merely incremental technological changes which need to be upgraded by cooperation with foreign MNEs (for further details see Colson, 1985; Woog, 1986; Inter-American Development Bank, 1988; Evans and Tigre, 1989).

Mexico

Although Table 6.7 does not enable a sectoral comparison to be made of inward and outward direct investment in Mexico, studies have shown that Mexican MNEs have been notable in non-metallic mineral products. For example, in 1989 Cementos Mexicanos invested in excess of US $800 million in cement companies not only in Mexico but also in Texas and California, which enabled the firm to become the largest cement manufacturer in the North American market and the fourth largest in the world. In addition, in the same year Vitro, which is a technologically dynamic, vertically integrated glass group in Mexico with extensive experience in the export of glass products, glassmaking equipment and process technology, purchased the second largest US glassmaker, Anchor Glass Container Corporation. Vitro initially acquired the process technology and production design in glass production through licensing and technical assistance contracts with a US firm and adapted them to suit a smaller market. The adapted process technology combined with the glassmaking equipment provided by Fama, a local capital goods producer with design and production capabilities, has enabled Vitro to expand its product line and considerably to improve the productivity of the originally imported process technology. Fama's design and production of glassmaking equipment were prompted by the unavailability of glassmaking equipment from the USA, Germany and the UK during the Second World War (Baker, 1989).

The lower-income developing countries

A narrower range of technological specialisation is seen in indigenous firms from the lower-income developing countries such as Colombia and Peru in

Table 6.7 The industrial distribution of foreign direct capital stock in Mexico, 1989, US $ million

Industry	Inward investment[1]	Outward investment
Primary	418.9	
Agriculture	28.0	
Mining and quarrying	390.0	
Oil	NA	
Secondary	17,700.8	
Food, drink and tobacco		
Textiles, leather and clothing		
Paper products		
Chemical products		Not
Coal and petroleum products		available
Rubber products		
Non-metallic mineral products		
Metal products		
Mechanical engineering		
Electrical equipment		
Motor vehicles		
Other transportation equipment		
Other manufacturing		
Tertiary	8,467.4	
Distributive trade	1,888.5	
Transport, storage and communication		
Finance and insurance		
Property		
Other services	6,578.9	
Total	26,587.1	

Source: Secretaria de Comercio y Fomento Industrial (SECOFI), Mexico.
Note: 1 Represents authorised foreign direct investments.

Latin America and China, India, Malaysia, the Philippines and Thailand in Asia. Most of the firms from these countries have engaged in the overseas production of less technologically intensive products compared to MNEs from the higher-income NICs. For example, the data in Table 6.8 suggest that outward investments by Colombian MNEs have been most notable on food, drink and tobacco which account for 86 per cent of their investments in the manufacturing sector. The second and third most important manufacturing sectors for Colombian MNEs are chemical and paper products which together account for only 9 per cent of their investments in the manufacturing sector. These sectors are also important areas of manufacturing investment by foreign MNEs in Colombia. By contrast, although Table 6.9 does not give an accurate sectoral distribution of the outward investment of Peruvian firms in the manufacturing sector, studies have shown that Peruvian firms are actively

Table 6.8 The industrial distribution of foreign direct capital stock in Colombia, 1987, US $ million

Industry	Inward investment	Outward investment
Primary	1,449.41	13.29
Agriculture	16.77	10.97
Mining and quarrying	1,432.64	2.32
Oil	NA	NA
Secondary	1,188.92	66.24
Food, drink and tobacco	178.13	57.03
Textiles, leather and clothing	51.14	0.37
Paper products	152.96	2.35
Chemical products	527.18	3.81
Coal and petroleum products	NSA	NSA
Rubber products	NA	NA
Non-metallic mineral products	74.52	0.75
Metal products	14.33	0.25
Mechanical engineering	172.43	1.63
Electrical equipment ⎫	Not	Not
Motor vehicles ⎬	available	available
Other transportation equipment ⎭		
Other manufacturing[1]	18.23	0.05
Tertiary	350.68	252.11
Construction	−1.31	0.46
Distributive trade	144.46	12.51
Transport, storage and communication	30.99	7.65
Finance and insurance	163.64	228.30
Property	9.93	0.68
Other services	2.97	2.51
Total[2]	2,992.26	334.54

Source: Banco de la República de Colombia, Oficina de Cambios, Sección Análisis Económico.
Notes: 1 Includes wood and furniture.
2 Includes unclassified sectors, US $3.25 million in the case of inward investment and US $2.90 in the case of outward investment.

engaged in international investment in the mineral and food products sectors (see, for example, Diaz-Alejandro, 1977, and White, 1981). Although further studies still have to be undertaken on the ways in which Colombian and Peruvian firms have accumulated their technological advantages, the hypothesis is that the presence of foreign MNEs may have played a tutorial role for industrialisation in the way that Kojima (1975, 1978) predicted and which may have led to the establishment of competitive indigenous firms.

Table 6.10 suggests that MNEs from Thailand have successfully engaged in the international production of food, drink and tobacco, textiles, leather and clothing and metal products, which are also important sectors of inward

Table 6.9   The industrial distribution of foreign direct capital stock in Peru, 1987, US $ million

Industry	Inward investment	Outward investment
Primary	423.63	0.74
Agriculture	6.93 ⎫	*Not*
Mining and quarrying	411.45 ⎬	*separately*
Oil	5.25 ⎭	*available*
Secondary	407.41	5.32
Food, drink and tobacco	99.13 ⎫	
Textiles, leather and clothing	26.39	
Paper products	10.09	
Chemical products	94.07	
Coal and petroleum products	1.97	*Not*
Rubber products	25.53	*separately*
Non-metallic mineral products	10.90 ⎬	*available*
Metal products	22.58	
Mechanical engineering	18.84	
Electrical equipment	60.14	
Motor vehicles	26.26	
Other transportation equipment	4.06	
Other manufacturing	7.45 ⎭	
Tertiary	342.19	53.88
Construction	1.44 ⎫	
Distributive trade	173.13	*Not*
Transport, storage and communication	6.59	*separately*
Finance and insurance	84.02 ⎬	*available*
Property	14.90	
Other services	62.11 ⎭	
Total	1,173.23	59.94

Source: Comisión Nacional de Inversiones y Tecnologías Extranjeras (CONITE), Ministerio de Economía y Finanzás de Peru.

foreign direct investment in Thailand. However, the low level of indigenous technological capabilities of Thai firms in other higher technologically intensive industrial sectors such as petroleum refining, tin smelting and electronic equipment has resulted in the total control of these sectors by foreign MNEs (see Phiphatseritham, 1982). For example, the absence of local firms equipped with the technology needed in the production of local materials and parts for the electrical products and equipment industries has resulted in the importation of these materials and parts by foreign MNEs (see also Tambunlertchai and McGovern, 1984). The high import dependence of this industry has meant that few or no linkages have been established by foreign MNEs with local firms, which has led to a dependent state of industrial development in Thailand (Prasartset, 1988). Perhaps a major reason for the limited transfer of foreign technology in Thailand is the lack of implementation of government policies which stipulate the use, transfer and regulation of imported technologies (Patarsuk, 1989).

Chinese MNEs are also becoming more important international investors in the fields of metal products, textiles, leather and clothing, transportation equipment, electrical equipment and non-metallic mineral products, as shown in Table 6.11. The establishment of joint ventures with foreign firms has been an important source of new foreign technology and management skills since 1979 when China promulgated the Law on Joint Ventures (Tai, 1987; Huasheng and Kerkhofs, 1988). The expansion of Shenzhen Electronics Group, a Chinese MNE which consists of a group of 117 small electrical/electronics firms in China in the production and eventual marketing of high technology

Table 6.10 The industrial distribution of foreign direct capital stock in Thailand, 1987, Thailand Baht million

Industry	Inward investment[1]	Outward investment[2]
Primary	11,091.6	9.0
Agriculture	928.4	3.0
Mining and quarrying	1,108.9	6.0
Oil	9,054.3	*NA*
Secondary	24,134.2	96.2
Food, drink and tobacco	2,003.6	54.2
Textiles, leather and clothing	4,040.5	27.0
Paper products	*NA*	
Chemical products	3,619.5	*Not*
Coal and petroleum products	2,262.4	*available*
Rubber products	*NA*	
Non-metallic mineral products	1,884.6	0.6
Metal products	*NA*	14.4
Mechanical engineering	1,495.9	
Electrical equipment	6,360.1	*Not*
Motor vehicles	*NA*	*available*
Other transportation equipment	*NA*	
Other manufacturing	2,467.6	
Tertiary	34,037.5	4,704.7
Construction	10,666.6	8.5
Distributive trade	12,758.7	93.6
Transport, storage and communication	2,731.8	0.2
Finance and insurance	3,183.5	4,571.0
Property	1,168.0	2.8
Other services	3,528.9	28.6
Total	69,263.3	4,809.9

Source: Bank of Thailand, *Monthly Bulletin*, various issues.
Notes: 1 Represents cumulative flows of inward foreign direct investment as from 1970.
      2 Represents cumulative flows of outward foreign direct investment as from 1978.

Table 6.11 The industrial distribution of foreign direct capital stock in China, 1987, China Yuan million[1]

Industry	Inward investment	Outward investment
Primary	4,384.3	322.6
Agriculture	1,268.7	176.9
Mining and quarrying	1,060.9	145.7
Oil	2,054.7	NA
Secondary	20,364.1	516.7
Food, drink and tobacco		19.4
Textiles, leather and clothing		80.2
Paper products	11.9	
Chemical products		21.0
Coal and petroleum products	Not	3.1
Rubber products	separately	30.2
Non-metallic mineral products	available	7.2
Metal products		209.6
Mechanical engineering		3.1
Electrical equipment		75.6
Motor vehicles		0.7
Other transportation equipment		15.9
Other manufacturing		38.8
Tertiary	24,660.9	371.8
Construction	773.1	46.4
Distributive trade	2,222.1	97.4
Transport, storage and communication	800.2	64.1
Finance and insurance	184.9	26.6
Property	20,680.6[2]	NA
Other services	NSA	137.3
Total[3]	52,625.8	1,731.5

Source: China Ministry of Foreign Economic Relations and Trade, unpublished data.
Notes: 1 Based on approved investment data.
2 Includes other services.
3 Includes unclassified sectors.

intensive products such as integrated circuits, is fundamentally dependent on links with large electronics firms based in Japan, North America, Europe and, to a lesser extent, Hong Kong (Forestier, 1989).

Perhaps the major exception to the stylised fact made about the range of technological specialisation and the stage of countries' development is that of firms from India and Malaysia. Table 6.12 shows the diverse range of sectors that characterises Indian outward investment and, as Chapters 9 and 10 will show, such a phenomenon may be attributed to earlier protectionist policies by the Indian government, which emphasised the development of indigenously generated technological effort aided in part by educational expansion in scientific and technological training and research, combined with restrictive policies towards inflows of foreign technology (see, for example, Eisemon, 1984). For example, the Indian government achieved its objective of self-suf-ficiency in the computer industry through participation in the ownership and control of subsidiaries of foreign-based computer firms and progress was made in fostering local sources of computer supplies and in fabricating technologi-cally sophisticated computer systems. However, although the technological lag in the adoption of the latest and most advanced computer technology in India has declined and cost-competitiveness has been achieved, foreign com-puter firms continue to supply large computer systems that represent very advanced technology that local firms cannot yet handle (Negandhi and Palia, 1989). Similar results have also occurred in the petrochemicals industry where

Table 6.12 The industrial distribution of foreign direct capital stock in India, 1980–2, Indian Rupee million

Industry	Inward investment, 1980	Outward investment, 1982
Primary	831.0	22.9
Agriculture	385.0	0.9
Mining and quarrying	78.0	22.0
Oil	368.0	0.0
Secondary	8,116.0	994.1
Food, drink and tobacco	391.0	99.6
Textiles, leather and clothing	320.0	236.5
Paper products	NA	131.6
Chemical products	3,018.0	225.9
Coal and petroleum products	NA	NA
Rubber products	463.0	9.4
Non-metallic mineral products	NA	48.4
Metal products	1,187.0	85.2
Mechanical engineering	710.0	106.3
Electrical equipment	975.0	17.4
Motor vehicles	480.0	NA
Other transportation equipment	35.0	22.6
Other manufacturing	537.0	11.2[1]
Tertiary	385.0	197.1
Construction	64.0	40.8
Distributive trade	209.0	5.7
Transport, storage and communication	NA	3.4
Finance and insurance	47.0	70.7
Property	NA	1.0
Other services	65.0	75.5[2]
Total	9,332.0	1,214.1

Sources: Reserve Bank of India, *Bulletin*, various issues; Indian Investment Centre, unpublished statement on 'Indian Joint Ventures Abroad as of 30/9/82'.
Notes: 1 Represents bottle capping.
      2 Includes hotels and restaurants, Indian Rupee 68,859.0 million.

local firms were more successful in the production of basic and intermediate chemicals than in speciality chemicals in which large foreign-based MNEs continued to dominate *inter alia* because of numerous restrictions they had imposed on technology transfer, and the lack of direction and focus of Indian R&D institutions (Khanna, 1984).

Although the Indian government's interventionist policies, aimed at achieving technological self-reliance, have enabled the country to build up a diverse and fairly sophisticated base in industrial technologies, these policies have also led large sectors of domestic industry to technological obsolescence and inefficiency (Lall, 1982a, 1983b, 1983d, 1984d, 1985a). Studies have shown that in cases where Indian firms have imported technology, their indigenous R&D activities have been encouraged or enhanced (see, for example, Katrak, 1989). For example, the Indian–Indonesian joint venture that operates the largest rayon (synthetic fibre) producing plant in Indonesia uses state-of-the-art technology comparable to any Japanese or other rayon-producing plant, which was made possible through a tie-up with an Austrian company prominent in the manufacture of rayon-producing equipment. In addition, Tata Engineering and Locomotive Company has the technological capability to design special machine tools for truck manufacture, aided initially by the importation of advanced foreign technology from Daimler Benz (Lall, 1982a, 1983b, 1985a; ESCAP/UNCTC, 1988).

Since the liberalisation of foreign investment policies in the mid-1980s, inward investment as a modality for the transfer of imported foreign technology has taken on a far more significant role. The industrial potential of Indian firms in the production of more highly technology intensive goods such as electronics, food-processing equipment and pharmaceuticals has been increasingly influenced by inward foreign investment, particularly from the USA (Debes, Holstein and Javetski, 1987; Rayner, 1989). In addition, India's industrial development in motor vehicle production and related industries, electrical and electronic products, mechanical engineering and chemical products has been enhanced by joint ventures established with Japanese firms which have become important modalities of technology transfer (Joseph, 1989).

Similarly, although Table 6.13 fails to reflect the sectoral distribution of Malaysian outward investment adequately, the case of Malaysian MNEs is unique to Third World outward investment, as will be shown in Chapters 9 and 10. Some indication of the sectoral distribution of Malaysian outward direct investment activity can also be seen in Seaward (1987). A substantial proportion of Malaysian MNEs were formerly foreign owned but have become nationalised as part of the foreign investment policy of the Malaysian government (Lim, 1984). The relatively high degree of indigenous Malaysian firms' technological capabilities has enabled them successfully to take over the operations of foreign MNEs and to broaden the range of their technological specialisation. During the 1980s Malaysia's comparative advantage evolved in the electronics industry, enabling the country to become the world's third largest exporter of semiconductor devices after Japan and the United States

(Goldstein, 1989).[17] This was sustained through the ability of Malaysia to attract investment by US and Japanese semiconductor firms in line with its New Economic Policy of 1970 to industrialise the economy, based at least initially on the attraction of foreign capital (Ehrlich, 1988). Foreign technology in Malaysia has mainly been accessed through technical assistance and joint venture agreements as well as management contracts, mainly from Japan, the USA and the UK (Jegathesan, 1990). Nevertheless, the modality of foreign direct investment has transferred technology mostly through foreign and on-the-job training and the presence of resident expatriate consultants and visiting foreign expatriates. Local firms have demonstrated fairly high capabilities to absorb and adapt new technology, including new management

Table 6.13 The industrial distribution of foreign direct capital stock in Malaysia, 1987, Malaysian $ million

Industry	Inward investment	Outward investment
Primary		142.2
Agriculture	Not	7.8
Mining and quarrying	available	134.4
Oil		NA
Secondary	4,960.7	−129.8
Food, drink and tobacco	1,154.3	
Textiles, leather and clothing	526.1	
Paper products	72.5	
Chemical products	392.0	
Coal and petroleum products	516.8	Not
Rubber products	232.8	separately
Non-metallic mineral products	483.6	available
Metal products	550.8	
Mechanical engineering	97.5	
Electrical equipment	534.8	
Motor vehicles	210.9	
Other transportation equipment NA		
Other manufacturing	188.6[1]	
Tertiary		1,967.2
Construction		−34.0
Distributive trade	Not	−6.0
Transport, storage and communication	available	NA
Finance and insurance		1,912.1
Property		NA
Other services		95.1
Total	4,960.7	2,005.3[2]

Source: Malaysian Department of Statistics, *Report on the Financial Survey of Limited Companies*, various issues.

Notes: 1 Data include wood and wood products, furniture and fixtures, scientific and measuring equipment and miscellaneous manufacturing.
2 Includes unspecified sectors.

systems, because of their relatively abundant supplies of skills and technical knowledge (Ariff and Lim, 1984; Lee, 1989).

Although Table 6.14 does not give an accurate sectoral distribution of outward investment in the manufacturing sector undertaken by MNEs from the Philippines, information gathered from field research suggests that Philippine firms have some degree of indigenous technological capabilities in such industrial sectors as food and beverages, textiles, leather and clothing, wood

Table 6.14 The industrial distribution of foreign direct capital stock in the Philippines, 1987–8, US $ million

Industry	Inward investment, 1987[1]	Outward investment, 1988
Primary	816.12	
Agriculture	45.75	*Not*
Mining and quarrying	121.58	*available*
Oil	648.79	
Secondary	1,373.56	7.81
Food, drink and tobacco	287.64	*NA*
Textiles, leather and clothing	59.48	7.81
Paper products	*NA*	
Chemical products	385.67	
Coal and petroleum products	86.64	
Rubber products	*NA*	
Non-metallic mineral products	*NA*	*Not*
Metal products	169.28[2]	*available*
Mechanical engineering	*NA*	
Electrical equipment	*NA*	
Motor vehicles	108.17	
Other transportation equipment	*NA*	
Other manufacturing	276.68	
Tertiary	639.96	31.66
Construction	21.57	*NA*
Distributive trade	84.90	0.18
Transport, storage and communication	16.37	*NA*
Finance and insurance	355.11	26.87
Property	25.82	*NA*
Other services	136.19[3]	4.61
Total	2,829.64	39.47

Source: Central Bank of the Philippines, Management of External Debt, Investment and Aid Department (MEDIAD).

Notes: 1 Figures represent stock of actual inward remittances of foreign direct equity investments net of cancellation and adjustments starting from 21 February, 1970 when the outstanding level of accumulated foreign equity investments was first monitored, until the first quarter of 1983.

2 Represents basic metal products and metal products except machinery and transport equipment.

3 Represents other services and other unclassified sectors.

and non-metallic minerals which has been strengthened in part by the presence of foreign MNEs and enhanced through purely technical collaboration agreements (or licensing agreements) with foreign firms. However, technology transfer through joint ventures is often associated with new business ventures involving fairly complex technology, requiring close supervision and control and/or marketing tie-ups. These technology transfer arrangements were mainly concluded with firms from the USA, Japan, the United Kingdom, Switzerland and Germany (Bautista, 1990). The next two chapters analyse the growth of Philippine MNEs, which constitutes one of the major case studies of this book.

CONCLUSIONS AND POLICY IMPLICATIONS

This final section of this chapter will attempt to summarise the major issues pertaining to technology development in developing countries. These issues pertain to the presence of a technological gap in the world economy and the way in which foreign technology may potentially assist in narrowing such a gap provided certain favourable conditions exist. Such conditions are entrenched in both the demand for and supply of foreign technology in developing countries in which government policies, the bargaining capacities of local firms, the idea of regional integration and technical cooperation between developing countries and other factors play very important roles. The main thrust of the argument is that if conditions are fulfilled, foreign technology may assist in narrowing the technological gap by enabling developing countries to catch up in sectors that offer greater opportunities for profitable investment and growth.

The presence of a technological gap and the role of foreign technology

There are several factors that would tend to maintain the presence of a technological gap in the world economy for fairly long periods. They include the quality and scale of commitment to R&D, the clustering of technical innovations and dynamic economies of scale (Posner, 1961). Evidence to support this view is shown by the fact that changes in technological leadership between countries only occur during changes in the techno-economic paradigm every forty to sixty years (Freeman, 1988a, 1988b). This helps to explain long waves in the overall rate of technological innovation that occurred at the end of the nineteenth century and in the immediate post-Second World War period (Cantwell, 1991a). Although opportunities for technological catching up or closing the technological gap are associated with upswings in the long wave cycle, the reopening of the gap between world technology and developing country technology is associated with downswings in the cycle with the diffusion of the major heartland technology to industries in the developing countries (Kaplinsky, 1984).

Similar views on the continuous existence of technological gaps between firms and countries are provided by Dosi and Soete (1988). In their view, the conditions of convergence or divergence in inter-firm and inter-national technological capabilities are explained by the evolutionary process of innovation which is determined by the interplay of science-related opportunities, country-specific and technology-specific institutions and the nature and intensity of economic stimuli. These reasons may help to explain why, although there has been rapid development and diffusion of advanced technology between industrialised countries, a significant technological lag continues to exist in developing countries. In recent years, the introduction of advanced technology in developing countries has been hampered because *inter alia* these countries lack favourable locational advantages for innovation, such as large size and growth of markets, and the infrastructural, human and financial resources and other factors intrinsic to countries that have a higher level of development and that are fundamental to the achievement of technological interdependence.

The work of Hufbauer (1965) on the longer imitation lags in firms from large developing countries as opposed to large-, medium- or small-sized firms from the industrialised countries has shown that the level of development is a much more crucial determinant of technological progress. The ability of firms and countries to engage in higher technologically intensive industries whose growth is dependent on R&D crucially determines their position in technological competition not only in international trade, as shown by Soete (1981b), but also in a whole range of international economic activities. Since technology determines cross-country variations in international economic performance in international industries, firms and countries that are more likely to succeed in world markets are those able to engage in the development and design of new products as well as the improvement of old products and the manufacturing technology by which such products are made. By contrast, firms and countries that remain in industries typified by abundant natural resource endowment or technological maturity and low research intensity are in danger of being locked in to a low growth development pattern and at the same time locked out of technologies that offer greater potential for higher growth. This fact raises a number of policy issues for developing countries, including the need to keep pace with the continuous process of attaining indigenous technological learning capacities to prevent a further widening of the technological gap.

As shown in this chapter by the lessons of Japan and the NICs, the role of foreign technology transfer should lie in assisting developing countries to realise their future comparative advantage through the development of their indigenous technological capabilities. Foreign technology is desirable to the extent that a dynamic growth path of changing comparative advantage is initiated (Ozawa, 1981; Lall, 1984a) and rapid industrialisation is achieved.

The findings of this chapter show that a process of technological change accompanied by a rapid growth in innovation has been occurring in developing countries in recent years (for related views see Dahlman, 1989). Foreign technology has a greater chance of perpetuating a virtuous circle of technological development by increasing the slope of the indigenous technical progress function in countries where innovative domestic industries exist. The competitive stimulus to the innovation activities of local firms offered by the introduction of foreign technology, especially through the modality of inward direct investment, are especially significant where local industry has a high technological capability, as in the NICs. Such a beneficial impact is enhanced when some form of applied research facilities is established in the host developing country. However, joint ventures and strategic alliances and other forms of foreign collaborative arrangements have helped to upgrade the production capabilities of local firms that were technologically weak but had capabilities in closely related types of activity. In the NICs these capabilities were sustained in part through efforts in R&D initiated by the state that sought to increase the technological capabilities of local firms. For the more advanced of the developing countries the theories of development have provided a more useful framework to describe the positive role of foreign technology in their technological development.

As a result, in those sectors where foreign technology has fulfilled a developmental role, technological progress in developing countries has been achieved in terms of narrowing the gap between the indigenous level of technology and that of the technologically more advanced developed countries. Local firms in these countries and especially in the NICs have been able to catch up with technological developments in the West and as a result increase their international competitiveness in a wide range of international economic activities, of which one important component is direct investment abroad. In particular, the modality of foreign direct investment offers these firms the opportunity to expand their network of productive and technological activities.

However, in the case of lower-income developing countries with a narrower range of technological specialisation, local firms are generally weak in the research-intensive sectors that offer greater opportunities for growth. Where foreign direct investment has been the main source of advanced technology, the capabilities of local firms have been undermined because the internationally competitive foreign firms rely on research carried out in their home country or elsewhere in their international network. Foreign firms had little incentive to locate research facilities in areas where the technological capabilities of local firms were weak and as a result these local firms have been undermined by the local expansion of foreign firms. Theories of underdevelopment will continue to be relevant to explain the more negative impact of foreign technology on indigenous technological development in research-intensive sectors in lower-income developing countries where firms lack competitive, indigenously generated technological capacities.

Supply and demand for foreign technology

Given that foreign technology will continue to be of fundamental importance for progress in developing countries, of which the growth of Third World MNEs is only one part, the supply of, and demand for, such technology becomes the pivotal point for discussion. The crucial question then is how the supply of and demand for foreign technology in the late twentieth century and beyond will affect future technological development of Third World firms.

On the supply side, the question that needs to be raised is how policies of developed countries towards outflows of technology will be fashioned in the future given the increasing trend towards reluctance as regards international technology transfer among the major sources of technology. For example, the declining technological competitiveness and adverse reactions to technology transfer among US MNEs and trades unions' reactions to direct investments abroad by US firms may well contribute to the declining role of the United States as a transferrer of technology in the future (see, for example, Johnstone 1990b). Japan is also increasingly reluctant to transfer advanced forms of technology for fear of having a Frankenstein competitor (Slaybaugh, 1981) or experiencing a 'boomerang effect' whereby firms which have received foreign investment capital and technology eventually become competing rivals (Matsuura, 1989). The reluctance of developed countries to transfer technology also stems from the risk of leakage of confidential data and the use of valuable company assets that have high opportunity costs (Slaybaugh, 1981). The increasing emergence of technological protectionism in the advanced industrialised countries has prompted the Asian NICs in particular to hasten the domestic development of technical, engineering and scientific manpower, in part through the enactment and enforcement of legislation on the protection of intellectual property rights, the promotion of inward investment in high technology industries and the acquisition of foreign technology through licensing and other contractual routes of foreign technological transfer. These technological efforts are in addition to efforts in research-linked investment in centres of innovation in the industrialised countries (Chia, 1989).

On the other hand, one could envisage a school of thought which holds that, apart from the financial gains that firms from developed countries can experience from the sale of technologies, which could help subsidise the high costs of R&D, the promotion of technological development in developing countries could work in the best interests of developed countries, as these developing countries represent areas of potential market growth for their products in the future. This is shown in the views of some writers such as Kirkland (1988) who contend that given the fast growth of GNP per capita and the large share of developing countries as destinations for exports from the developed countries, the developing countries may be among the only sustainable markets in terms of potential for growth. Since the trend in recent times is towards increasing global competition, innovators in the developed countries may want to provide new technology to firms in developing countries that have the capacity to implement their technology, as part of their overall global strategy to gain stronger footholds in world markets. At present, and in the foreseeable future, these countries continue to represent an important source of competitive inputs which can assist directly in maintaining the worldwide competitiveness of globally integrated MNEs. An interesting scenario would be the extent to which the growth and development that are now evident in the NICs would promote increased trade, licensing and foreign direct investment in most if not all developing countries in the future in response to the growth of local market demand. In the area of foreign direct investment, the trend has been rising foreign direct inflows in developing countries since 1984. A survey of US MNEs suggests that of those firms that have increased their Third World investments, 68 per cent have attributed it to their strategy of positioning themselves for the future in the face of the growing globalisation of the world economy and the need to gain access to potentially strategic and expanding markets, as well as to production capabilities and lower manufacturing costs in these countries (Wallace, 1989). With the creation of new markets, developing countries may be drawn further into the arena of technological competition between multinational enterprises based in the advanced industrialised countries as these enterprises seek to strengthen their market position with increasing locational advantages in these markets.

Such expectations of the increased participation of developing countries in international business and access to more advanced foreign technology must be assessed in the context of demand factors for foreign technology. Developing countries have to realise that technology transfer from abroad does not constitute the only solution to their problems of technological underdevelopment. Developing countries themselves have to initiate the process of technological transformation by creating effective demand for foreign technologies through their inherent knowledge and capacities to select, adopt, transplant, diffuse and develop foreign technology according to their unique requirements (UNCTAD, 1984).

The lesson of Japan suggests that technological requirements may not be restricted to those for which a country currently has a comparative advantage. Indeed, the analysis of the post-war development of Japan and the current newly industrialising developing countries provides support for the view that the sectors in which a country enjoys the greatest potential for innovation and in which investment is most beneficial are not necessarily those in which the country currently has a comparative advantage (Pasinetti, 1981). As part of their strategies to catch up with the West, Japan and the NICs have not only emphasised the development of labour-intensive technologies but also more modern capital- and research-intensive technologies, or what Blumenthal and Teubal (1975) refer to as 'future-oriented technologies', whose use may be unjustified by existing factor proportions but which may prove to be more efficient in view of future requirements.

Role of government policies and regional integration

Since technological intensity is increasingly becoming an important determinant of growth, the establishment and growth of infrastructure for R&D and the effective linkages to production have to be given emphasis in the development of technological policies for developing countries. These considerations ought to be of paramount importance to national governments in their quest to develop indigenous forms of competitive technology as part of their industrial development goals. Although foreign technology may play a vital role, policies that set in motion a dynamic process of technological accumulation, consistent with the effective and cumulative development of an indigenous technological capacity, have to be nationally defined and oriented (Bienefeld, 1984). Given that foreign technology is desirable to the extent that a dynamic growth path of changing comparative advantage and rapid industrialisation is achieved, the correct strategy for development must be dependent on having the required conditions to undertake the difficult and complex process involved in the effective blending and assimilation of foreign technology with local technological efforts to achieve production efficiency and the competitive technologies necessary for world competition. These require technology planning, including the development of human resources and infrastructure to meet projected technological requirements, and the implementation of policies that ensure adequate inflow, absorption and adaptation of foreign technology. Such careful management and control over the process of importation and assimilation of foreign technology, which complements the development of indigenous technological capabilities, is consistent with the pursuit of an active strategy for technological transfer and development (UNCTC, 1987, 1990a).

Foreign technological efforts are required to enable developing countries to acquire deeper and more sophisticated capabilities in know-why, such as those required in basic design and new product/process development (Lall, 1984a) and there is, therefore, a need for a selective policy towards the import of foreign technology (Stewart, 1984). Although active technological strategies through government intervention to reduce access to readily available foreign technologies are favoured, especially where domestic know-why can be competitively developed, the further development of know-why requires keeping pace with rapid technological developments in the advanced industrialised countries. Even the ways in which firms from the more advanced countries design their technological strategies may be described more in terms of interdependence than independence. These firms rely on international as well as domestic research investments to catch up with the advancement of the frontiers of technology.

Policies directed towards the national system of innovation are the essential foundation for a catching up strategy in development (Freeman, 1988a). Domestic technology policy is bounded by scientific, industrial and foreign trade policies and other macroeconomic and institutional policies and therefore needs to be generally integrated with these policies (Prasad, 1979; Dahlman, 1984; Stewart, 1984; Freeman, 1988). The development of indigenous technological capacity is constantly affected by the political economy of the country and region, and by the world economy (King, 1984). The role of public-sector participation in the development of national systems of innovation is crucial since long-term and consistent policies are needed to stimulate the necessary continuous interaction between producers, users and external sources of technology (Gregersen, 1988).[18]

For example, since governments and the suppliers of capital and technology, such as the MNEs, are the predominant external influences on technological change in the early stages of technological development, governments in developing countries have an important role in fostering the international competitiveness of local firms (see Kim, 1980; Rohlwink, 1987). The role of government-directed comparative advantage in the internationalisation of developing country firms through trade and foreign direct investment has been analysed by Aggarwal and Agmon (1990). This may occur *inter alia* through the implementation of measures to ensure local market protection, through import substitution and controls on foreign investment, etc., to provide incentives for technological change. At later stages of development, although customers and competitors become the more important external forces that influence technological change, governments could still fulfil a supportive role through the implementation of measures to ensure that bureaucratic systems are gradually replaced by market forces and that competition is enforced through policies of export promotion and anti-trust.

Government policies therefore play a crucial role in determining technological development and the role of foreign technology in that process.[19] While some policy interventions are necessary for sustained industrial development, technological dynamism has been achieved through the combination of incentives with adequate capabilities and institutions, each supported by a proper balance of selective and functional policy interventions (UNCTAD, 1991).

For example, the control and co-ordination of governments in South Korea and Taiwan played an important role in the rapid development of high technology industries in those countries. Their governments orchestrated the whole process of technological advancement, from finding economical sources of finance and locating potential markets for their new products to explicitly defining the role of foreign technology in hastening the pace of technological catching up. The co-ordinating role of the government in the heavily specialised new industries is especially relevant in Taiwan which has few large firms comparable in size to the Korean conglomerates or *chaebols*. Through the co-ordination of the industry, the government of Taiwan enabled the country to benefit from forward and backward integration characteristic of the operations of large firms. Government intervention in support of the infant industries is perhaps the single most important reason behind Taiwan's dramatic achievements in the semiconductor industry; South Korea's global success in steel, personal computers, automobiles and other heavy industries; and Brazil's accomplishments in the development of a globally competitive capital goods industry (Saghafi and Davidson, 1989).

For developing countries with smaller national markets, government policies to emphasise the growth of national systems of innovation may be supplemented by regional integration and technical cooperation activities among developing countries which may also potentially fulfil important roles in the technological development of those countries (see, for example, Unger, 1988). Regional policies to foster integration may not only promote the rationalisation of industries which otherwise cannot reach optimum size within national frontiers but may also strengthen local innovative activities through regional technological cooperation. These technological cooperation activities may in part encourage development of research in science and technology, enable better access to information about experience in technology acquisition and foster greater co-ordination of policies for the international transfer of technology (Germidis, 1977; UNCTAD, 1984; Singh, 1986). An excellent case in point emerges from the channels of technology transfer in Europe which have been assisted by the formation of the EC (Elliott and Wood, 1981).

However, unlike the current integration plans of developed countries such as those in the EC and despite the interdependence in trade and investment among some regional groups such as Asia and Latin America, the regional integration schemes of developing countries have been stagnant owing to economic, political and structural obstacles. This has led to delays in implementation or the inability of these groupings to agree on trade liberalisation programmes, regional industrial programmes or the formulation of other common regional policies. Among the regional and integration groupings in developing countries, only the Andean Pact has adopted a common policy towards FDI and technology transfer, embodied in Decision 24. The technology transfer provisions embodied in this policy were intended to ensure that FDI made an effective contribution to these economies. The restrictive requirements of Decision 24, from the point of view of foreign investors, led to its replacement in May 1987 by Decision 220 which grants policy autonomy to member countries, although within a common set of guidelines (UNCTC, 1990b).

Although regional efforts at integration within developing countries may potentially provide a larger market for trade in products and technologies suited to the requirements of these markets, local technological development may become counterproductive if these efforts are taken to extremes, condemning industry to progressive backwardness and further widening instead of narrowing of the technological gap (Lall, 1984c). Global competition requires that even when the technological capabilities of developing countries have graduated to internationally competitive levels, exports of technologies from advanced industrialised countries to developing countries should be treated as complementary to exports of technologies between developing countries.

Opportunities for technological catch-up: increased bargaining position of developing countries

At the present stage of the world economy, where the ability to research and develop is becoming of paramount importance, the capabilities of developing countries to catch up is increasingly determined by their capacity to participate in the generation and improvement of rapidly growing new technologies. This means being able to enter either as early imitators or as innovators of new products or processes (Perez and Soete, 1988; Perez, 1988). The expectations are that since developing countries are less entrenched in the use of an outdated technology in terms of actual production, investment and skills, their potential to enter new areas of technology, which currently offer investment and growth opportunities, are higher than those of other countries with a much greater commitment to the maintenance of the prevailing technology. The lessons of this chapter suggest that the realisation of this expectation, with the assistance of foreign technology, is fundamentally dependent on high levels of externalities and scientific and technical knowledge which many developing countries do not yet possess, and which may represent considerable barriers to entry (see also Akashah, 1987, and Unger, 1988). The difficult and costly process of catching up is exacerbated by the nature of technology as a constantly changing force.

Perhaps many developing countries can learn something from Japan, where the process of industrial structural change was enhanced by the existence of a dual industrial structure during the 1950s. The process of technological catching up initially focused on both the pre-war originated modern industries characterised by capital intensity such as iron and steel, chemicals and shipbuilding, automobiles and some engineering sectors and on the post-war, science-based modern industries such as petrochemicals, synthetics and electronics. However, after pre-war based activities had caught up in the period after 1966, technological imports were concentrated on bringing in new technology for the science-based industries (Ohkawa and Rosovsky, 1973). The co-existence of a capital-intensive, modern sector composed of Japan's industrially leading firms, the *zaibatsu* group, with the labour-intensive, traditional sector composed of a large number of small firms enabled modern foreign technology to be borrowed despite the inconsistency between the required factor proportions and those available in Japan (Blumenthal and Teubal, 1975). The traditional sector functioned to sustain Japan's exports, earning precious foreign exchange to purchase much-needed industrial raw materials and fuels for the modern sector which served as a spearhead of technological and economic growth for the rest of the economy. As this strategy of dualistic development succeeded, rapid capital accumulation occurred in the modern sector, particularly in the 1960s, and the overall factor proportions of Japanese industry quickly became relatively capital abundant and labour scarce (Ozawa, 1981).

This brings us back to the question raised at the beginning of the chapter.

Table 6.15   Some foreign collaborative arrangements of South Korean MNEs with foreign firms

South Korean MNE	Foreign firm	Imported technology	Form of technology transfer
Borneo International Korea Furniture	NA	Furniture designs from Italy and furniture manufacture using the most sophisticated German technology and US assembly line techniques	NA
Sanchok Industrial	Monsanto	Production of silicon wafers	Joint venture
Samsung	Hewlett-Packard & AT&T(USA)	Production of semi-conductors	Joint venture
	Nippon Electric Corporation (Japan) Sharp (Japan)		Technical agreement Technical links
	Intel (USA)		NA
	IBM (USA)		Cross-licensing
	Micro Technology, Hewlett-Packard (USA), Hitachi (Japan), Bell Telephone Manufacturing (Belgium)		Licensing

Does foreign technology fulfil a beneficial role in the technological development of Third World firms? We sought answers through our review of the theories of development and underdevelopment and found that if capitalist development and associated integration with the world economy were considered a favourable strategy for development, then foreign technology would generally fulfil a positive role in the attainment of such an objective, provided that certain conditions in recipient countries exist to enable a technology transfer process to be successfully fulfilled. Such conditions are entrenched in the growth of national technological or innovative activities which together with technology imports and investments determine the process of economic growth (Fagerberg, 1988).

Certain writers, such as Kobrin (1979) and Warren (1980), argue that political independence makes a difference in increasing the relative bargaining strength of developing countries in their dealings with more developed countries. This chapter has shown that among other factors the bargaining strength

Table 6.15 (continued)

South Korean MNE	Foreign firm	Imported technology	Form of technology transfer
Goldstar	AT&T(USA)	Production of semi-conductors	Joint venture
	Honeywell (USA)		Technical links
	Hitachi (Japan)	Production of 1 MB DRAM chips	Turnkey agreements
Samsung, Hyundai, Daewoo, Goldstar	Tolerant (USA)	Production of super minicomputers	*NA*
Daewoo	Northern Telecom (Canada)	Designing and producing semi-conductors	Technology exchange agreement
Hyundai	Mitsubishi (Japan)	Finance and international marketing of motor vehicles	Joint venture (15 per cent equity)
	Oplex (Japan)	Production of the STN-grade liquid crystal displays for televisions, word processors and office automation equipment	Technical link

Sources: UNCTC, 1986; Kwon and Ryans, 1987; Oman, 1988; Clifford, 1989; Evans and Tigre, 1989; *Global Trade*, 1989; Saghafi and Davidson, 1989; *Korea Business World*, April 1990; Mody, 1990.

of such countries is enhanced by their indigenous capacity to catch up with technological developments in the West, which assures that perpetual under-development and economic backwardness will increasingly become an irrelevant phenomen. The enhancement of bargaining capacities through in-digenous technological development is one important way of lowering the costs and increasing the benefits of integration into the international economic system associated with capitalist development. As Dunning (1983b) has quite rightly pointed out, strong indigenous competition within a particular country and government encouragement to foreign affiliates to engage in activities consistent with the country's long-term comparative advantage, particularly with respect to the creation of technological capacity, form the most effective countervailing power that countries may have against foreign-based MNEs.

Given the strengthened bargaining position of developing countries and the attainment of a certain minimum level of technological development such that gaps may be narrowed, their role in the new international division of labour may have to be redefined. Indeed the expectations of Germidis (1977) may be realised as the dividing line in the international division of labour between developed and developing countries is redrawn not *between* industrial activities but *within* industrial activities, i.e. between the work of designing products and techniques and the work of making the goods. Perhaps with the growth and enhancement of the national innovative capabilities of developing countries, especially in the area of research and development, we can even venture to predict that the tasks undertaken by developing countries will encompass much more than the competitive production of existing products by moving towards the competitive design of new products and processes.

The dynamic interdependence
between inward and outward
investment in the Philippines

INTRODUCTION

The last chapter reinterpreted Dunning's investment development cycle in a microeconomic framework by introducing the theory of technological competence in analysing the dynamic interaction between technologically innovative indigenous firms in developing countries and foreign technology. Such interaction has enabled foreign technology and expertise to be diffused within the local economy and to have important implications for indigenous technological development and the stimulation of local enterprise. This chapter focuses on the extent to which foreign direct investment as a modality for the transfer of advanced foreign technology has played a complementary role in the development of the indigenous technological capacity of Philippine firms which has enabled these firms to engage in outward direct investment.

The analysis of the developmental role of foreign MNEs in Philippine firms focuses on three main periods: the colonial period (1521–1946); the period of import-substituting industrialisation (1950–70) and the period of export-oriented industrialisation (1970 to the present).

FOREIGN INVESTMENT IN THE COLONIAL PERIOD (1521–1946)

The history of foreign investment in the Philippines can be traced back to the colonial era. The Philippines had the longest colonial history of all the ASEAN nations. Three centuries of Spanish rule resulted in considerable amounts of foreign investment from Spain and from neighbouring traders, particularly China. The Philippines became an independent republic in 1896 but independence was short lived as the victory of the USA in the Spanish–American war of 1898 subjugated the Philippines to fifty years of American rule.

The United States had the following broad objectives for the Philippines: to secure the country as an additional market for US manufactured products and a source of raw materials for US industry; to function as a gateway for US trade in the East, given the strategic commercial location of the Philippines and the increasingly important commercial roles in the area of Japan, France, Germany, Russia and other rival countries; and to secure the country as a military and naval base in the Pacific (Lodge, 1898; Beveridge, 1900). As a result, the Philippine economy was largely reoriented towards trade with and subsequent foreign direct investment from the USA. Exports of raw materials from the Philippines to the United States, such as sugar, hemp, tobacco and coconut oil, were facilitated largely by the enforcement of the Payne-Aldrich tariff law from 1909 to 1929 which guaranteed the US market tariff-free access to these products. Similar free trade arrangements existed for exports of industrial products from the USA to the Philippines (Schirmer, 1975).

Direct investments were the predominant form of foreign investment during this period. Table 7.1 gives an indication of the high share of direct investment to total foreign investment in the periods 1914, 1930 and 1935. The stock of foreign direct investment in the Philippines was estimated to be US $100 million in 1914 and further increased to US $300 million in 1930 and US $315 million in 1935. By 1940, the stock of FDI amounted to US $372 million. Table 7.2 shows that the main sources of these foreign investments in 1940 were the USA, China, Spain, the UK, Japan and other European countries such as Sweden and Switzerland.

The data contained in Tables 7.3 to 7.7 show the industrial distribution of foreign portfolio and direct investments in the Philippines by firms from the USA, China, Spain, the UK and Japan respectively. These foreign investments were primarily concentrated on the production and export of agricultural commodities and on the establishment and operation of public utilities. These investments can largely be explained by locational theories of trade and capital movements in which international production is taken to be an extension of domestic economic activity and an instrument of colonial policies. Foreign firms from the USA, Spain, the UK and Japan were engaged in backward vertically integrated investments in resource-based extractive activities in the Philippines to supply their home countries with raw materials.

After the onset of American rule in the Philippines in the early twentieth century, US capital flowed into the primary sector and in particular into the extraction, processing and export of minerals, sugar, coconut and abaca (hemp) industries and in lumber operations. Table 7.3 shows that a total of 54 per cent of US direct investments in the Philippines in 1940 was directed to the primary sector. The growth of such investments had been enhanced by the free trading relationship between the Philippines and the United States from 1909 to 1929 under the Payne-Aldrich Law and from 1935 to 1940 under the Tydings-McDuffie Act. In the period 1940–6, Philippine products receiving preferential treatment were subject to an export tax, increasing annually by 5 per cent to 25 per cent in 1946 and 100 per cent after 1946 (Jenkins, 1954). Similar preferential trading arrangements were also made with trading firms under the auspices of the Japanese military government during the war years (Virata, 1972).

Nevertheless, although US FDI was concentrated in key industries, the share of the Philippines in the overall outward FDI stock of the USA is considered to be very small, accounting for between 0.6 and 1.3 per cent in

Table 7.1 Total foreign investment in the Philippines, 1914, 1930 and 1935, US
$ million

Year	Direct investment	Portfolio investment	Total
1914	100.0	12.0	112.0
1930	300.0	85.0	385.0
1935	315.0	61.0	376.0

Source: Callis, 1942.

Table 7.2 The geographical distribution of foreign direct capital stock in the
Philippines, 1940, US $ million

Country	Value	Percentage distribution
USA	218.0	58.6
China	55.8	15.0
UK	37.2	10.0
Japan, Spain, Switzerland, Sweden and other Europe	61.0	16.4
Total	372.0	100.0

Source: Richards, 1944.

selected years since 1929, with a declining trend. These investments were not
considered crucial or strategic to the USA, largely due to the fact that the main
thrust of US direct investment, especially in the period prior to 1914, was
market oriented since the USA was rich in most natural resources. Moreover,
the Philippines offered several locational disadvantages: the distance between
the USA and the Philippines; the lack of incentives offered to US investors;
and the restrictive government policies with respect to land ownership, repatriation of profits and taxation. These location disadvantages plus the
additional political factor of a possible US withdrawal from the Philippines
explain the low level of US investment in the Philippines (Callis, 1942).

Similarly, Spanish, UK and Japanese investments in the Philippines in 1940
were directed to the primary sector, in particular sugar, coconut, tobacco,
abaca, mining and fishery. By contrast, as shown in Table 7.4, Chinese investments in the Philippines in 1940 were largely directed to the merchandising
and rice milling sectors. Prior to 1932, Chinese investors accounted for 85 per
cent of the retail trade and a considerable proportion of the wholesale trade
sector in the Philippines.

This process of essentially dependent development was supported by investments in public utilities, especially by firms from the UK and the USA.
Table 7.3 shows that 12 per cent of US direct investments in 1940 was geared
to public utilities such as the development of electrical power and other public

Table 7.3 The industrial distribution of US investments in the Philippines, 1940, US $ million

Industry	Amount	Percentage distribution
Rentier investments		
Government bonds	32.40	
Government guaranteed corporate issue	2.70	
Private issues	1.50	
	36.60	14.0
Direct investments		
Sugar industry		
Centrals	30.58	
Land and improvements	5.44	
Crop loans	3.32	
	39.34	15.0
Coconut industry		
Mills and refineries	5.55	
Land and improvements	8.37	
	13.92	5.0
Abaca (hemp) industry		
Mills and equipment	3.93	
Land and improvements	5.50	9.43	4.0
Mining industry	70.00	27.0
Public utilities	31.85	12.0
Transportation	19.10	7.0
General merchandising	15.00	6.0
Lumber and logging operations	6.40	3.0
Minor investments		
General manufacturing	4.10	
Engineering	3.85	
Oil merchandising	2.66	
Hotels	1.50	
Banking	1.42	
Pineapple industry	1.00	
Exporting	0.32	
Pearl buttons	0.10	
Miscellaneous	1.97	
	16.92	7.0
Total	258.56	100.0

Source: Richards, 1944.

facilities and another 7 per cent was geared towards transportation. In addition, 26 per cent of UK investments was directed towards the development of railway systems. These investments in public utilities as well as trade facilitated the development of production and markets in the Philippines.

These early forms of foreign investment were largely in the form of joint

Table 7.4 The industrial distribution of Chinese investments in the Philippines, 1940, US $ million

Industry	Amount	Percentage distribution
Coconut oil	0.60	1.4
Soap	0.40	0.9
Abaca (bailing)	0.10	0.2
Hats	0.20	0.5
Corn mills	0.25	0.6
Cordage	0.10	0.2
Tobacco	0.10	0.2
Distilleries	0.20	0.5
Rice mills	8.20	19.6
Confectionery and baking	0.90	2.2
Lumber	1.54	3.7
Utilities	0.25	0.6
Merchandising	25.00	59.8
Banking	1.00	2.4
Hotels	2.00	4.8
Others	1.00	2.4
Total	41.84	100.0

Source: Richards, 1944.

Table 7.5 The industrial distribution of Spanish investments in the Philippines, 1940, US $ million

Industry	Amount	Percentage distribution
Sugar	20.12	49.0
Coconut	4.71	11.5
Tobacco	6.46	15.7
Brewery	0.50	1.2
Lumber	0.25	0.6
Mining	5.00	12.2
Merchandising	2.00	4.9
Others	2.00	4.9
Total	41.04	100.0

Source: Richards, 1944.

ventures with local firms since the foreign investment policy of the Philippine government at that time required that foreign investment with respect to the exploitation of natural resources, land-holding and public utility enterprises must have a minimum Filipino equity participation. The growth of a powerful class of Filipino entrepreneurs led to an increase in Filipino ownership of non-agricultural lands and real estate from less than 15 per cent of the total assets in 1935 to 45 per cent in 1938. In contrast, the combined American and

Table 7.6 The industrial distribution of UK investments in the Philippines, 1940, US $ million

Industry	Amount	Percentage distribution
Abaca	1.41	3.6
Coconut	3.50	8.8
Molasses	0.18	0.5
Confectionery	0.05	0.1
Lumber	0.75	1.9
Banks	2.00	5.0
Insurance	5.00	12.5
Sugar loans	5.00	12.5
Mining	5.00	12.5
Merchandising	6.00	15.0
Manila railroad bonds	10.00	25.1
Others	1.00	2.5
Total	39.89	100.0

Source: Richards, 1944.

Table 7.7 The industrial distribution of Japanese investments in the Philippines, 1940, US $ million

Industry	Amount	Percentage distribution
Abaca and land improvements	16.50	51.3
Other land and improvements	0.50	1.6
Abaca bailing and dessicating	0.75	2.3
Dessicated coconut	0.20	0.6
Confectionery	0.20	0.6
Cotton mill	0.30	0.9
Rubber shoes	0.15	0.5
Fish canning	0.15	0.5
Fishing	0.50	1.6
Lumber	0.75	2.3
Public utilities	0.18	0.5
Brewery	1.00	3.1
Merchandising	7.50	23.3
Banks	1.00	3.1
Mining	1.50	4.7
Others	1.00	3.1
Total	32.18	100.0

Source: Richards, 1944.

Chinese interests accounted for 25 per cent and 14 per cent respectively (Espenshade, 1955). Such joint venture arrangements have had favourable developmental impacts on indigenous Philippine firms which have acquired

Table 7.8 The distribution of the 324 largest foreign affiliates in the Philippines in 1976 by period of establishment and trade orientation

Period	Trade orientation			Total
	Import-substitution[1]	Export[2]	Non-trade goods[3]	
1934–9	16	20	20	56
1945–8	7	5	5	17
1949–52	24	2	2	28
1953–6	28	4	4	36
1957–60	22	4	4	30
1961–4	36	3	9	48
1965–8	22	4	10	36
1969–72	9	2	3	14
1973–6	16	1	5	22
Subtotal	180	45	62	287
Year of establishment not available	22	6	9	37
Total	202	51	71	324
Percentage of total	62.4	15.7	21.9	100.0

Source: Tsuda *et al.*, 1978.
Notes: 1 Including beverages, dairy products, flour milling, tobacco, textiles (excluding cordage), paper and paper products, rubber tyres, chemicals and chemical products (including drugs and cosmetics), glass, glass products, metal products, basic metals, machinery, electrical machinery, appliances, transport equipment, petroleum products and miscellaneous manufactures.
2 Including metal ore, sugar, coconut oil, cordage, fruit farming and canning, plywood and veneer.
3 Goods and services which have not been imported or exported by the Philippines in the past to any significant degree and do not normally enter international trade owing to their non-tradable nature, i.e. electricity, construction and real estate.

some technological expertise in the exploitation and exploration of several natural resources with which the Philippines is relatively well endowed.

Such a close link in the sectoral pattern of trade and investment that characterised the early forms of foreign investment in the Philippines persisted throughout the colonial period. However, in the period from 1934 onwards, the emergence of FDI by modern MNEs in the Philippines, which was distinct from trading company and resource-based investments, was increasingly becoming more important. Table 7.8 shows the trade orientation and period of establishment of the 324 foreign affiliates identified as being among the top 1,000 firms in the country by 1976. The table indicates that in the period 1934–9, a total of 16 import-substituting manufacturing affiliates, 20 export-oriented manufacturing affiliates and 20 foreign affiliates in the services sector were established by the 324 largest foreign affiliates.

THE PERIOD OF POLITICAL INDEPENDENCE (1946–50)

The Philippines gained formal political independence from the USA on 4 July 1946. The granting of independence was in part precipitated by specific interest groups such as the American Federation of Labor, which voiced concern over the supposed threat of competition from cheap labour in Asia and from Oriental immigration to the United States. However, the major pressure for Philippine independence emerged from the American farm lobby which viewed continued trade preferences to the Philippines as a threat to domestic agricultural production (Jenkins, 1954).

However, despite formal political independence, the USA continued to retain control over strategic military bases and to enjoy substantial economic privileges, giving rise to the period of neocolonialism – the exercise of indirect and informal control or dominance by foreign imperial powers over subject nations (see, for example, Schirmer and Shalom, 1987). In particular, the neocolonial policies were considered to be a necessary part of US capital investments in the Philippines. As a result, a US-dependent state of Philippine development was ushered in as the United States became not only an important market for Philippine exports but also a significant source of manufactured products and investment capital (Schirmer, 1975).

Such neocolonial policies associated with dependent development were manifested in the Philippine Trade Act (Bell Trade Act) of 1946 which defined economic relations between the United States and the Philippines after political independence. This Act established a system of preferential trading arrangements including free trade between the USA and the Philippines for a period of eight years except in such export products as sugar, cordage, rice, cigars, scrap tobacco, coconut oil and buttons of pearl or shell, where absolute quotas were imposed; and amended the Philippine constitution to include a parity clause which gave US citizens equal rights with Filipino citizens in the exploitation, development and utilisation of natural resources and in the operation of public utilities. The Trade Act also imposed various infringements on the exercise of Philippine sovereignty, with stipulations forbidding the imposition of a Philippine export tariff or the export of major primary products to countries other than the United States and vesting power and control in the US president to restrict foreign exchange and capital movements in the Philippines. The acceptance of the Trade Act was partly facilitated by its being linked to the US rehabilitation aid that is fundamental not only for material rehabilitation and for the maintenance of an expensive governmental system in the Philippines but also as a source of American political and military support (Diokno, 1946; Shalom, 1980).

The Laurel-Langley Agreement replaced the Bell Trade Act upon the expiration of the latter in 1954. This agreement, which governed US–Philippine economic relations until 1974, was to a considerable extent less restrictive, with the removal of various restrictions on local government control of the Philippine economy and the modification of the parity clause to allow for reciprocity and enable Philippine investors to enjoy the same rights to invest in the United States as US investors had in the Philippines. In addition, free trade between the USA and the Philippines was abolished and replaced by gradually diminishing mutual trade preferences. However, the Laurel-Langley Agreement extended the protection accorded US capital in the Philippines by introducing a new clause that guaranteed US investors equal treatment with Filipino investors not just in the natural resources and public utilities sectors as stipulated in the Bell Trade Act of 1946 but in all areas of the economy. Colonial legislation therefore continued to influence foreign investment in the Philippines for almost thirty years after political independence from the USA.[1]

Nevertheless, the policy of economic self-sufficiency and nationalism in the period immediately after political independence spurred the enactment of various statutes to grant incentives for the development of the industrial sector. For example, Republic Act 35, called the New and Necessary Industries Act, was enacted on 30 September 1946 to initiate and encourage investment in new and necessary industries through 100 per cent tax exemptions for a period of four years. The question of necessity was rather vaguely defined so that the effect of the Act was to encourage the establishment of assembly or processing type operations.

This Act was amended in 1953 by Republic Act 901 which continued to provide manufacturing firms with 100 per cent tax exemptions for a period of five years, after which gradually diminishing exemptions were enforced. These tax exemptions were given to product lines so that there was no discrimination between local and foreign firms. However, the roles of domestic and foreign investment were harmonised in such a way that pioneering industries which required heavy capital investment and modern technology were open to foreign investment. The role of foreign investment in terms of transfer of technology, employment and training of personnel was considered a valuable contribution to the Philippine economy, especially in those industries that had not been operated on a commercial scale and where foreign firms could fulfil pioneering roles. A provision of fading out of foreign firms in these industries was introduced to allow for a programmed transfer of 60 per cent of the voting stock of a firm to Philippine nationals over a period of twenty or forty years. Among these pioneering industries were the smelting of ores, refining of metals, and rolling and extrusion of metals as well as petroleum and salt-based chemicals, integrated pulp and paper mills and much mechanical and electrical components manufacturing. On the other hand, non-pioneering industries, along with public utilities, the primary sector and the development of public land, were reserved for domestic investors although joint ventures with foreign firms were also encouraged to supplement domestic capital. The major industrial goal of the Act was therefore to link the processing industries to the extractive industries through the transformation of ores, agricultural products and wood products into finished consumer and industrial products. The Act was considered instrumental in the establishment of new industries and by the end of the 1950s, a hundred product lines had been exempted. The growth of some Philippine MNEs in resource-based industries can be attributed to the successful implementation of this Act.

As a direct result of the incentives offered by these policies and the limitations imposed on foreign holdings in extractive industries, there was a shift in the sectoral distribution of FDI in the post-Second World War period towards the manufacturing sector and away from the primary sector and towards industrial restructuring, such as the rebuilding of sugar and coconut mills, the rehabilitation of gold mining and public utilities such as electricity, communications and transportation. By 1948, foreign investment accounted for 51 per cent of the assets in the manufacturing sector (Friedman and Kalmanoff, 1961). The importance of ownership advantages in the internationalisation of the modern, technologically innovative MNE became a far more significant determinant of the international production of firms in this period than in the colonial period when location advantages played a much more important role.

THE PERIOD OF IMPORT-SUBSTITUTING INDUSTRIALISATION (ISI) (1950–67)

The last half of the 1940s saw the emergence of a balance-of-payments crisis traceable to a large extent to the inability of exports to recover rapidly from the effects of war; the heavy reliance on imports of finished consumption goods and the decline in the inflow of foreign exchange in the form of foreign aid (primarily from the USA) for post-war reconstruction and rehabilitation. The highly protective tariff and import and exchange controls, as well as the devaluation of the Philippine peso implemented in 1949, gave powerful impetus to the import-substituting industrialisation (ISI) enforced during the period 1950–67. The period of ISI attracted a large volume of foreign investment geared to replace imports of final consumption goods that had already found market acceptance (see also Lindsey, 1983, 1985). A number of wholly owned foreign subsidiaries were established to assemble, package, fabricate or undertake light manufacturing in the Philippines in such sectors as textiles, pharmaceuticals, cigarettes, household appliances, petroleum refineries, car assembly, rubber tyres, containers, food products, flour mills, milk canneries, confectionary, biscuits and soft drinks (Virata, 1972). Foreign direct investment continued to be the principal vehicle of international capital movements in the Philippines.

The Basic Industries Act passed by the Philippine Congress in 1961 was an attempt to strengthen the New and Necessary Industries Act (Republic Act 35). The basic industries identified by this Act were granted exemptions from payments of tax and customs duties on their importation of equipment, spare parts and machinery. Full exemptions were given for the first four years from mid-1961–5, gradually diminishing to 75 per cent in the period 1966–8 and 50 per cent in the period 1969–70. The Basic Industries Law of 1964, otherwise known as Republic Act 4095, was passed by Congress specifically to amend the Basic Industries Act of 1961. The list of basic industries expanded to include the manufacture of basic chemicals, basic iron, steel, nickel and aluminium; the processing of food products using local raw materials; the manufacture of cigars and tobacco; and the mining and exploration of base metals and crude oil.

As a direct result of these policies of industrialisation, foreign participation in the manufacturing sector dramatically increased. With the parity provision in the Philippine Trade Act and later in the Laurel-Langley Agreement, the United States accounted for some 80 per cent of the total foreign-owned equity in the Philippines in 1970, a share much higher than that in the colonial period (Schirmer and Shalom, 1987). A total of forty-seven US-owned firms ranked among the 200 largest firms in the Philippines in 1971 accounting for 30.4 per cent, 29.4 per cent and 28.0 per cent respectively of the total equity, sales and assets of these 200 firms in all sectors. Their significance is even more apparent in the manufacturing sector where 37.0 per cent, 33.6 per cent and 34.5 per cent of the total equity, sales and assets of the 110 largest Philippine manufacturing firms were accounted for by thirty-five US firms (Corporate Information Center, 1973).

Table 7.9 shows the foreign and domestic shares of assets in major non-financial firms in the Philippines in 1965. Foreign firms accounted for 66.1 per cent of the total assets in the manufacturing sector. An analysis of the sub-sectors shows that foreign firms accounted for a majority of the assets in paper and paper products (91.7 per cent), chemical products (90.9 per cent), machinery equipment (90.6 per cent), petroleum (82.3 per cent), wood products (78.3 per cent), household durables (73.2 per cent), metal products (63.5 per cent), non-metallic mineral products (58.8 per cent) and food and beverages (51.9 per cent). In addition, foreign firms accounted for 100 per cent of the total assets in the transportation equipment sector. Apart from considerable asset and equity ownership obtained through foreign direct investment, the considerable influence of foreign firms in the Philippine economy is also manifested in the licensing of technology (Lindsey, 1983, 1985).

On the other hand, the presence of Philippine firms in resource-based industries such as textiles and cement that have accumulated production experience and the capacity for localised technological innovation has enabled those firms to account for a majority of the total assets in these sectors. Other resource-based sectors in which local firms have positively responded to the presence of foreign firms are food and beverages, metal products and non-metallic mineral products. Philippine firms accounted for 48 per cent, 37 per cent and 41 per cent respectively of the total assets in these sectors.

Apart from the manufacturing sector, indigenous firms in the services sector have also benefited from the competitive stimulus offered by foreign MNEs. In addition, the stipulation of the Philippine Constitution that investments in public utilities, along with natural resources and the development of public

Table 7.9 Foreign and domestic shares of assets in major private non-financial firms in the Philippines, 1965, Philippine Peso million

Industry	Foreign assets	%	Domestic assets	%
All sectors	2,010.0	65.6	1,052.04	34.4
Manufacturing	1,397.0	66.1	718.0	33.9
Petroleum	487.0	82.3	105.0	17.7
Food and beverages	286.0	51.9	265.0	48.1
Chemical products	120.0	90.9	12.0	9.1
Textile products	55.0	34.8	103.0	65.2
Lumber, veneer, plywood, furniture	54.0	78.3	15.0	21.7
Metal products	47.0	63.5	27.0	36.5
Automotive, agricultural and transportation equipment	37.0	100.0	0.0	0.0
Household durables except furniture	30.0	73.2	11.0	26.8
Machinery equipment, apparatus and parts (excl. office equipment)	29.0	90.6	3.0	9.4
Paper and paper products	22.0	91.7	2.0	8.3
Cement	17.0	32.1	36.0	67.9
Glass, ceramics and clay products	10.0	58.8	7.0	41.2
Others	203.0	60.6	132.0	39.4
Mining, ore processing, quarrying and mineral exploration	16.0	37.2	27.0	62.8
Agriculture, fisheries and forestry	13.0	99.7	0.04	0.3
Commerce, warehousing, customs brokerage, etc.	377.0	86.5	59.0	13.5
Utilities	97.0	40.9	140.0	59.1
Construction	37.0	84.1	7.0	15.9
Services	52.0	34.0	101.0	66.0
Real estate	21.0	100.0	0.0	0.0

Source: Poblador, 1969.

lands, should be undertaken by firms with 60 per cent of their capital stock owned by Filipino citizens, resulted in indigenous firms accounting for 59 per cent and 66 per cent of total assets in the utilities and services sectors respectively by the 1960s. The Manila Electric Company (MERALCO) which represents a merger between Manila Electric Railroad and Lighting Company and Manila Suburban Railways Company had been sold by the American parent company, General Public Utilities Corporation, to a local holding company as a result *inter alia* of the anti-monopoly law in the United States, the restriction on profit remittances by the Central Bank of the Philippines and the declining profitability of the company. Similarly, American equity participation in the Philippine Long Distance Telephone Company (originally known as the Philippine Islands Telephone & Telegraph Company), the largest telephone and telecommunications company in the Philippines, with a book value of US $15 million and representing 28 per cent of the total equity of the firm, had been sold to a local group in response to rising economic nationalism in the country.

Although foreign firms had been instrumental in the establishment of the infrastructure of the Philippines, in the provision of transport and communication systems and other public utilities, such foreign participation gradually diminished as indigenous firms with gradually increasing ownership advantages were increasingly able to replace foreign firms as the government adopted policies of economic nationalism. This development broadly paralleled events that had taken place in the developed countries in an earlier period when foreign firms had also been instrumental in building the infrastructure but whose contribution had slowly diminished as local firms took over their operations (Cantwell, 1987a).

In addition, although Philippine construction firms accounted for only 16 per cent of the total assets in the construction sector at this time, these firms displayed the capacity to combine advanced construction technology introduced by foreign firms with their own localised innovation. The accumulated technology was applied to their domestic construction activities and to foreign construction activities at a later stage.

That import-substituting investments increase the scope for the transfer of advanced industrial technology has been confirmed by Reuber *et al.* (1973), and stems largely from the greater bargaining position of the host country on the one hand and the long-term commitment of foreign firms to local demand and cost conditions on the other. Such a neoclassical view of the role of MNEs, analysed in the last chapter in the context of theories of development, regards them as transferrers of technology, as shown by their capacity to provide training and management programmes to local personnel in order to increase both their technological and economic capabilities.

On the other hand, the theories of underdevelopment put forward a more critical view by arguing that far less technology is transferred by MNEs than conventional arguments claim because most R&D undertaken by foreign MNEs is concentrated in the home country located in the more industrially advanced world. Such products of R&D in the home countries become the major source of ownership advantages of foreign MNEs and have to be protected from diffusion to potential indigenous competitors. Indigenous firms are therefore forced into a state of technological dependence as a result of the inhibited development of their technological and scientific capabilities.

The theory of technological competence assumes an intermediate stance between the two schools of thought on the role of FDI to explain the variable impact of foreign investment on the development of indigenous technological capacity and subsequent outward investment by local firms. Those who argue for the developmental role of FDI assert that inward foreign investment at an aggregate level perpetuates a virtuous cycle as regards indigenous firms; while those in the market power, dependency, world-system and neo-Marxist schools argue that inward foreign investment at an aggregate level inhibits the growth of indigenous technological capacity by promoting technological dependence. On the contrary, such dependence only arises when a large proportion of the country's technology is sourced from abroad (Stewart, 1977). Moreover, a system of double dependence arises when both the element of technical knowledge, i.e. know-how, and the capacity to use that knowledge in investment and production, i.e. know-why, have to be transferred (Cooper and Sercovich, 1970).

An analysis of the dynamic role of inward foreign investment in outward investment by Philippine firms provides support for the theory of technological competence and for the presence of technological interdependence between foreign investment and outward investment by Philippine firms in certain sectors in which local firms have acquired some production experience and technological capabilities. The technology and capital brought into the Philippines, particularly in the 1950s, by foreign MNEs in response to the ISI created linkages with Philippine firms in particular sectors which are largely resource based, such as textile products, cement and, to a lesser extent, food and beverages, metal products and non-metallic mineral products. Philippine firms have developed some inherent capacity for indigenous innovation in these sectors and have largely gained from the technological, pecuniary, marketing or entrepreneurial initiatives provided by foreign firms.

Studies on the impact of foreign technology transfer on selected industries in the Philippines confirm the presence of some indigenous technological capacity which enabled Philippine firms to make an independent response to the presence of foreign MNEs. The pharmaceutical industry is a case in point: Philippine firms benefited from foreign process technology in re-packing and dosage formulation. Although the capabilities of local firms in the production of chemical raw materials are still weak, as shown by their high import intensity for these products, some Philippine pharmaceutical firms such as United Laboratories, Pharma Industries and Zodiac Pharmaceuticals have become competitive on the basis of comparative advantage in low cost and locally available packing materials and labour. As a result, these firms have been able to match the efficiency of foreign affiliates in the Philippines and to engage in exports or direct investment on the basis of meeting international standards of quality control and the adoption of production, management, marketing and distribution strategies based on western models. The success of these Philippine drug firms is also manifested by their capacity to export technology abroad in the form of personnel or to train foreign drug technicians in their laboratories (Clemente and Bautista, 1978).

The domestic motor vehicle industry is another example in which indigenous firms have gradually increased their initial know-how in assembly to manufacturing through the presence of a number of American, European and Japanese automobile firms. This growth was enhanced by the Progressive Car Manufacturing Program launched by the Philippine government in 1973 to promote the local manufacture of cars geared for the domestic market. There were five major participants in this programme: Ford Philippines, Incorporated; General Motors Philippines, Incorporated; Chrysler Philippines Corporation; Delta Motors Corporation; and DMG Incorporated. The last two are domestic firms while the first three are affiliates of foreign MNEs.

While the capability of the Philippine motor vehicle industry in assembly is well developed, its capability in manufacturing is still at an early stage and as a result R&D activities related to car manufacture are fundamentally carried out by parent firms or other foreign firms that supply various components. In particular, sustained investment had to be made by the five major participants in assembly and manufacturing facilities which resulted in the development of a suitable utility vehicle for the countryside which entails the use of unsophisticated, locally stamped bodies and engines from the industry's major firms.

However, the sustained economic viability of the locally produced vehicle was threatened by a situation of market fragmentation owing to the limited size of the domestic market, which resulted in plant operating at levels far below their annual rated capacities. The ASEAN Regional Complementation Scheme, established to alleviate the problem of market fragmentation, yielded less than favourable results, owing to the large number of models produced relative to the size of the ASEAN market. But despite the lack of economic viability local firms have benefited from the training given to local managers, supervisors and rank and file employees, intended to reduce their technological dependence on foreign firms. The growth of the motor vehicle industry also produced developmental links with local firms in related industries that supply the motor vehicle industry with such vital inputs as steel, glass, rubber, electrical parts and plastics. The car companies have assisted these ancillary firms in upgrading their technical know-how (Laxa, Cardenas, Federizon and Gesmundo, 1978).[2]

The extent to which the technological development of indigenous Philippine firms was favourably affected by the presence of foreign MNEs during the period of ISI was to some extent constrained by the absence of government policies to promote further local technological development in local capital goods industries. The establishment of local capital goods industries is instrumental in the promotion of local technological development, in the enhancement of the bargaining power of local firms in relation to the importation of foreign technology and in the improvement of localised technological change (Stewart, 1977, 1979; Mitra, 1979; Pack, 1981).

As a result of the low level of technological development of local firms in such capital goods industries as mechanical engineering, the presence of foreign MNEs in these sectors perpetuated a vicious circle of cumulative decline by decreasing the slope of the indigenous technical progress function. These foreign MNEs stifled domestic competition through their increased market power, as predicated by Hymer, and reinforced the underdevelopment of the Philippines as predicated by the theories of underdevelopment.

Hirschman (1968) described the ISI process as consisting of tightly separated although highly sequential stages which began with the establishment of consumer goods industries and at a later stage progressed to the establishment of intermediate and capital goods industries. However, the establishment of consumer goods industries may be pushed to the maximum possible extent, leading to a premature widening of the productive structure, i.e. the production of increasingly sophisticated, high-income durable consumer goods rather than the backward integration into intermediate, investment and capital goods (Felix, 1964).

However, contrary to Hirschman's view, the replacement of imports of intermediate goods by domestic production tended to be highly capital intensive and subject to important economies of scale achieved only through a large minimum-efficient plant size which is far beyond the domestic needs of most developing countries whose margin of processing is relatively small and whose organisational and technical inefficiencies may lead to high costs. The domestic manufacture of motor vehicles in the Philippines is an excellent case in point: market fragmentation led to constrained growth of the domestic motor vehicle industry. Parry (1981) has suggested that technological adaptation undertaken by foreign MNEs through a two-stage technology transfer via an intermediate economy (i.e. an economy that has characteristics in common with both developed and developing countries) may lead to more appropriate technology in developing countries and thus relieve the problems of inappropriate factor proportions and large-scale plant that result in market fragmentation. But Lall (1975) notes that in the modern, technologically advanced industries, particularly those geared to export markets, there is no scope for an intermediate technology because the most advanced technique is the most appropriate.

Similarly, the production of producer and consumer durables is also subject to economies of scale which relate not so much to plant size, as with intermediate goods, as to horizontal and vertical specialisation which entails reduction in product variety and the manufacture of parts, components and accessories on an efficient scale in separate plant. Such products require skilled and technical labour and, to a lesser extent, the application of sophisticated technology.

Developing countries that have small national markets may, however, be inefficient manufacturers of highly physical- and capital-intensive intermediate goods as well as skill-intensive producer and consumer goods because of the limited size of their national markets, which constrains the possibilities for exploitation of economies of scale. Moreover, the need to import raw materials and machinery limits the scope for savings in foreign exchange.

Despite the rather limited impact of foreign firms on technology transfer to local firms, ISI laid the foundations for Philippine economic growth. The economic policies adopted during that time under a system of exchange and import controls were basically directed towards creating and encouraging light and intermediate industries and to processing, assembly, packaging and final consumer goods. Table 7.8 shows that 74 per cent of foreign affiliates established during the period 1949–68 were in import-substituting industries. The highly protective tariffs biased in favour of consumer goods and the adoption of fixed multiple exchange rates alleviated the balance-of-payments crisis significantly through a reduction in imports of final consumer goods but with a corresponding increase in imports of capital goods such as machinery and intermediate products. A large quantity of consumer goods was locally manufactured and production in many product lines was considered to be at a world standard of efficiency (Castro, 1969).

That ISI was incapable of accomplishing a high degree of industrialisation is evident from an examination of the average annual growth rate of the manufacturing sector as well as the share of this sector in the Net Domestic Product (NDP) during the period. The average annual growth rate of the manufacturing sector declined from 14 per cent between 1949 and 1953 to 5 per cent between 1961 and 1965. As a proportion of NDP, the share of manufacturing grew marginally from 10 per cent in 1950 to 13 per cent in 1955 and 15 per cent in 1960. On the other hand, the share of agriculture in the NDP declined. Moreover, the market-determined ISI strategy adopted by the Philippines led to heavy dependence on a variety of foreign inputs in the form of intermediate goods, i.e. raw materials, technology, etc., required in the manufacture of consumer goods.

ISI was also the main industrialisation strategy adopted by some of the larger South and Southeast Asian economies such as Pakistan and India, and also by some of the larger Latin American economies such as Brazil, Argentina and Mexico in the 1950s and 1960s. In addition, ISI began to be adopted in a number of the sub-Saharan African economies such as Ghana, Kenya, Nigeria and Zambia, and in smaller Latin American and Southeast Asian countries in the early to mid-1960s. On the other hand, ISI and export-oriented industrialisation (EOI) – an infant, exporter-based, protected export promotion strategy – have been successfully combined in the Republic of Korea and other East Asian NICs where import protection has provided the basis for the development of local manufacturing capacity and subsequent exporting activities (Singer, 1984; Liang, 1990).

The Latin American countries adopted a strategy of ISI as a response to the ideas of Raul Prebisch who viewed export production as a constraint on the economic growth of developing countries because of the adverse foreign market conditions for their primary product exports as well as the lack of international competitiveness of their manufactured product exports. As a result of the constraint of exports on growth, developing countries were advised to orient their manufacturing industries towards domestic markets through industrial protection. Similar ideas expressed by Gunnar Myrdal influenced the policies followed by India, the Soviet Union and the European socialist countries. The presence of determined government intervention in some of the larger developing countries such as Brazil, Mexico, Argentina, Malaysia, India, Pakistan and South Korea, and some of the European socialist countries, led to the establishment of both intermediate and capital goods industries in these countries under an ISI regime. These industries would not have been established with unplanned, spontaneous and market-determined ISI (Colman and Nixson, 1986).

As a result of government policies to reduce technological dependence through the promotion of local technological development, particularly in the intermediate and capital goods industries, and sector-specific policies towards FDI indigenous firms in these countries have been able to adapt and modify foreign technologies to suit their own technological innovation and to produce sophisticated products. The high capabilities of firms in these countries to assimilate foreign technology to suit their particular technological needs have in part led Warren (1973) to argue that technological dependence does not arise from imperialist monopoly or other forms of domination but from the lack of capabilities of indigenous firms to assimilate foreign technology. These capabilities increase with the advance of commercialisation and industrialisation and with the acquisition of education and experience, including bargaining experience (Warren, 1980). The introduction of such technical change, with innovative activities occurring not only among the developed but also the developing countries, is taken by Soete (1981a) to be irreconcilable with the primarily static analysis of technological dependency.

Similarly, Lall (1975) has shown that the concept of technological dependence is irrelevant in analysing the underdevelopment of developing countries since a number of developed countries also rely heavily on foreign technology and therefore the issue lies not in the presence or absence of dependence but in the degree of such dependence. Reasoning along these lines leads to the proposition that higher levels of technological capabilities of local firms would enable some alleviation from technological dependence and foster technological interdependence on foreign technology.

THE PERIOD OF EXPORT-ORIENTED INDUSTRIALISATION (EOI) (1970 TO THE PRESENT)

Although considerable amounts of foreign investment were established in the Philippines during the 1950s in response to the policy of import-substituting industrialisation, by the 1960s and early 1970s new inflows of foreign investment, from the USA in particular, were reduced as the policy reached its limits, with the rising tide of nationalism and the expiration of the Laurel-Langley Agreement marking the end of the preferential access of US capital in the Philippines (Lindsey, 1983, 1985). Several measures were nevertheless enacted to counteract the forces that led to a declining trend of foreign investment in the Philippines in response to the limitations on further expansion through import-substituting industrialisation policies. First, emphasis was gradually shifted to the promotion of export industries which could contribute substantially to the foreign exchange earnings of the country. Second, martial law was declared to curb the rising tide of nationalism and to assure foreign investment of a receptive and favourable investment climate.

Republic Act 6135, otherwise known as the Export Incentives Act, was passed by the Philippine Congress and approved by the president on 31 August 1970. As with the Foreign Business Regulations Act, the Export Incentives Act complemented the earlier Investment Incentives Act with regard to the foreign ownership of registered enterprises and embodied the two-pronged policy of the state, namely the active encouragement, promotion and diversification of non-traditional exports, consisting of services and manufactured products utilising domestic raw materials to the fullest extent; and the development of new markets for Philippine products.

The Export Incentives Act granted various incentives to export producers both in pioneer and non-pioneer areas of investment; and also to export traders and service exporters registered with the Board of Investments (BOI) which exports products listed in the Export Priorities Plan (EPP) as well as to other firms exporting at least 50 per cent of their products. These incentives, consisting *inter alia* of tax exemptions and deductions, tariff protection and liberty in the employment of foreign nationals, are given either to a Philippine national or a Philippine firm with at least 60 per cent of its capital stock owned by Philippine nationals, or a pioneer enterprise whether foreign or domestic.

The strategies for export promotion, especially in manufactured products, had been further enhanced by Republic Act 5490 which became law on on 21 June 1969 and led to the creation of export processing zones (EPZ). The Philippines created a number of EPZ which are particularly suited to production conforming to the new international division of labour such as textiles and clothing and electrical machinery. The first of these EPZ was established in Mariveles at the tip of the Bataan Peninsula. However, despite the designation of Mariveles as principal port of entry where merchandise could be brought in without being subject to the customs and internal revenue laws and regulations, no appropriation was made for the estimated 500 million Philippine pesos (US $71 million) necessary for the actual physical development of the zone. During the next three years, the development of the zone was supported by the transfer of savings from other governmental agencies totalling 25 million Philippine pesos (US $4 million) (Ken, 1977). As a result of the poor infrastructural facilities of the EPZ, and other locational disadvantages of the Philippines compared to other Asian countries, few export-oriented foreign investments were encouraged.

As in the previous investment legislation enacted during the period of import-substituting industrialisation, discrimination against foreign investors is inherent in the Export Incentives Act of the export-oriented industrialisation period. Apart from pioneer firms, wholly owned foreign firms are not entitled to incentives granted by the Act even if such firms are potential exporters. Thus, this policy did not encourage foreign import-substituting firms to reorient their operations towards fulfilling both domestic and foreign demand. Both the older import-substitution and the newer export-oriented foreign investment in the Philippines were highly import dependent (Lindsey, 1983, 1985).

The era of export-oriented industrialisation therefore had only a minor impact on the level and pattern of foreign investment even though the value of exports, especially manufactured exports, from the Philippines had actually increased. By 1970, the value of exports amounted to 6,183 million Philippine pesos (US $961 million) compared to an average value of 2,071 million Philippine pesos (US $610 million) during the period of ISI from 1950–69. The share of manufactured exports in the total exports of the Philippines rose from 4 per cent in 1960 to 17 per cent in 1975 and 35 per cent in 1979. Such an increase in the value of exports can be explained by the *de facto* devaluation of the Philippine peso *vis-à-vis* the US dollar from 3.9 pesos to 6.4 pesos to the US dollar in the period 1969–70 as a result of which the Philippines was placed in a much more favourable position as far as labour-intensive export products were concerned.

Although manufactured exports accounted for a significant proportion of total exports in the Philippines, the absolute value of such exports amounted to only US $34.2 per capita in 1979. Table 7.10 gives a comparison of the absolute value of manufactured exports per capita for a number of developing countries in South and East Asia. This table shows that exports of manufactured products per capita in the Philippines are very small compared to those of the Asian NICs such as South Korea, Hong Kong and Singapore which are at a more advanced stage of development. These NICs accounted for 35 per cent of the total manufacturing exports of the developing Asian region in 1979, owing largely to their longer experience in export-oriented industrialisation and more favourable locational advantages, compared to the Philippines, which enabled these countries to attract more foreign MNEs which were almost exclusively confined to export manufacturing. By comparison, Table 7.8 suggests that FDI in the Philippines continued to have a more significant role in import-substituting investment.

Such a high concentration of FDI in export-oriented manufacturing in the NICs and the capacity to supply inputs, technology and market access have enabled local firms in these countries to engage in competitive international expansion through exports and outward investment. The effect on local firms of the success of foreign MNEs in export markets has been an important force behind the growth of entrepreneurship in Hong Kong firms and their majority share in the export of labour-intensive manufactured goods in the 1960s

Table 7.10 Exports of manufactured products of developed countries in South and East Asia, 1979

Country	Manufactured exports, US $ million	Population, million	Manufactured exports per capita, US $
Total South Asia[1]	5,484.0	890.5	6.2
Nepal	28.0	14.0	2.0
Bangladesh	437.0	88.9	4.9
India	3,729.0	659.2	5.7
Pakistan	1,140.0	79.7	14.3
Sri Lanka	122.0	14.5	8.4
Burma	28.0	32.9	0.9
Total East Asia[2]	50,410.0	308.6	163.4
Indonesia	488.0	142.9	3.4
Thailand	1,327.0	45.5	29.2
Philippines	1,596.0	46.7	34.2
Malaysia	1,966.0	13.1	150.1
Republic of Korea	13,299.0	37.8	351.8
Hong Kong	10,804.0	5.0	2,160.8
Singapore	7,372.0	2.4	3,071.7
Totally centrally planned economies of Asia[3]	5,883.0	1,036.5	5.7
China	5,311.0	964.5	5.5
Vietnam	297.0	52.9	5.6
Total South, East and CPE Asia	61,777.0	2,235.6	27.6

Sources: Maex, 1983, from data obtained from *World Development Report*, 1981 and 1982 and *Far Eastern Economic Review's Asia 1981 Yearbook*.

Notes: 1 Including Bhutan and the Maldives.
2 Including Brunei, Laos, Kampuchea, Macau, etc.
3 Including Mongolia and the Democratic Republic of Korea.

(Riedel, 1975). More importantly, the constraint on further export growth as a result of the quantitative restrictions imposed by the export markets in the developed countries has been a primary determinant of outward investment at a later stage.

By contrast, although export-oriented foreign MNEs established operations in the Philippines in response to the export-oriented industrialisation strategy in 1970, these investments were few and far between. Table 7.8 shows that only three of the 324 largest foreign affiliates in the Philippines established export-oriented manufacturing in such resource-based sectors as metal ore, sugar, coconut oil, cordage, fruit farming and processing, plywood and veneer in the period 1969–76. Hence, the role of FDI in promoting the envisaged structural change in Philippine exports from primary products to manufactured products has been limited compared to other countries. FDI played a far more significant role in South Korea, Taiwan, Mexico, Colombia and Singapore where the proportion of manufactured exports attributed to foreign MNEs were 17 per cent, 20 per cent, 33 per cent, 58 per cent and 88 per cent respectively (de la Torre, 1974; Cohen, 1975; Lall, 1977; Jenkins, 1979; Westphal, Rhee and Pursell, 1979).

Reuber *et al.* (1973), de la Torre (1974), Helleiner (1975), Keesing (1979) and Parry (1980) have argued that FDI can play a crucial role in the rapid growth of manufactured exports from developing countries through greater access to inputs, technology, market information, distribution channels and marketing skills. However, export processing zones in the Philippines designed to attract export-oriented foreign MNEs only accounted for 12.3 per cent of the total exports of manufactured products in the Philippines in 1976. This means that the production of manufactured exports is still undertaken largely by Philippine firms and only to a much lesser extent by the few export-oriented foreign MNEs engaged in a large variety of standardised commodity-type goods such as textiles, apparel, shoes and leather products. The far more significant role in the expansion of Philippine exports played by local firms has broad parallelism with the period in the 1960s when local firms in the East Asian NICs provided the major impetus for export growth with the help of marketing services provided by foreign buyers such as Japanese trading houses and the largest retail buying groups in the developed countries. The role played by MNEs only became more significant when the prospects for exports from these countries became clear and when the composition of exports started to shift to more sophisticated product lines (Westphal *et al.* 1979; Lall, 1981). Until such time as these countries had become attractive locations for export-oriented foreign direct investment, local firms continued to export a significantly higher proportion of their output compared to foreign firms in the same industry (Cohen, 1975; Riedel 1975; Jenkins, 1979).

This was especially the case in the Philippines where the original investment of foreign MNEs was in import-substituting industries and in which the policies stipulated conditions limiting the subsequent export of output. Foreign MNEs' low share in the manufactured exports of the Philippines supports the conclusion made by Nayyar (1978) that the share of foreign MNEs, excluding foreign buying groups, in manufactured exports of low-income countries is lower than is widely believed and was no larger than 15 per cent in the period around 1974.

However, a more important reason for the less significant role played by FDI in the rapid growth of manufactured exports and outward investment by the Philippines compared to the NICs and other lower-income Asian developing countries was the Philippines' lack of adequate locational advantages as a host country – in particular, the undesirable investment climate, a result of the rising trends in nationalism, social unrest, etc., and the existing foreign investment policy which provided incentives but placed restrictions on the degree of foreign equity participation in preferred non-pioneer areas of investment and allowed for nationalisation in pioneer areas. As a result of the declining locational advantages and the more conducive investment climate offered by other Southeast Asian countries such as South Korea, Thailand and Taiwan, the Philippines experienced substantial outflows of FDI while those countries experienced substantial inflows from the latter part of the 1960s and early 1970s. The experience of the Philippines with respect to the provision of investment incentives reflects their limited effectiveness and hence the competition between different developing countries in the provision of investment incentives, for footloose investment would only result in the attraction of FDI when other more important political and economic determinants of the investment climate, such as the potential for market growth, suitability of local forms of work organisation, the local capacity for organisational and related innovation and the existence of an appropriate infrastructure, were rendered conducive. The provision of fiscal incentives, therefore, is not necessary to encourage a higher level of FDI in developing countries; more important are the presence of natural resources and a proven record of economic performance (Root and Ahmed, 1978; Lim, 1983; Cantwell, 1987b).

The low level of technological capabilities of local firms associated with the early stage of industrial development of the Philippines has been shown to be a particularly important factor in the limited success of government local-content programmes in the capital- and technology-intensive sectors such as motor vehicles and electrical appliances that were intended to foster the development of small industries through the establishment of subcontracting networks with large foreign firms (Hill, 1985). The crucial factor in the difficult process of transfer of skills to local personnel by foreign firms in such industries as electrical appliances and iron foundries has been the limited level of Philippine capabilities which are confined, for example, in areas of manual dexterity as in the operation of machine tools, sheet-metal work or welding as opposed to more sophisticated operations such as foundries or annealing. In addition, the Filipino work ethic has placed quality control and pride of workmanship low down in order of importance (Odaka, 1984).

A strong indication of the existence of local firms which are technologically weak and experience difficulties in catching up, and whose technological capabilities are undermined, is the local expansion of internationally competitive foreign firms that rely on research carried out in their home country or elsewhere in their international network. A recent survey conducted among the fourteen subsidiaries and affiliates of foreign MNEs in the Philippines that count among the 1,000 largest firms in the country suggests that the 36 per cent of foreign affiliates that conducted basic research limited their activities mostly to applied research such as the adaptation of products to local market demand. More extensive fundamental research was conducted in their home country. The most important contribution by foreign firms to the development of local firms in the Philippines centred largely on the provision of employee training which had a generally beneficial impact on the country's labour force. This training is manifested in the practice of sending employees abroad and in the visits of expatriates (Miranda, 1989). In the banking sector, foreign banks such as Citibank pride themselves on being the training ground for the majority of Philippine bank presidents (Tiglao, 1990c).

Recent reforms by the government and recommendations by the World Bank have been aimed at making growth in the industrial sector more self-sustaining through the harnessing of local technological capabilities. They include the implementation of rules to facilitate backward linkages in the key export sectors of garments and semiconductors (Galang, 1988). In addition, industrial estates are being established that emphasise the formation of advanced technology and skills. However, unlike the Asian NICs in which industrial estates are mostly established by the government, the two Philippine estates that are to be established in the southern provinces adjacent to Manila are being financed by private firms. These are the 344-hectare Ayala Laguna Technology Park, to be established by the Ayala Group of Companies, and the 143-hectare Science Park which is being established by a group of firms including the Philippine American Life Insurance Company, Bechtel Investments of the USA and the Investment and Capital Corporation of the Philippines, a joint venture of the Philippine-based Far East Bank and Trust Company and American Express. The expectation is that these Philippine conglomerate firms will constitute an important vanguard for Philippine economic development in the same way that the *keiretsu* and *chaebol* have fulfilled crucial roles in the growth of Japan and South Korea (Tiglao, 1990b).

Evidence of the structural change occurring in the Philippine industrial sector is found in the increased share of manufactured products in total exports, up from 22 per cent in 1976 to 61 per cent in 1986. Data from the National Statistical Office of the Republic of the Philippines suggest that by 1989 these products accounted for 74 per cent of the total value of the ten leading exports of the Philippines, which include such high value-added products as semiconductor devices, electronic microcircuits and consigned finished electronic and electrical machinery equipment and parts. An increasing number of semiconductor exports is accounted for by the concentration of both local and foreign-owned subcontract plant and small firms that perform assembly and specialised tasks for technologically simpler semiconductors (Henderson, 1989). However, in recent years the attainment of more advanced technological capacities is increasingly apparent in some local firms such as the Ayala Corporation and the Solid Corporation, which have emerged as manufacturers of integrated circuits and electronics products respectively (Tiglao, 1990a, c).

An interesting scenario would be the extent to which the Philippines can achieve economic growth and increase their locational advantages in the wide arena of international economic activities. In addition to an English-speaking and very easily trained labour force, the Philippines is blessed with a steady stream of entrepreneurs and management executives helped by a well-developed network of business schools that count as among the best in the world. However, there are several other locational factors that are fundamental to this process, of which the attainment of political stability and the balance of payments are crucial. The recently concluded aid programme, Multilateral Assistance Initiative, involving nineteen donor countries and seven international aid organisations, is helping to create the conditions for economic growth in the Philippines through private sector investment. The goal of the programme is to create an environment attractive to domestic and foreign investors through infrastructure development, the establishment of economic development zones and local government projects, and the provision of pre-investment studies – market analyses, environmental assessments, full feasibility studies, and natural resource identification or confirmation.

CONCLUSIONS AND POLICY IMPLICATIONS

This chapter focused on the extent to which foreign direct investment as a modality for the transfer of advanced foreign technology has played a complementary role in the development of the indigenous technological capacity of Philippine firms, which enabled these firms to engage in outward direct investment.

The early forms of foreign investment in the primary sector in the colonial period, which represented joint venture arrangements with Philippine firms, exerted favourable developmental influences on local firms which acquired some production experience in the exploitation and exploration of several natural resources with which the country is relatively well endowed. The era of import-substituting industrialisation (ISI) in the period 1950–67 witnessed an increase in foreign participation in the manufacturing sector and enabled some technologically innovative firms in such sectors as textiles, cement, food and beverages, pharmaceuticals, metal products, non-metallic mineral products and construction, which have accumulated production experience and demonstrated the capacity for localised innovation, to respond positively to the presence of foreign firms. On the other hand, the low level of technological development of Philippine firms in capital goods industries such as *inter alia* mechanical engineering, chemical products, petroleum, paper and paper products and metal products has enabled foreign MNEs to be dominant in these sectors and to perpetuate a vicious circle of cumulative decline. Such a variable impact of foreign direct investment on the development of indigenous technological capacity was analysed within the framework of the theory of technological competence, suitably modified to take account of the distinctive nature of innovative capacities of Third World firms.

The extent to which the technological development of the Philippines was favourably affected by the presence of MNEs during the period of ISI was constrained by the absence of government policies to promote further technological development in the capital goods industries. By contrast, in some of the larger developing countries such as Brazil, Mexico, Argentina, Malaysia, India, Pakistan and South Korea the presence of determined government intervention led to the establishment of both intermediate and capital goods industries which enabled local firms to reduce their technological dependence on and increase their technological interdependence with foreign MNEs. In recent years, as the last chapter has shown, government intervention in some of these countries has fostered the growth of more technologically intensive industries.

The era of export-oriented industrialisation from 1970 onwards has had a minor impact on the level and pattern of foreign investment in the Philippines even though the value of manufactured exports actually increased. Foreign direct investment accounted for a very small proportion of the manufactured exports of the Philippines owing to the lack of adequate locational advantages. By contrast, the higher proportion of FDI in export-oriented manufacturing in the NICs enabled local firms in those countries to catch up with the older and more established investors by engaging in competitive international expansion through exports and outward investment.

The extent to which FDI will favourably affect the development of the indigenous technological capacity of Philippine firms in the future fundamentally depends on the extent to which important economic, social and political determinants of the investment climate improve. The foreign investment policies of the Philippines since the end of the martial law regime in 1982 and the newly ratified constitution of 1987 have been increasingly geared towards identifying the specific sectors in which foreign investment can fulfil development objectives. Those sectors in which foreign firms could foster greater integration and, consequently, developmental spill-over effects with indigenous firms are encouraged, while in those sectors with limited capacity for such developmental effects foreign investment policies are geared to preventing the abuse of monopoly power and to achieving greater bargaining power as far as the distribution of benefits is concerned.

A significant although neglected aspect in the formulation of foreign investment policy in the Philippines is the important distinction that has to be made between policies geared to existing investments and policies geared to attract new investment. Policies geared to existing investments should ensure greater interaction with local firms and have a more positive impact on industrialisation through *inter alia* greater local downstream processing activities. On the other hand, policies which are geared to attracting new export-oriented or footloose investment that would facilitate greater integration and linkages with local firms should offer incentives in the form of tax *et al.* benefits once the investment climate has been improved.

The growth of some Philippine MNEs which developed sufficient ownership-specific advantages to be transferred and exploited in foreign markets may be attributed in part to the beneficial impact of inward FDI on their indigenous technological capabilities. The phenomenon of Philippine MNEs is further discussed in the next chapter.

8 A profile of Philippine multinationals

INTRODUCTION

This chapter reports on the findings of extensive and original empirical research on the emergence and growth of multinational investment activities by Philippine firms. The research was undertaken during a critical period in the country's political history. The survey period started in October 1985, a few months before the overthrow of President Ferdinand Marcos by the People Power revolution led by the incumbent president, Corazon C. Aquino, in February 1986, and ended in December 1986. A considerable amount of these outward direct and portfolio investments by Philippine firms may have been prompted by the need to diversify risks owing to the political and economic instability in the Philippines.

Background data on the estimated value of foreign direct investment by Philippine firms, their geographical and industrial distribution and performance overseas were collected from Wells (1983) and Lim (1984). Preliminary information reveals that there are at least thirty Philippine firms with direct investment overseas. A consensus based on the different sources of information shows that, by number of firms, Philippine MNEs are predominantly in the following areas of economic activity: banking and finance (53.4 per cent), construction (23.3 per cent), manufacturing (16.7 per cent), and others, which include resource-based and service sector investments (6.6 per cent).

Owing to the small number of direct investor firms identified, all thirty Philippine parent firms were selected for the survey. The only criterion imposed on the selection was the continued international operation of the firm after the overthrow of the regime of President Marcos in February 1986. Such a criterion would ensure that the foreign operations of the firm represented a commercial venture in the real sense. Moreover, it would eliminate companies closely associated with the former Marcos administration which were subsequently sequestered after the overthrow of the regime. These firms were identified as mere conduits for fund transfers abroad and are more correctly referred to as portfolio rather than direct investments.

Some background information on some selected Philippine MNEs is summarised in Tables 8.1 and 8.2. The tables suggest that the average age of the

Table 8.1 Background information on some Philippine MNEs

Name of firm	Date of establishment	Main activity	Foreign investment		
			Country	Activity	% of total equity
San Miguel Corporation	21 September 1890	manufacturing: beer, soft drinks, dairy products, animal feeds and livestock, packaging materials, agribusiness, aquaculture	India	dairy products	joint venture
			Singapore[1]	dairy products	joint venture
			USA[1]	dairy products	joint venture
			Guam[2]	beer	100
			Papua New Guinea[3]	beer	100
			Hong Kong	beer	100
			Indonesia	beer	100
			Spain	beer	100
Atlantic, Gulf & Pacific Company of Manila (AG&P)	1900	construction	contracts in Saudi Arabia, Kuwait, Qatar, Iraq, Algeria, Sudan, Congo, Sri Lanka, Papua New Guinea, Indonesia, Vietnam, Malaysia, Okinawa	construction	
			Saudi Arabia	construction	50
			Malaysia	construction	45
			Indonesia	construction	49

Table 8.1 (continued)

Name of firm	Date of establishment	Main activity	Foreign investment		
			Country	Activity	% of total equity
Construction & Development Corporation of the Philippines	2 November 1966	construction	contracts in several Middle East countries	construction	
Engineering Equipment Incorporated	1931	industrial machinery distribution, foundry, domestic and overseas construction, steel fabrication	contracts in several Middle East countries	construction	
			Middle East	construction	joint venture
Erectors Incorporated	6 November 1957	general construction	contracts in Iraq, Saudi Arabia and other Middle East countries	construction	

Table 8.1 (continued)

Name of firm	Date of establishment	Main activity	Foreign investment		
			Country	Activity	% of total equity
Ayala International (Phils.) Incorporated	1983	exporter and distributer of marine products, food items, steel products, coffee/cacao beans, nectar fruit juices and general merchandise	Hong Kong Singapore Thailand Brunei USA Japan Malaysia Spain Several ASEAN countries	trade and distribution	100
			Malaysia	construction contracts insurance plantation real estate	30 –
			Thailand	trading insurance real estate plantations and ranches	–
			Singapore	trading investment management	
			USA	real estate hotels condominiums freight forwarding reservations ticketing for People's Airways	50

Table 8.1 (continued)

Name of firm	Date of establishment	Main activity	Foreign investment		
			Country	Activity	% of total equity
Landoil Resources Corporation	7 February 1973	holding management, real estate, oil exploration, development, construction	USA	health services provision	joint venture
			40 offices in Asia	market development centre, promotion of product lines and services	100
			France Singapore Switzerland Bahrain	construction management services construction	joint venture joint venture joint venture
			Panama	provision of manpower services oil exploration	joint venture joint venture
			contracts in Middle East	construction	
D. M. Consunji	1954	general engineering and general building	Brunei Iraq Saudi Arabia	construction	100 joint venture joint venture

Table 8.1 (continued)

Name of firm	Date of establishment	Main activity	Foreign investment		
			Country	Activity	% of total equity
Allied Banking Corporation	2 June 1977	banking	Hong Kong	deposit-taking subsidiary	100
			Bahrain	offshore banking unit	100
			London Singapore Sydney Tokyo	representative offices	
Bank of the Philippine Islands	1 August 1851	universal banking	Hong Kong	deposit-taking subsidiary	100
China Banking Corporation	20 July 1920	commercial banking	Hong Kong	deposit-taking subsidiary	–
			New York, USA	banking	–
Equitable Banking Corporation	26 September 1950	commerical banking	Hong Kong	banking	100

Table 8.1 (continued)

Name of firm	Date of establishment	Main activity	Foreign investment		
			Country	Activity	% of total equity
Manila Banking Corporation	23 January 1961	expanded commercial banking	Hong Kong USA	deposit-taking subsidiary banking	100 100
Metropolitan Bank & Trust Company	6 April 1962	universal banking	USA Hong Kong	banking banking	100 100
Philippine Commercial International Bank	8 July 1938	banking	New York, USA	banking	100
			Los Angeles, USA	banking	100
			Houston, USA Madrid, Spain Frankfurt, Germany London, England[4]	representative office	100
			Hong Kong	deposit-taking subsidiary	100

Table 8.1 (continued)

Name of firm	Date of establishment	Main activity	Foreign investment		
			Country	Activity	% of total equity
Philippine National Bank	4 February 1916	banking	Singapore		100
			New York, USA		100
			Houston, Texas, USA	banking	100
			London, England		100
			Beijing, People's Republic of China		100

Source: Questionnaires, interviews, annual reports.
Notes: 1 These ventures were divested in the late 1950s. In the case of Singapore, the divestment was due to political uncertainties and in the case of the US ventures, the divestment was due to severe competition due to the emergence of large national breweries.
2 This brewery was sold when the market dried up after the end of the Vietnam War.
3 This brewery was sold to Malayan Breweries, a Singapore MNE, after facing massive losses. It continues to have a technical licensing agreement to produce San Miguel beer.
4 This representative office was closed in early 1987.

Table 8.2 Revenues of some Philippine MNEs, 1988, US $ million[1]

Sector	Revenue
Industrial sector	
San Miguel Corporation	970.5
United Laboratories	123.7
Universal Robina Corporation	79.6
Banking and finance	
Philippine National Bank	1,816.6
Bank of the Philippine Islands	1,240.1
Metropolitan Bank and Trust Company	1,209.2
Philippine Commercial International Bank	924.7
Allied Banking Corporation	420.1
Equitable Banking Corporation	383.9
The Manila Banking Corporation	360.1
China Banking Corporation	278.6
Construction and engineering	
Benguet Corporation	198.9
Atlantic, Gulf & Pacific Company of Manila	85.9
Engineering Equipment Incorporated	48.9
Erectors Incorporated	32.2
Philippine National Construction Corporation	14.4
Asian Construction and Development Corporation	12.5
Hydro Resources Contractors Corporation	6.3

Source: Annual reports.
Note: 1 The data on banks represent assets. The data on Erectors Incorporated also represent total assets as of 1987.

parent MNEs is about fifty years. Even excluding the Bank of the Philippine Islands and San Miguel Corporation, which were established in 1851 and 1890 respectively, the average age of Philippine parent MNEs is about forty years. The industrial distribution of parent MNEs within the Philippines leans heavily towards the services sector, and primarily the construction, banking and trading sectors. However, the manufacturing sector is particularly important for some large conglomerate companies such as the San Miguel Corporation and the Gokongwei Group. There are at least four main conglomerate groups among Philippine parent MNEs, namely San Miguel Corporation, Gokongwei Group, Landoil Resources Corporation and Ayala International (Phils.) Incorporated.

The geographical distribution of foreign investment is concentrated in the developing countries of Asia and Australasia. However, some foreign investments are located in Europe, the USA, Australia and Japan. The Middle East countries feature as particularly significant for Philippine firms in the construction sector. Most of these foreign affiliates are predominantly in the same sector of activity as the parent MNE although there is some evidence for vertically integrated foreign investment. In contrast, there is limited evidence for horizontally diversified foreign investment. The only exception for horizontally diversified foreign investments were those of affiliates of Ayala International (Phils.) which has ventured into sectors such as construction, real estate and finance which is outside the scope of the principal activity of the parent firm, i.e. export and distribution of food products, steel products and general merchandise.[1]

Responses were obtained from twenty Philippine MNEs. The industrial distribution of this sample by number of firms is as follows: banking and finance (50.0 per cent), construction (25.0 per cent), manufacturing (15.0 per cent) and others (10 per cent). The survey was conducted in two stages. The first stage consisted of sending questionnaires to the chief operating officers of the Philippine parent firms. The questionnaire sought to find out basic company information, the firms' perceived competitive assets, the reasons for their foreign operations, the mode of their foreign operations, their reasons for internalisation, their major criteria in selecting the location of their foreign operations and other relevant information.

The second stage of the survey consisted of personal interviews with company executives involved in the international operations of the firm to clarify the responses given in the questionnaire and to gather analytical information. Third-party research was undertaken on occasions where the information could not be gathered through the first two stages. (This was the case with San Miguel Corporation, the largest Philippine parent firm in manufacturing which was sequestered and subjected to government investigation during the survey.) The next section of this chapter discusses the value of Philippine direct investment, comparing the wide discrepancies between official values recorded in the books of the Central Bank of the Philippines and values recorded in the official foreign investment statistics of the different host countries. Subsequent sections analyse the industrial and geographical distribution of Philippine direct investment abroad in the context of a stages approach to Third World outward investment which is determined largely by the pattern of domestic industrial development of home countries and the emergence of ownership advantages of local firms. The survey encompasses the primary, secondary and construction sectors whose growth may be explained by the application of the concept of technological competence.[2]

The trend in the sectoral distribution of Philippine MNEs is seen as generally following the developmental path of some of the more traditional investor countries as well as the newer investor countries from the developed countries and the more industrialised developing countries. Generally these outward investments are predominant in the primary sector at an early stage, and then lead to simple manufacturing activities and service activities at a later stage as their capacity for technological accumulation expands and their experience in international production broadens.

The central questions addressed in this and subsequent chapters conform to the two interrelated issues raised in this book. We first address the issue pertaining to the industrial distribution of Philippine MNEs as they relate to the nature of ownership advantages of Third World multinationals as described by Wells' product cycle model and Lall's theory of localised technological change. Next we examine the second issue pertaining to the geographical scope of Philippine MNEs as a result of increasing complexity in the sectoral distribution of their outward investment as their capacity for technological change expands and their multinational investment experience broadens.

THE VALUE OF PHILIPPINE DIRECT INVESTMENT ABROAD

Measuring the actual value of direct investment abroad by Philippine firms is a Herculean task. The complexity arises mainly because the only available official statistics on outward investment are those officially approved and recorded by the Central Bank of the Philippines. Such official data refer to annual outflows of investments abroad as from 1980 when the Central Bank started collecting data on a systematic basis. In principle, the Central Bank required that all existing investments abroad of resident corporations, entities or individuals, including those not previously approved, had to be registered with its Management of External Debt Department and Foreign Exchange Operations and Investments Department. Tables 8.3 and 8.4 present the Central Bank's actual outwardly remitted investments by sector and country respectively for the period 1980 to 1988.[3]

The data provided by the Central Bank of the Philippines reflect only a fraction of the total amount of approved outflows of foreign investments from the Philippines. Overall, total accumulated flows of actual outwardly remitted investments during the period 1980–8 amounted to US $77.51 million. The official data show that the banking and finance sector is the most important for Philippine foreign direct investments in such countries as the USA and Hong Kong. Total outflows of US $42.63 million were directed to this sector in the period 1980–4 alone, representing a share of 64.5 per cent of the total flows of actual outwardly remitted investment during the period. A large proportion of outward remittances during the early 1980s were channelled to financial institutions established abroad, mostly by government or financial institutions closely associated with government authorities.

The manufacturing sector is second in importance with outward remitted investments amounting to US $11.03 million by 1988, or a share of 14 per cent of the total flows of outward remitted investments. The data suggest that US $9.98 million was outwardly remitted in the textile sector in Panama in 1984. Such investment may be presumed to have been held in trust and not channelled to direct investment in the manufacturing sector. The services sector, other than banking and finance, is the least important for foreign investments according to the statistics provided by the Central Bank. Flows of outward remitted foreign investment to this sector amounted to only US $1.02 million during the 1980s. The early stage of the outward investment of Philippine MNEs and the initial concentration on resource-based investment by these

Table 8.3 Central Bank outwardly remitted investment flows by industry, 1980–8, US $ million[1]

Industry Group	1980	1981	1983	1984	1987	1988
Banks and other Financial institutions	5.72	4.74	6.56	25.61		1.41
Banks	4.72	4.74	6.56	25.61		
Other financial institutions	1.00	NA	NA	NA		
Manufacturing	NA	1.05	NA	9.98		
Textiles	NA	NA	NA	9.98		
Food	NA	1.00	NA	NA		
Furniture and fixtures		Not available				
Machinery, except electrical	NA	0.05	NA	NA		
Paper and paper products		Not available				
Rubber		Not available				
Others		Not available				
Public utility	NA	0.01	NA	NA		
Water transport	NA	0.01	NA	NA		
Communications		Not available				
Commerce	1.00	NA	NA	NA		
Wholesale	1.00	NA	NA	NA		
Real Estate		Not available				
Others		Not available				
Services	NA	NA	0.01	NA		
Business		Not available				
Personal	NA	NA	0.01	NA		
Construction		Not available				
Building		Not available				
Transportation facilities		Not available				
Infrastructure projects		Not available				
Agriculture, fishery and forestry		Not available				
Livestock and poultry		Not available				
Others	0.09	NA	NA	NA		4.63
Total	6.81	5.80	6.57	35.59	16.70	6.04

Source: Central Bank of the Philippines, Management of External Debt Department and Foreign Exchange Operations and Investments Department.
Note: 1 No data available for 1982, 1985 and 1986.

Table 8.4 Central Bank outwardly remitted investment flows by country, 1980–8, US $ million[1]

Country	1980	1981	1983	1984	1987	1988
Hong Kong	6.81	4.74	1.52	2.32	NA	NA
Brunei		Not available		1.35		4.63
USA	NA	1.00	5.00	1.75		1.41
Singapore	NA	0.05	0.05		Not available	
Saudi Arabia	NA	0.01	NA	NA	NA	NA
Libya		Not available		19.19	NA	NA
Panama			NA	9.98	NA	NA
Bahamas			NA	1.00	NA	NA
Bermuda						0.005
Total	6.81	5.80	6.57	35.59	16.70	6.04

Source: Central Bank of the Philippines, Management of External Debt Department and Foreign Exchange Operations and Investments Department.
Note: 1 No data available for 1982, 1985 and 1986

firms are therefore not reflected in the officially reported data. The author's estimate of resource-based investments by Philippine MNEs, published in the Dunning and Cantwell (1987) volume of statistics on international investment, is that the share of such investments in total Philippine outward FDI in 1981 was approximately 19.2 per cent.

The geographical distribution of Central Bank-approved outward investment shows that Libya is the most important destination for Philippine foreign investment with flows of remitted investment amounting to at least US $19.2 million during the period 1980–8, followed by Hong Kong with at least US $15.4 million. Panama and the USA are the next most important countries with US $10 million and US $9.2 million during the period 1980–8. Finally, Brunei is also a significant destination for Philippine outward investment with accumulated flows of US $6 million during the period.

Such official statistics on outflows of foreign investment suffer from a gross underestimation. Policy restraints on foreign exchange mean that a large proportion of outward investments is unofficially remitted and therefore not officially approved and recorded with the Central Bank. Any realistic quantitative assessment of the value of foreign direct investment undertaken by Philippine firms abroad has to rely on foreign direct investment statistics recorded by host countries. Table 8.5 presents the estimated stock of foreign direct investment by Philippine firms in several host countries for which data were available. The author estimates that at least US $315.7 million were invested by Philippine firms as at 1988.

THE OWNERSHIP ADVANTAGES OF PHILIPPINE MNES

The nature of ownership advantages of Third World MNEs has been examined in two competing theories in Chapter 4. The product cycle model formulated

Table 8.5 Estimated stock of Philippine direct investment in selected host countries, 1988, US $ million[1]

Country	Investment
USA	140.0
China	70.6
Malaysia	39.0
Hong Kong	34.0
Singapore	18.3
Indonesia	12.9
Thailand	0.6
Republic of Korea	0.2
Sri Lanka	0.08[2]
Total	315.68

Sources: Bank of Indonesia, *Report of Financial Year 1987/88*; Government of the Republic of Singapore, Department of Statistics, unpublished data and Economic Development Board, *Yearbook*, and *Annual Report*, 1987/1988, 1988/1989; Korean Ministry of Finance, International Bureau, Overseas Investment Division, *The Status of Inward Foreign Investment*, 1988; Industry Department, Government of Territory of Hong Kong, *Report on the Survey of Overseas Investment in Hong Kong's Manufacturing Industries 1989*; UNCTC (1992) from primary data obtained from US Department of Commerce, *Survey of Current Business*, August 1990; China Ministry of Foreign Economic Relations and Trade, unpublished data; Bank of Thailand, unpublished data; Malaysian Department of Statistics, *Report on the Financial Survey of Limited Companies*, various issues; Sri Lankan Foreign Investment Advisory Committee and Greater Colombo Economic Commission, unpublished data.

Notes: 1 The data on China represent cumulative approved inflows since 1979. The data on Malaysia has been estimated based on the share of the Philippines in total approved paid-up capital in the manufacturing sector only. The data on Hong Kong refer to fixed assets and working capital in the manufacturing sector only. The data on Singapore refer to the amount of paid-up capital contributed by foreign investors and the amount of reserves attributable to these foreign investors, including a small amount of portfolio investment (less than 10 per cent). The data on Indonesia represent cumulative approved inflows from June 1967 to end December 1988, taking into account the cancellations and shifting of projects from foreign to domestic investment. The data on Thailand refer to cumulative net inflows of foreign direct investment since 1970. This consists of foreign equity investments and intercompany loans. The data on Korea refer to cumulative net inflows of foreign direct investment since 1962. This consists of equity and reinvested profits. The data on Sri Lanka are based on cumulative inflows of foreign investment approved by the Foreign Investment Advisory Committee since 1977 and amounts contracted by the Greater Colombo Economic Commission since 1978.

2 Represents 1987.

by Vernon and applied to Third World multinationals by Wells suggests that the competitive advantages of firms from developing countries are predicated to derive from the peculiar nature of their home markets. Firms from developing countries are taken to follow the technological course of developed countries through the imitation and adaptation of foreign technology in accordance with the requirements of Third World markets and production conditions. As a result of such imitation and adaptation of imported technology, the advantages of Third World MNEs derive from the following: first, predominance in industries characterised by low expenditure on R&D and low product differentiation; second, development of small-scale, labour-intensive processes and products and a higher tendency to source inputs locally; third, development of technology for manufacturing in small volumes; and fourth, greater flexibility.

Perhaps the major limitation of Wells' product cycle model is the narrower scope for innovation by Third World MNEs whose technological course is limited to the imitation and adaptation of foreign technology in accordance with the requirements of Third World markets and production conditions (Wells, 1986). The model fails to consider when firms from developing countries follow an independent or localised technological trajectory or evolutionary pattern in which foreign technology is integrated with their indigenously generated technology. The theory of localised technological change advanced by Lall shows that the advantages of Third World MNEs derive from their ability to innovate on essentially different lines from those of the more advanced countries, i.e. innovations that are based on lower levels of research, size, technological experience and skills and that achieve improvements by modernising an older technique, including foreign outdated technology. The capability of a firm from a developing country to innovate for a unique proprietary advantage which does not merely represent an imitation or adaptation of foreign technology conforms to Nelson and Winter's (1977, 1982) evolutionary theory of technical change which explains that technical change is localised at the firm level and is irreversible. Similar views on the localisation of technical progress where firms comprehend only a limited range of technologies and hence any shifts in the production function are located around known techniques are shared by Atkinson and Stiglitz (1969) and Stiglitz (1987).

A further analysis of the important country-, industry- and firm-specific factors that generate and sustain the particular ownership advantages of Philippine firms in certain sectors is needed in order to distinguish the nature and extent of the international economic involvement of these firms *vis-à-vis* firms of other nationalities.

Resource-based investments

The outward FDI of Philippine firms in the primary sector can be related to the resource-based investments undertaken by foreign-based MNEs such as those of the USA in the Philippines during the colonial and neocolonial period as well as the more recent Japanese investments in the Philippines in the early 1970s. The last chapter showed that these resource-based investments are linked to the vertically integrated export of raw materials to assist expansion of the industrial sector of the home country. Such investments in the Philippines acquired much more significance for Japan than for the USA *inter alia* due to the fact that the main thrust of US direct investment, especially in the period prior to 1914, was market oriented as the USA was rich in most natural resources and also because of certain locational disadvantages of the Philippines such as geographical distance. On the other hand, resource-based investments by Japanese investors in neighbouring Asian countries were crucial to Japanese industrialisation. The Ricardian trap of industrialisation brought about by the shortage of natural resources in the home country explains the major motive for resource-based investment of resource-poor industrial countries (or Ricardian economies), such as Japan, in resource-rich countries (Ozawa, 1982). The resource-based investments of Japanese MNEs in neighbouring Asian countries can be compared to the outward FDI by resource-poor newly industrialising countries (NICs) as well as other developing countries that face a Ricardian trap of industrialisation as a result of shortage of vital natural resources needed for industrial expansion in the home country.

The forward integration of American and Japanese investors into the secondary processing of minerals, timber and wood products and other primary products as a result of Philippine foreign investment policy, designed to stimulate former export-oriented investors to undertake downstream processing activities and to have greater integration with local firms through tax *et al.* incentives, has exerted a developmental influence on some Philippine firms. These firms have gradually developed unique ownership advantages, specifically, the indigenous capacity for technological innovation enhanced through foreign technology, which was later exploited in foreign markets. These innovative activities are manifested initially in the form of backward vertical integration into the acquisition of essential raw materials and at a later stage in forward vertical integration into the secondary processing of raw materials in host countries.

The resource-based investments undertaken by Philippine firms and specifically the backward vertical integration into the acquisition of raw materials such as timber and oil were needed not so much to supplement inadequate local resources and establish reliable sources of raw materials for the home country as to gain advantages over competitors through the control of supply of inputs, product or production strategy and access to markets through the use of their ownership-specific competitive assets. Several Philippine firms in the timber industry, undoubtedly spurred by deforestation in the Philippines, have made fuller use of their access to large timber resources in Indonesia and, in addition, have opened new markets other than that in the Philippines and exported to such third countries as Japan, Korea and Taiwan. Some of these raw materials were also used in the manufacture of end-products to which the Philippines is particularly suited. The internalisation of activities, owing to the presence of raw materials in Indonesia and the efficiency of the Philippines as a location for extraction and downstream processing, enabled Philippine timber firms to take advantage of the gains associated with vertical integration.

The availability of abundant timber resources in the Philippines and the accumulated experience acquired in exploitation *inter alia* as a result of the developmental impact of foreign MNEs have been important country-specific factors that have favoured the generation of some ownership-specific advantages for Philippine firms engaged in the agricultural and forestry sector of Indonesia. Access to knowledge about natural resources is an important ownership-specific advantage of these firms, which have clearly specified that their most important competitive assets are their familiarity with working in virgin tropical forests and in conditions where there may be difficulty in the utilisation of heavy equipment. Another principal competitive asset of these firms is their access to trained labour. Hence, in the purely resource-based investments of Philippine MNEs, the evidence generally supports the product cycle model in which the capacity of Philippine firms to accumulate ownership advantages derives from their acquisition of production experience, in part through the imitation and adaptation of foreign technology suitably modified to take account of Third World markets and production conditions.

Table 8.6 shows that about one-third of the total stock of Philippine investment in Indonesia, representing US $20.3 million, was directed to the agricultural and forestry sector by 1982. An indirect form of government intervention in the host country, as exemplified in the open-arms policy of the Indonesian government towards foreign investment and the ban on the export of logs since 1980, has favourably influenced the creation of distinct ownership-specific advantages which affect and enhance other ownership-specific endowments of Philippine firms. One of the first companies to take advantage of the opportunities for foreign investment in Indonesia was Gonzalo Puyat and Sons, which formed a joint venture with the Indonesian firm D. V. Djatikenbang. With an initial capitalisation of US $4.0 million, the main purpose of the joint venture was to exploit 370,000 acres (150,000 hectares) of woodland in East Kalimantan (Lim, 1984). A further eight Philippine and American firms based in the Philippines entered into contracts with the Indonesian government in later years.

By contrast, the backward vertical integration into petroleum exploration was prompted by the lack of petroleum resources in the Philippines. The government has overseas investments in energy-intensive projects under the ASEAN Industrial Complementation Programme. In the private sector, firms such as the Landoil Group have entered into agreements with several members of the Organisation of Petroleum Exporting Countries (OPEC) in order to engage in the exploration for oil. The parent company, Landoil Resources Corporation (LORC), and its affiliate, Basic Petroleum and Minerals (BASIC), initiated oil exploration activities in the Philippines and internationalised their operations in the Middle East in 1981. The firm was the first oil and energy group in Southeast Asia to acquire oil and gas concessions in the Middle East. In the early 1980s, the firm was named operator of the Arab–Canadian–Australian consortium that has concession rights over the entire offshore and onshore areas of Ajman, United Arab Emirates. Gulf Consolidated Company for Services and Industries EC (GCSI), a Bahrain-

Table 8.6 Industrial distribution of Philippine direct investment in selected host countries, 1980–2, US $ million[1]

Industry	Indonesia (1982)	Hong Kong (1982)	USA (1980)	Malaysia (1980)	Thailand (1981)
Agriculture and forestry	20.3	Not available		Not available	Not available
Mining	NA		Not available		0.5
Manufacturing	35.7	8.4		0.06	NA
Food and beverages	Not separately available	6.9		0.04	0.1
Textiles and leather		0.1	available	0.01	0.2
Non-metal		NA		NA	0.2
Electrical and electronic		0.2		neg.	NA
Other manufacturing		1.2		NA	NA
Construction	0.8	Not available		Not available	NA
Real estate	0.5				NA
Other services	3.0				0.5
Banking	NA		36.0		NA
Total	60.3	8.4	73.0	0.06	1.0

Sources: Bank of Indonesia, *Report of Financial Year 1981/82* and *Indonesian Financial Statistics*; Industry Department, Government of Territory of Hong Kong; US Department of Commerce, *Survey of Current Business*, August 1985; Malaysian Industrial Development Authority; and Board of Investment, Thailand.

Note: 1 The data on Indonesia refer to equity investment and exclude investments in the petroleum sector.
The data on Hong Kong refer to fixed assets and working capital in the manufacturing sector only.
The data on Malaysia refer to paid-up capital in the manufacturing sector only.
The data on Thailand refer to paid-up capital.
The total data on the USA include unspecified sectors.

based conglomerate and an associate company of the firm, is financing the costs of the Ajman drilling operations.

Apart from access to finance, associations of Philippine and foreign firms are formed to gain access to foreign technology. For example, the Philippine conglomerate Landoil Resources Corporation obtained an equity interest in the Canadian energy company Charriot Resource in order to supplement the firm's limited technological expertise in petroleum exploration. Similarly, the 80 per cent Philippine-owned joint venture P. T. Taliaban Timber, formed between the Philippine firm R. J. Gulanes and Company and an Indonesian firm in July 1969, entered into an agreement with Mitsubishi Shoji Koisha Limited of Japan. The agreement stipulated that Mitsubishi Shoji Koisha would provide machinery, technology and finance to supplement and/or complement the production machinery, materials and labour coming from the Philippines.

Manufacturing sector investments

At a later stage, the relative importance of purely resource-based, extractive investment declines while that of forward vertically integrated manufacturing investment which represents the secondary processing of natural resources increases. The latter type of investment is prompted by the need to internalise new firm-specific ownership advantages that have emerged from an initial advantage based on the knowledge of natural resources, their access and the best way in which they can be used. The increasing capacity for localised technological innovation first becomes apparent at this stage.

Both the increasing technological advantages of Philippine firms and the 1980 Indonesian host government policy to prevent the export of logs have contributed to the shift in emphasis of Philippine outward investment from purely extractive investments towards the establishment of wood-processing plant in Indonesia. In 1970, ten Philippine firms with a total investment of US $275.5 million operated fifteen logging projects which aimed to exploit some 2.2 million hectares of Indonesian forest land. This made the Philippines the third largest logging concessionaire in Indonesia at that time. The concentration on purely extractive investments continued until the late 1970s but by the 1980s the processing of extracted timber in the host country had become an equally important production activity. By 1987, the wood-processing industry and construction constituted the most important sectors of investment in Indonesia for Philippine firms (see Table 8.7).

Such forward vertical expansion into secondary processing in host countries is also reflected in countries other than Indonesia. In 1978, a wood-based industrial complex was built in Tawan, Sabah, Malaysia through a joint venture agreement between the Sabah Foundation, Santa Ines Medale Forest Products Corporation and the Construction and Development Corporation of the Philippines (CDCP).[4] The joint venture had a capitalisation of US $12.0 million with the Sabah Foundation holding 51 per cent equity interest and

Table 8.7   Industrial distribution of Philippine direct investment in selected host countries, 1984–90, US $ million

Industry	Bangladesh (1987)[1]	Indonesia (1987)[2]	Thailand (1984)[3]	USA (1990)
Primary sector				−1.0
Petroleum				−1.0
Manufacturing sector		9.9	0.5	11.0[5]
Food, beverages and tobacco		2.3	0.04	
Textiles		0.5	0.007	
Wood and wood products		4.3		
Metal products		2.8	0.45[4]	
Plastic, PVC and rubber products	0.43			
Electrical products			0.004	
Mechanical equipment				1.0
Services sector		5.1	0.09	51.0
Construction		4.0		
Wholesale trade				8.0
Real estate				−1.0
Banking and finance		1.1	0.09	44.0
Other services				
Unclassified				16.06[6]
Total	0.43	15.0	0.59	77.06

Sources: Directorate of Industries, Government of Bangladesh; Capital Investment Coordinating Board, Indonesia (Badar Koordinasi Penanaman Modal); Bank of Thailand; and US Department of Commerce.

Notes: 1   Refers to sanctioned investment with multinational enterprises since 1974–5.
2   Refers to approved investments from 1967 until end March 1987.
3   Refers to cumulative net foreign direct investment inflow by source from 1978.
4   Includes non-metallic mineral products.
5   Data on manufacturing sub-sectors are suppressed in the original data source to avoid disclosure of data of individual companies. These sectors are *inter alia* food and kindred products, and chemicals and allied products. In the latter sector, however, less than $500,000 has been invested.
6   Includes insurance and other primary sectors.

Santa Ines and CDCP holding 24.5 per cent each. The joint venture agreement stipulated that the Sabah Foundation would ensure the supply of logs for the wood-based industrial complex from the 181,622 hectare Rana forest, while Santa Ines and CDCP would provide the necessary technological, managerial, marketing and financial requirements to ensure the efficient operation and management of the wood-based industrial complex.

Although Philippine firms engage in purely resource-based extractive investments in resource-rich, neighbouring developing countries in the early stage, domestic activity in the host countries is upgraded in the later stages, largely as a result of the gradual process of technological accumulation from mere production experience towards the capacity to generate localised technological innovation. Downstream processing activities in the host countries assume greater importance over time, resulting in a shift in the industrial and geographical pattern of investment from an initial concentration on the purely resource-based kind of investment in neighbouring developing countries or developing countries that have an abundance of the required natural resources. Such upgrading of the domestic activity of Philippine firms in host countries is also enhanced by the foreign investment policy of the host country government which encourages former export-oriented investors to undertake downstream processing activities and to have greater integration with local firms through tax *et al.* incentives. In the case of investments in the timber sector in Indonesia in particular, the 1980 host government policy to prevent the export of logs undoubtedly provided the spur and dictated the timing for the actual shift in the sectoral pattern of investment by Philippine timber firms.

International production activities by Philippine firms in other manufacturing sectors also demonstrate their localised technological capacity for innovation. In general, however, Philippine direct investments in the manufacturing sector have not been as large as those of other developing countries, partly because very few firms in the manufacturing sector, apart from large conglomerate firms such as the San Miguel Corporation and the Gokongwei Group, have developed the technological and organisational capacity required for effective competition in foreign markets. The motives for internationalisation of these conglomerate MNEs seem to conform to the dynamic view of the growth of production and the distinctive nature of the process of technological accumulation. These MNEs have engaged in international production primarily to exploit indigenously created ownership advantages within the firm which can be utilised in a foreign country as part of the firm's growth strategy.

In the food and beverage sector, the San Miguel Corporation, in particular, stressed the need to protect its property rights (trademark) by ensuring the product quality of its world-famous San Miguel beer, as the firm's most important motive for undertaking the establishment of breweries in several foreign markets. The desire to control the product and the associated production strategy to gain profits as well as to control access to certain export markets have also been mentioned as secondary factors in a firm's drive towards internationalisation. Similar reasons have compelled the Gokongwei Group, San Miguel's direct competitor, to invest in ASEAN markets.

Tables 8.6 and 8.7 show that the manufacturing sector accounts for a significant proportion of Philippine direct investments in Indonesia, Hong Kong, Malaysia and Thailand. Total manufacturing investments in these four countries amounted to at least US $45 million in the 1980–2 period. In Indonesia in particular, the manufacturing sector accounted for almost three-fifths of Philippine direct investments in 1982, and by 1987 the share of the sector had increased further to two-thirds. On the other hand, half of Philippine direct investment in Thailand in 1981 was channelled into the manufacturing sector, particularly the non-metal and electrical and electronic industries. By 1984, the manufacturing sector accounted for almost 85 per cent of the total amount of Philippine outward investment in Thailand, primarily in the metal products sector. Almost all Philippine direct investments in Hong Kong and Malaysia are involved in the food and beverage and the textile and leather sectors. However, small amounts of direct investment are also present in the electrical and electronic and other manufacturing sectors, especially in Hong Kong. The previous chapter showed that these sectors may be those in which inward FDI exerted a favourable influence on the indigenous technological capability of Philippine firms.

Philippine manufacturing MNEs involved in the production of textile and leather products, non-metallic mineral products, and electrical and electronic goods indicated that the particular kind of technology involved in small-scale production makes the contractual alternative less feasible when compared to the kind of large-scale technology provided by MNEs from the developed countries. It is certainly true in the case of Philippine manufacturing firms that the technology involved is disembodied and cannot be codified and sold explicitly. Often the competitive assets of these firms lie in their managers who are trained in labour-intensive means of production and have the maximum flexibility and adaptation. The methods used in exploiting machinery are not codifiable and therefore not easily transferable between independent firms. Production experience as a result of learning by doing and learning by using creates difficulties in the accurate assessment of time and costs required in contractual arrangements. Moreover, where a less developed country has no indigenous capital goods production, the technology exports may be based upon elementary learning which is best exported in the form of direct investment (Lall, 1982a).

In sum, Philippine direct investment in the manufacturing sector in neighbouring Asian countries in the early 1980s was concentrated in the food and beverage, electrical and electronic, and non-metallic mineral product sectors. By the late 1980s, wood products had gained importance as a sector of Philippine outward investment, followed by metal products and food, beverages and tobacco. The industrial pattern of Philippine foreign investment in the manufacturing sector does not seem to conform wholly to the product cycle generalisation of Lecraw (1977, 1981) and Wells (1983, 1986) about Third World multinationals, which indicates that the industrial pattern and ownership advantages of MNEs from developing countries result from the imitation and adaptation of foreign technology leading to mature and relatively simple, labour-intensive, small-scale technologies with limited capacity for product differentiation. The case of the famous Philippine brewer and leading manufacturer of consumer and producer goods, the San Miguel Corporation, is an excellent demonstration of the wider technological capacities of Philippine firms to engage in localised technological innovation, as envisaged by Lall. This firm has developed a brand image in the domestic market which was transferred successfully to other countries through the establishment of a chain of breweries in Hong Kong, Guam, Papua New Guinea, Indonesia and Spain, which were previously export markets. Marketing networks were also intensified with the establishment of distribution agencies in Australia, Canada, Malaysia and Singapore in order to promote beer exports. The main thrust of the firm's foreign operations is to protect and expand export markets by promoting sales of the internationally known San Miguel beer. The acceptance of advertising and other persuasive marketing methods, as well as the continued adherence to quality standards in the domestic market concomitant with the development of consumer tastes and culture, are country-specific factors that have enabled the San Miguel Corporation to produce a differentiated consumer product. To date, the firm has engaged in the production of more than five different brands, and also product packaging, feed and livestock, dairy products manufacture, coconut oil milling and refining, product exportation, investments and fast-food franchising, as well as various other agribusiness projects. Similarly, the Gokongwei Group, a direct competitor of San Miguel involved primarily in food products, has successfully invested in other ASEAN countries since the 1970s on the basis of a strong indigenous capacity for technological innovation. The firm has similarly competed on the basis of high product quality standards and product differentiation.

Construction sector investments

The gradual shift in the structure of the Philippine economy from an initial reliance on the primary sector towards the development of the manufacturing sector has led to the increased significance of other forms of service investments in response to the development of new manufacturing activities. In particular, the increasing importance of the manufacturing sector in the Philippine economy prompted the establishment of the construction sector.

Perhaps one important factor in the rapid growth of overseas construction investments by Philippine MNEs in particular and Third World MNEs in general is the fast growth of construction activities in their home countries to support the growth of the domestic industrial sector. Construction experience allied to technological advantages accumulated by these firms in their domestic construction activities, enhanced by the presence of foreign construction firms, was exploited in foreign markets at a later stage. The major drive behind the international operations of a majority of Philippine construction firms is to expand the narrow home market relative to the firm's production/service capacity through the opening of new construction opportunities in foreign markets. The compelling need to make a firm's presence felt in the local market has also been an important secondary drive to establish international construction activities. Such local presence is sometimes greatly influenced by host government policy.

The role of direct government intervention is an important home country-specific variable which has generated and sustained the growth of ownership-specific advantages of Philippine construction firms. The vigorous promotion of the export of Filipino manpower as a priority in the government's 1974 labour policy was formalised through the establishment of three government structures: the Overseas Employment Development Board (OEDB), the state-owned recruiting agency for land-based workers; the National Seaman Board (NSB), the government-run placement agency for seamen; and the Bureau of Employment Services (BES), the government bureau that authorises and regulates the operations of privately owned recruiting agencies. The OEDB is vested with the authority to enter into recruitment agreements with foreign government ministries, agencies and entities. However, the Philippine Overseas Construction Board was specifically organised as the official government entity vested with the legal right to export Filipino construction labour. Apart from these government structures, established in 1975, a Middle East Labour Centre was organised by the Ministry of Philippine Labour and Employment (MOLE) in Jeddah, Saudi Arabia, in 1982 to open new markets for Filipino skills and entrepreneurship in nine Middle Eastern countries (briefing paper, MOLE, 11 February 1982).

The main ownership advantages of Philippine construction firms, deriving from their construction experience, are based on their capacity to generate localised technological change in construction, in some cases complemented by foreign technology. These advantages, which are harnessed by the direct support of the Philippine government, conform to Lall's theory of localised technological change which predicates that firms from developing countries have the capacity to engage in genuine and unique innovation.

Apart from their expertise in working in developing countries, the fundamental home country-specific factor that has favourably influenced the propensity of Philippine construction firms in particular and Third World construction firms in general to acquire some degree of competitiveness in foreign markets is their access to a plentiful and relatively low-cost labour supply with a reasonable level of skill in construction and engineering. These factors have provided Philippine construction firms with a competitive edge, at least against competitors from the industrialised countries in developing host countries. In recent years, construction firms from lower-income developing countries such as the Philippines have sustained a competitive cost edge against construction firms based in the NICs whose labour costs have significantly increased.

In addition, the experience of Philippine construction firms in developing countries has enabled them to cope with problems intrinsic to environments similar to their own, and hence introduce appropriate technology. Moreover, their expertise enables these firms to design small-scale plant or simple buildings more suitable for the lower incomes and tropical climates of their host countries. These firms are more inclined to substitute imported for locally available construction materials and therefore create domestic linkages. Finally, their construction experience in developing countries enables Philippine firms to estimate costs more accurately and they are thus more likely to offer the lowest bid, as a result of fewer uncertainties, than local and foreign firms from an advanced country. Philippine construction firms have also pointed out the transaction advantages that arise out of their operations in several Middle Eastern markets. The cyclical nature of construction firms is effectively synchronised when operating subsidiaries are established in several host countries.

The relative importance of the perceived competitive assets of Philippine construction firms *vis-à-vis* local construction firms and construction firms from developed countries is tabulated in Tables 8.8 and 8.9 respectively. These firms were asked to rank several potential competitive assets according to their relative declining order of importance with one being the most important. The ranks achieved by the different competitive assets were then averaged to provide an overall indication of the relative importance of each competitive asset for all the construction firms included in the survey. The major ownership advantages of Philippine firms *vis-à-vis* indigenous construction firms in the host country derive from their access to a large supply of cheaper factors of production such as skilled construction labour and engineering expertise in the home country; superior technical knowledge and management expertise; longer experience in construction in developing countries; and established reputation. Perhaps the major competitive advantage of Philippine firms *vis-à-vis* construction firms from other developing countries such as India, Korea and Latin America is their increasing capacity to provide superior construction and engineering services at lower costs. Korean construction firms, in particular, have indicated that business prospects in the overseas construction industry were not promising in the 1980s, *inter alia* because of their declining competitiveness due to construction firms from developing countries *vis-à-vis* Korea's higher labour costs and aging labour pool (Korean Development Bank, 1985; Mi-Yong, 1990). The comparison between Philippine construction firms and those from other developing countries is further discussed in Chapter 9.

Apart from lower factor costs the major ownership advantages of Philippine firms with respect to contractors from developed countries in the host country are greater flexibility and adaptability in construction in developing countries and more appropriate technology for local host country conditions. In general, however, the main ownership advantages of firms from developing countries in general, and those from the Philippines in particular, derive from their access to lower cost skilled labour. By contrast, the ownership advantages of construction firms from the developed countries derive more from their access to superior engineering, management and financial resources, and their overall strength in overseas markets.

The limited amount of indigenous technology has, to some extent, compelled Philippine firms to source sophisticated technology abroad. Thus, access to foreign technology constitutes an important modality through which Philippine construction firms strengthen their ownership-specific advantages. For example, the Landoil Group has successfully gained access to more

Table 8.8 Perceived competitive assets of Philippine construction firms with respect to indigenous construction firms in the host country

Perceived asset	Mean
Lower factor costs	1.00
Superior technical knowledge and management expertise	1.66
Longer experience in construction in developing countries	2.00
Established reputation	2.00
Access to home government assistance	2.50
Ability to take advantage of different markets owing to operation in several host countries	2.50

Source: Questionnaires.

Table 8.9 Perceived competitive assets of Philippine construction firms with respect to contractors from developed countries in the host country

Perceived asset	Mean
Lower costs for factors of production	1.00
Greater flexibility and adaptability in construction in developing countries	1.50
More appropriate technology for local host country conditions	1.50
Host government policies preferring foreign firms from developing countries to those from developed countries	2.00
Access to home government assistance	2.50

Source: Questionnaires.

sophisticated construction technology through equity participation in Siframe SA, a French construction and engineering company. The Landoil construction group was reported by the Philippine Overseas Construction Board as being the largest overseas contractor with contracts amounting to US $374 million during 1982.

Internalisation theory, as applied to the international construction industry, predicates that certain forms of non-equity investment in which maximum benefits could be achieved in terms of the effective maintenance of supervision over product quality may be favourable. In addition, the risks associated with arms-length transactions in foreign markets could be minimised. For example, in a joint venture arrangement, the pooling of competitive assets and financial resources resulted in higher chances of success in obtaining bids for construction projects. In addition, the instability of exchange rates and interest rates since the early 1970s led most construction firms to reappraise the risk-sharing opportunities offered by their joint ventures or contractual agreements abroad. Empirical research undertaken on the international construction industry has shown that joint ventures and management contracts were the most widely used forms of non-equity foreign investment undertaken by construction firms (Seymour, 1987). A survey conducted in 1981 showed that 98 per cent of the 142 largest international non-American contractors and 91 per cent of the 80 largest US contractors predicted that joint venture arrangements in construction would increase. A similar survey conducted in 1980 revealed significantly lower expectations at 80 per cent and 75 per cent respectively (*Engineering News Record*, 1981).

The main motive for partial internalisation derives from construction firms' need to develop and maintain comparative advantages in tasks (Buckley and Enderwick, 1985). Where the ownership advantages are linked to country-specific factors, the incentive for internalisation is greater since construction firms from other countries may not possess the advantage themselves. For example, while construction firms from the Philippines in particular, and developing countries in general, would choose to internalise their access to a skilled labour force, construction firms from the more developed countries would choose to internalise their access to management, engineering and technical expertise, and finance. As a result, contractors from developing countries would strive for geographical diversification into several sub-markets to maximise the utilisation of their country-specific advantage in skilled labour and aim for technology transfer from contractors from developed countries in order to increase their technological competitiveness. By contrast, contractors from the developed countries would strive to introduce higher forms of technological skills in their product in order to maximise the utilisation of their country-specific advantage which lies in technological expertise (Seymour, 1987).

A majority of Philippine construction MNEs with international construction activities, such as Atlantic Gulf & Pacific Company of Manila (AG & P), Erectors Incorporated, Construction & Development Corporation of the Philippines Incorporated (CDCP), Engineering Equipment Incorporated (EEI), Landoil Resources Corporation and D. M. Consunji, undertake management contract arrangements and joint ventures. However, a few of these firms, such as AG & P, Ayala International (Phils.) and D. M. Consunji also operate wholly owned or at least majority-owned subsidiaries.

Although the formation of joint ventures with local partners is a prerequisite for obtaining a construction project in Middle Eastern countries, many Philippine construction firms have indicated that the assistance provided by the local partner as far as introduction to the regional characteristics of the construction site is concerned has been invaluable. Apart from providing their Philippine partners with valuable regional insights, the increased joint reputation of the partners as a result of pooling their ownership-specific advantages, as well as financial resources, has augmented their chances of successful bids in construction projects. For example, the success of the joint venture established by a Philippine construction firm with an indigenous firm in Brunei has been attributed to the complementarity of assets between the partners.

However, in a number of cases Philippine construction firms have indicated that some of their joint ventures represent merely nominal arrangements in which the Philippine firm provides both the capital and technical expertise, while the local partner only provides the use of its name which is a major firm-specific ownership advantage. The success of these nominal joint ventures in construction contracts in the Middle East is therefore highly dependent on the management and technical expertise of the major Philippine partner. In fact, Philippine construction firms perceive their highly skilled technical personnel and superior management know-how as the major competitive advantages that have induced foreign companies to seek collaborative arrangements.

On the other hand, joint ventures with foreign companies from developed countries have fulfilled the need of Philippine construction firms for advanced construction technology which needs to be integrated with the independent and unique technological advantages of the firm. The Philippine conglomerate MNE Landoil Resources Corporation, which purchased an equity interest in Siframe SA, a French construction and engineering company, in order to gain access to more sophisticated construction technology, is an excellent case in point. The desire to gain access to foreign technology through collaboration with foreign companies from developed countries is the underlying factor that motivates South Korean construction firms to seek joint venture arrangements (Westphal *et al.*, 1984). Some firm-specific evidence of the technological advantages of Philippine construction firms that have enabled them to become successful contractors in foreign markets is given below.

Engineering Equipment Incorporated (EEI), is involved in the international distribution of industrial machinery and equipment, and in providing engineering and construction services. The firm has accumulated technological expertise and achieved domestic and international competitiveness through a combination of indigenous effort and the importation and adaptation of foreign technology. In addition to being the third largest engineering firm in the Philippines, EEI has developed into a major industrial firm which owns and operates the largest and most modern commercial steel foundry and steel fabrication plant in the country.

Aggressive marketing efforts have enabled EEI to obtain high-revenue and often technology-intensive construction contracts both in domestic and international markets. The growth of the overseas construction division was enhanced by the acquisition of contracts for the construction of Petromin-Mobil's fuel refinery, a petrochemical complex in Yanbu, Saudi Arabia; Aramco's microwave and optic fibre communication systems, also in Saudi Arabia; and the Assir power plant for Hitachi Zosen of Japan in 1986. In the mid-1980s, the firm constructed the Mina Abdulla refinery for Kuwait National Petroleum Company. The firm is also pursuing another contract involving boiler erection and piping fabrication works for Azzour power station in Kuwait. EEI's orders and contracts amounted to 1.2 billion Philippine pesos (US $126.5 million) in 1982, which represents a significant increase over the previous year's figure: 868 million Philippine pesos (US $105.9 million).

Despite sluggish market conditions and escalating expenses in the early 1980s, the firm undertook major expansion and diversification projects. Among these were the 32 million Philippine peso (US $3.5 million) grinding ball venture, originally a joint venture with Cozino Rio Tinto of Australia and Midland Ross of the USA. However, because of the difficulties encountered in the registration of the planned 40 per cent foreign equity participation, EEI decided to pursue the project as a 100 per cent Philippine-owned undertaking. But, along with other Philippine construction firms, EEI has been experiencing difficulties in operating in Saudi Arabia because of the pressures of the competitive market environment and the deceleration of new construction opportunities in the region.

Similarly, the Philippine firm Atlantic, Gulf & Pacific Company Limited (AG & P) is considered one of the pioneers in commercial and heavy industrial plant construction in the country. The firm operates an overseas affiliate, AG & P Arabia, in Saudi Arabia and maintains a main overseas office in San Francisco, California. The firm has worked with reputable international contractors in the construction of industrial and petrochemical plant overseas and has undertaken the construction of onshore and offshore petroleum and gas production facilities through subcontracting arrangements in Saudi Arabia, Kuwait, Qatar, Iraq, Algeria, Sudan, The Congo, Sri Lanka, Papua New Guinea, Indonesia, Vietnam, Malaysia and Okinawa.

Another Philippine firm, Erectors Incorporated, was established in 1957 as a company specialising in structural steel fabrication but has since developed into a dynamic construction firm whose main product lines include industrial plant construction, commercial building construction, installations of power-generation transmission lines and, more recently, heavy civil works. It is currently considered one of the largest construction firms operating in the Middle East. The firm initiated overseas construction in 1977 with the acquisition of the Hartha power station construction contract in Basrah, Iraq. The firm's second major overseas contract was the construction of the Philsinports stevedore accommodation-II in Saudi Arabia. The company has completed over eighty construction projects both in the home country and overseas. Among the numerous overseas construction projects it has undertaken are the Baraidah sport complex in Al Gassin, Saudi Arabia; the Habbaniya House, Baghdad Sheraton Hotel and Baghdad Conference Palace in Baghdad, Iraq; and the Nasiriya hospital also in Iraq.

As early as 1973, the Construction & Development Corporation of the Philippines (CDCP) restored the Borobudur temples in Indonesia for US $9.0 million. The firm ranked fourth largest among Philippine MNEs in terms of total assets and gross revenues as early as 1982 and is still considered the largest construction firm in the industry with a proven track record in obtaining impressive contracts in highly competitive markets, both local and foreign. The firm's overseas projects include the construction of roads and drainage systems in Saudi Arabia. In 1982 alone, the company had 9 billion Philippine pesos (US $981.4 million) in construction contracts, of which more than 70 per cent was accounted for by overseas projects.

D.M. Consunji Incorporated is another of the leading firms in the construction industry. Its spectacular success and growth in the domestic construction market is the result of its high level of technological capability and skilled personnel, which are drawn from the cream of the country's managerial and technical talents. The firm capitalised on such assets in their foreign ventures and, as of 1986, had established a branch in Brunei and joint ventures in Iraq and Saudi Arabia.

Another Philippine-based MNE, Ayala International (Phils.) Incorporated, undertook the largest single construction project ever undertaken by an ASEAN firm. Ayala was responsible for the construction of Brunei's US $350-million national palace, *Istana*, which is said to be the largest edifice in the world. In addition, the firm has developed a 1,480-acre (600-hectare) residential complex in Klang, on the outskirts of Kuala Lumpur in Malaysia, and was responsible for the construction of the Centrepoint hotel-bus-terminal-commercial complex in Kota Kinabalu, Sabah.

Philippine construction firms have built strong technological advantages, not just through the simple scaling down or imitation of mature technologies acquired from the industrialised countries, as predicted by Wells' and Lecraw's product cycle model, but through efforts in technological change which have in turn enhanced the development of their indigenous technological advantages. In some cases efforts in technological change are manifested in terms of acquiring advanced technology through joint ventures. The ownership-specific advantage in skilled labour-intensive industries is also seen in non-construction service industries. Philippine-Singapore Ports Corporation, a joint venture between Filipino and Singapore partners and a subsidiary of the Landoil Group, is involved in the management of ports, while another associated company, the Philippine Hospital and Health Services, provides skilled personnel such as doctors, nurses, dentists and medical technicians for hospitals in the Middle East and North America. Landoil is also associated with International Caterer Incorporated which provides food and related services to Filipino workers in the Middle East.

THE GEOGRAPHICAL DEVELOPMENT OF PHILIPPINE MNES

The shift in the sectoral pattern of Philippine outward direct investment has developmental implications for the geographical pattern of this investment. Such industrial and geographical shifts may be explained in the context of the process of technological accumulation. For example, resource-based extractive investments were initially of particular importance in resource-rich, neighbouring countries with the closest psychic distance. However, the increased sophistication of investments from purely resource-based investments towards downstream processing of extracted natural resources in the host country in the first instance, followed by more complex kinds of international production at later stages, may increase the geographical scope of outward direct investment towards more sophisticated markets in other countries with increasingly further psychic distance. This section analyses the geographical shift in the outward direct investment of Philippine MNEs as a result of the emerging industrialisation of the Philippine economy which had developmental repercussions on the sectoral distribution of outward investments.

Investments in neighbouring and/or ethnically related territories

The first and generally the most important investments of new international investors are those directed towards neighbouring and/or ethnically related territories. Ethnic ties have been an essential factor in decisions involving foreign production for a number of firms, especially those of Asian origin (Wells, 1983). For example, the number of Chinese and Indian communities in foreign countries has encouraged exports that have led to investments at a subsequent stage or, in some cases, has led directly to investments without previous exports. This phenomenon has occurred even when specific ethnic products were not involved.

The cost of acquiring reliable information about foreign markets is believed to be considerably lower when ethnic ties are present in a potential foreign market because of the important element of trust. Apart from providing reliable market information which identifies foreign business opportunities, however, ethnic ties have exercised a significant influence in placing constraints on the ability of firms to manage scattered units of an enterprise. Most Asian firms, for example, have been small, family owned and family managed. The role of ethnic ties and, specifically, the role of family ties is important not only in identifying opportunities for trade and investment but in the operation and maintenance of international business activities. However, the evidence suggests that as firms gain experience in particular countries, the importance of ethnic factors declines. Therefore the reliance on ethnic contacts for initial foreign ventures would tend to diminish over time. But, for the smaller and more recent Third World investor firms, ethnic contacts continue to be of fundamental importance as regards the extent and form of their initial foreign involvement.

The data in Table 8.3 reveal that almost 30 per cent of Philippine foreign direct investment during the period 1980–8 was undertaken in neighbouring ASEAN member countries as well as Hong Kong. The estimated stock of Philippine FDI in Malaysia, Thailand and Indonesia stood at US $61.4 million at least in the period 1980–2 according to official data sources. Considerable direct investment by Philippine firms has also been found in other Asian countries such as China and Hong Kong. According to host country data shown in Table 8.5, Philippine direct investment in these countries was estimated to be in the region of US $105 million in 1988, representing a third of the estimated stock of Philippine FDI during that period. The data contained in Table 8.5 lead to the finding that at least US $176 million of Philippine foreign direct investment was located within the Asian region in 1988 which constitutes about 56 per cent of the estimated stock of Philippine FDI during the period. Empirical research suggests that direct investment by Philippine firms is also located in other Asian countries such as Japan and Taiwan. However, the inadequacy of published statistics and the confidentiality attached to quantitative information with regard to the value of foreign investment does not allow for any quantitative measurement of the level of FDI in these countries.

Interviews with some Philippine parent companies reveal that at least a part of locally held equity in the Philippine subsidiaries abroad is owned by parties of the same ethnic group. In some cases, the ethnic group can own more than 90 per cent of the local equity. Such an internationalisation process is described in Johanson's and Vahlne's (1977) model of knowledge development and increasing foreign commitments. The internationalisation process of the firm is regarded as following a stages approach in which the initial foreign investments would be directed to those markets with the greatest linkage in terms of psychic distance. The lack of market knowledge due to differences between countries with regard to language and culture, for example, is considered an important obstacle for decision making connected with the development of international operations. The model focuses on the impact of the process of gradual acquisition, integration and use of knowledge about foreign markets and operations on a firm's incrementally increasing commitments to foreign markets over time. The Johanson and Vahlnes model can therefore be considered a firm-level analogy to the product cycle model in which internationalisation is regarded as a process of incremental adjustments to the changing conditions of the firm and the environment.

A majority of the investments in neighbouring, ethnically related or other developing countries by Philippine MNEs was resource based, prompted not so much by considerations pertaining to psychic but to geographical distance in broad similarity to the concentration of Japanese resource-based investments in the early 1970s in Asian developing countries where natural resources were present in abundance. By contrast, resource-based investments by MNEs from the UK and the USA prior to the Second World War were undertaken in colonial territories in Africa and Latin America respectively.

The search for the raw material, timber, for the Philippine wood-processing industry was an important motive for the re-location of logging and wood-processing firms in neighbouring countries and particularly in Indonesia in the late 1960s. The processed timber was then exported to Japan, Korea and Taiwan. The influx of Philippine firms in the Indonesian timber industry was a result both of push factors associated with the home country and pull factors associated with the host country. The push factor was the devastating impact of deforestation on the expansion of the thriving Philippine wood-processing industry. The pull factor in the host country was the enticement offered by the rich and tempting timber resources in Indonesia and the open-arms foreign investment policy of the post-1965 Indonesian government towards foreign investment, especially in the form of joint ventures. The demand for another essential raw material, oil, spurred firms like the Landoil Group to enter into agreements with several members of the Organisation of Petroleum Exporting Countries in order to engage in the exploration for oil. Similarly, the Philippine government has overseas investments in energy-intensive projects under the ASEAN Industrial Complementation Programme.

Investments in neighbouring countries and/or ethnically related territories that have a lower level of industrialisation as the Philippines, or at least the same level, have been important for Philippine MNEs, at least in the early stages. Although official data on FDI from host countries suggest that at least 56 per cent of the estimated stock of Philippine overseas direct investment during 1988 was located within the Asian continent, the share could be higher if data on other Asian host countries were available.

As will be shown in Chapter 10, the relatively high share of intra-regional direct investment of developing country firms in Asia and Latin America lends support to the product life cycle hypothesis which predicates that Third World MNEs can best exploit their competitive advantages in other developing countries. The movement costs associated with investment in neighbouring and/or ethnically related territories are smaller than investment in other geographical areas. Transport costs are rendered minimal as the distance between the home country and the neighbouring host country is not great. The psychic distance involved in investment in these areas is also considerably reduced as cultures, customs, traditions and languages are similar. In addition, production costs are very much reduced in these areas as the nature of the production process adopted is similar to, if not the same as, that in the home country because factor proportions are similar. Labour costs and productivity would also be rendered comparable. Because of the reduced production and movement costs, there are increased possibilities for economies of scale that arise out of operating subsidiaries in several neighbouring and/or ethnically related territories.

In the context of Philippine MNEs in particular, and Third World MNEs in general, the geographical scope may be described as following three sequential stages. The first stage is production in neighbouring and/or ethnically related territories as a result of the reduced production and movement costs and close psychic distance. The second stage is production in several neighbouring and/or ethnically related territories or in other, non-ethnically related developing countries. The reduced production and movement costs and close psychic distance, as well as the increasing significance of other economic considerations such as the presence of favourable investment opportunities and the desire to make fuller use of regional economic cooperation and integration, seem to explain significant direct investments within distinct regional groupings.

The generally favourable climate for foreign investment in Asian countries in general, and ASEAN member countries in particular, can be attributed to the political and economic ties between the individual governments. Apart from restrictions on or prohibitions against FDI in some non-manufacturing industries, policies towards FDI in these host countries have been generally open. Barriers to FDI have been historically low or non-existent. Foreign firms that have wanted to invest were able to do so largely unhampered by host country policies (Lecraw, 1981; Kirchbach, 1981). These factors increase the possibilities for firms to exploit economies of scale that arise out of operating subsidiaries in several neighbouring and/or ethnically related territories.

At this second stage, firms are better able to benefit from internalisation advantages in terms of economies of externalities and interdependent activities. In the framework of analysis adopted by Dunning (1981a, 1988a), firms in the second stage are described as possessing both asset (*Oa*) and transaction advantages. The asset advantages and, in particular, the indigenous capacity of these firms for technological accumulation are increasingly becoming sophisticated and supported by the more important transaction cost advantages. The reliance on neighbouring and/or ethnically related territories as recipients of their outward direct investments is therefore reduced. However, in the dynamic context adopted in this book, such asset and transaction advantages are analysed in terms of the extent to which Third World MNEs in general and Philippine MNEs in particular engage in vertically integrated activities across national boundaries as a feature in the growth of production of firms. Each firm has distinct technological advantages which are difficult to transfer to and implement by other firms with different technological advantages, especially among firms in countries at different stages of development (Rosenberg, 1976). These views on the high cost of technology transfer are allied to Teece's (1990) distinct but interrelated features of internalisation, namely appropriation and co-ordination. The appropriation issue ensures greater efficiency when technology resources are transferred within the firm rather than between two independent firms. On the other hand, greater co-ordination between firms ensures lower costs of technology transfer since technology needs to be adapted. Finally, in the third stage, firms are able to exploit their advantages further away from home and from neighbouring developing countries, in territories with a further psychic distance and possibly also in developed countries.

Investments in developed countries

Philippine direct investments have been of some importance in developed countries. Substantiated information contained in Table 8.5 reveals that upstream investments have been undertaken by Philippine MNEs in such countries as the USA, and in earlier years in Spain and Australia. The collective stock of foreign direct investment by Philippine firms in these countries was estimated to be at least US $124.5 million in 1983–4, of which US $121.0 million was directed to the USA. By 1988, inward FDI to the USA from the Philippines had reached US $140 million. Empirical research shows that some direct investment has also been undertaken by Philippine MNEs in Japan, Canada and some West European countries. However, the magnitude of direct investment by Philippine firms in these other developed countries is difficult to assess, as the host countries do not often delineate the stock of inward foreign direct investment that is attributable to firms of Philippine origin.

Nevertheless, the motives for upstream investment by Philippine firms in developed countries are fundamentally industry and firm specific. The company history of San Miguel Corporation, which was established in 1890, shows that as early as 1937 the company expanded into the USA with the acquisition of George Muehleback Brewery Company in Kansas City. This initial US investment was followed in 1939 by the acquisition of another brewery, in San Antonio, Texas. These ventures were divested in the late 1950s owing to the severe competition posed by the emergence of large national breweries. In addition, the company established brewery plant in Guam during the Vietnam War, but these were sold when the war ended.

Similarly, service investments by Philippine firms in developed countries were initially export oriented and consisted of the establishment in 1979 of twelve accredited trading companies in connection with the government's export promotion drive which specifically aimed to boost exports not only of traditional primary products but also non-traditional products. However, the increased importance of the manufacturing sector for the Philippine economy has necessitated the growth of new forms of service investment to support the development of these new manufacturing activities. Substantial Philippine direct investments in developed countries such as the USA represent investments undertaken in the banking sector, since these countries are often the major export markets for Philippine products. The geographical distribution of overseas affiliates of banks from developing countries generally reflects the trade flows of home countries (Wells, 1983) and in this context Philippine banks are no exception.

Empirical research reveals that the motives of Philippine banks for foreign investment are a prime determinant of location. Banking investments were undertaken in developed countries primarily to serve the banking needs of Philippine companies involved in international trade and to serve the banking requirements of their valued clients involved in the large-scale transfer of their portfolio investments abroad. The provision of a full range of banking services has been a very significant motive for the establishment of overseas affiliates of Philippine banks. Access to low-cost funds in foreign countries such as the USA, and also in major Asian financial centres such as Hong Kong, to finance domestic banking operations in the Philippines is a second major reason for direct investment by Philippine banks in foreign markets. The Filipino-owned China Banking Corporation, for example, established a subsidiary, Manila CBC Finance (Hong Kong) Limited, in 1980 to meet the 500 million Philippine peso (US $65.8 million) capital requirement for universal banking status. Similar reasons have compelled Metropolitan Bank and Trust Company to establish an overseas subsidiary, the First Metro International Investment Company Limited, in Hong Kong.[5]

A third important motive for the establishment of overseas branches by Philippine banks in the major financial centres such as London and New York is to capture vital financial information quickly, and for effective fund management. Furthermore, loan syndication facilities and money and capital market infrastructures are more advanced in the major financial centres. Thus, the Philippine National Bank (PNB) established an overseas branch in New York as early as 1917 and another overseas branch in London in 1969. In the early 1980s, PNB opened another overseas branch in Houston, Texas, and two representative offices in Toronto and Vancouver. Other Philippine banks such as Philippine Commercial International Bank (PCI Bank) established overseas branches in New York and Los Angeles in the early 1980s and representative offices in Houston, Madrid, Frankfurt and London. However, in early 1987, PCIB decided to close its representative office in London.

Similarly, Metropolitan Bank & Trust Company established an overseas branch in 1979 in New York and another branch, the International Bank of California, was opened in Los Angeles. The Allied Banking Corporation established a branch in London in 1979 and operates representative offices in Sydney and Tokyo. The Bank of Philippine Islands opened an overseas branch in New York in 1985.

Another important motivation for the establishment of overseas branches in the developed countries is to provide local firms and financial institutions in the host country with information with regard to investment opportunities in the Philippines. The case of the Manila Banking Corporation establishing a wholly owned subsidiary commercial bank in Los Angeles is an excellent example. By contrast, the main motives of the Philippine Commercial International Bank in establishing overseas branches in developed countries such as the USA derive from their aggressive pursuit of the remittances of Filipino workers overseas.

There have been instances in which upstream investments were undertaken specifically for the purpose of acquiring foreign technology and management know-how. The Landoil Group, one of the largest Philippine MNEs, purchased an equity interest in a number of firms located in developed countries in order to complement and upgrade its limited indigenous technological and organisational expertise. Thus, joint venture agreements have been established between Landoil and the Consulan Group, a Swiss MNE specialising in hotel management, specifically to upgrade Landoil's hotel management capability. Similarly, joint venture agreements have been concluded between Landoil and Charriot Resource, a Canadian energy company, to upgrade Landoil's oil exploration technology. Finally, in order to gain access to more sophisticated construction technology, Landoil has concluded a joint venture agreement with Siframe SA, a French construction and engineering company.

Investments in construction sites in the Middle East and elsewhere

The emergence of Philippine firms in particular and Third World firms in general in overseas construction activities may be attributed to the fact that construction has been growing fastest in these countries in response to their development of new manufacturing activities. The emergence of Philippine firms with host country market-oriented construction activities abroad is a manifestation of the increasing importance of the manufacturing sector in the home country. The increased mobility of unskilled or semi-skilled labour, largely attributed to improvements in international transport, combined with the first oil price hike in the early 1970s have provided a powerful impetus to Philippine investment in the construction and service-related sectors in many countries, especially in the Middle East. The historic oil embargo associated with the first oil crisis in 1973–4 enabled Middle Eastern countries to enjoy increased earnings from their oil exports.

Increased oil export revenues encouraged these countries to embark on a massive infrastructure programme to hasten their goals of economic development. However, this large-scale infrastructure programme proved to be a formidable task, beyond the capabilities of local contractors. The inevitable consequence was the importation of foreign labour. Statistics reveal that 43 per cent of jobs in Saudi Arabia in the early 1980s were filled by migrant workers. The corresponding statistics for Kuwait, Qatar and the United Arab Emirates were even higher at 70 per cent, 81 per cent and 85 per cent respectively. Hence, the construction sector of the Middle Eastern countries has become an important area of international activity for Philippine firms. Such activities include the export of services, technology and skilled labour, and on occasion the establishment of fully fledged subsidiaries, and have been a significant potential source of at least short-term employment and foreign exchange earnings for the Philippine economy. During the seven-year period 1975–81, a total of 563,955 Filipino workers were sent to 111 countries. The Middle Eastern countries received 84 per cent of these workers and Saudi Arabia received 69 per cent (*Business Day*, 18 February 1982).

Data from the Philippine Ministry of Labour and Employment reveal that based on stipulated salaries in work contracts that had passed through the Overseas Employment Development Board, The Bureau of Employment Services and The National Seaman Board, the three government structures formed to vigorously promote the export of Philippine manpower, the estimated remittances of overseas workers amounted to a minimum of US $3,228.9 million during the period.

Overseas construction by Philippine firms continued to be favourable until the late 1980s despite the slowdown in spending of some Middle Eastern countries. Accredited Philippine contractor companies numbered 168 in 1984, of which the most significant were the Landoil Group, Atlantic Gulf & Pacific Company Limited (AG&P), Hydro Resources Company (an associate company of CDCP), D. M. Consunji Incorporated, Engineering Equipment

Incorporated (EEI) and Erectors Incorporated. The overseas contracts of all Philippine contractor firms had an outstanding value of US $2.3 billion in 1982 and, as of 1983, a total of 130 projects in thirteen countries had been undertaken by these firms.

By contrast, the construction activities of Ayala International are focused around the ASEAN region. The close business ties of Enrique Zobel, the President and Chairman of the Board, with Sultan Hassanal Bolkiah of Brunei enabled Ayala International to undertake the biggest single construction project completed by an ASEAN firm. The US $350 million project for the building of Brunei's national palace, *Istana*, has resulted in the palace's inclusion in the Guinness Book of Records as the largest edifice in the world. As was illustrated in the previous section, the firm also developed the 1,480-acre (600-hectare) residential complex in Klang, Malaysia, and the Centrepoint hotel-bus-terminal-commercial complex in Kota Kinabalu, Sabah.

Empirical evidence shows that many Philippine construction companies operating abroad penetrated the international construction market initially through contractual arrangements and subsequently established an office and/or a subsidiary in the local market. The establishment of a full-scale operating subsidiary or the purchase of equity interest in a design or construction project in a firm is preferred to contractual arrangements on some occasions, depending on the perception of the potential for market growth. The need to overcome pressures of competition from other Philippine construction firms, and other foreign construction firms which have established local operations, is also an explanatory factor in undertaking overseas direct investments. AG & P, EEI, Landoil Resources and D. M. Consunji are some of the Philippine construction firms that have established either wholly owned or majority-owned subsidiaries in foreign markets.

CONCLUSION

This chapter focused on the emergence and growth of multinationalisation of Philippine firms within the context of the stages approach to Third World outward investment adopted in this book, which is determined largely by the pattern of countries' domestic industrial development and the emergence of ownership advantages of local firms.

The industrial development of Philippine MNEs as related to their ownership advantages were analysed with specific reference to the nature of ownership advantages of Third World multinationals described by Wells' product cycle model and Lall's theory of localised technological change. The main hypothesis is that the importance of resource-based investments with the associated narrow scope for innovation at this stage, embodying mere production experience, declines at a later stage. Over time, the significance of simple manufacturing and service investments increases as a result of the gradual process of technological accumulation and, in particular, the increased capacity for localised technological change, which enables firms to undertake more significant and more complex forms of outward investment. Hence although the product cycle model may be useful as a conceptual framework for the early stages of outward investment, associated with the elementary forms of technological advantages, the acquisition of the capacity for localised technological change envisaged by Lall's model and the broader concept of technological competence become increasingly apparent in the later stages.

Such a shift in the sectoral pattern of Philippine outward investment, as a direct result of the process of technological accumulation and the gradual development of technological competence, from purely resource-based investments towards manufacturing and, consequently, service investments which support the growth of manufacturing sector investments, has developmental implications for the geographical distribution of such investments. The increasing complexity in the geographical development of Philippine outward investments in particular and Third World outward investments in general is postulated to follow a three-stage process from an initial concentration in neighbouring and/or ethnically related territories towards several neighbouring and/or ethnically related territories or developing countries and finally towards the developed countries. The next two chapters extend the analysis to other significant Third World investors in Asia and Latin America.

9 The industrial development of Third World outward investment

INTRODUCTION

This chapter seeks to extend the analysis of Philippine MNEs in the last chapter towards a broader perspective which will include significant investors from other developing countries based in Asia and Latin America. The chapter will pursue the major themes of this book, the first of which pertains to the gradual process of expanding technological capabilities of selected Third World firms associated with cumulative growth in their international direct investment. The second theme proposes that the emerging technological competence and growth of Third World firms is related to the pattern of their domestic industrial development and the cumulative process by which they build upon their unique technological experience. The existence and accumulation of technological capabilities are an important determinant as well as an effect of the pattern and growth of their international production activities. Taken together, these propositions imply that the sectoral distribution of the outward direct investment of a developing country changes gradually over time, in a way that to some extent can be predicted.

The theory of technological accumulation as firm specific, cumulative and differentiated advanced by Pavitt (1988) has been previously applied to explain the historical evolution of the sectoral pattern of comparative advantage in innovative activity of national groups of firms in industrialised countries (Cantwell, 1990) and in work on competition between rival technologies (Arthur, 1988, 1989). These innovative activities, especially in research and development, have been found to be statistically significant determinants of differences in export and productivity performance (Soete, 1981b; Fagerberg, 1988) and also of internationalisation of business among the major OECD countries (Vernon, 1974, 1979; Cantwell, 1989a).

Although the innovative activities of firms based in Third World countries have assumed a rather different form from those of firms in the more industrialised countries, Third World firms have followed a technological course that is to some extent independent and a function of their own unique learning experience. However, the unique technological course of Third World firms is not necessarily limited to descaling and increasing the labour intensity of sophisticated technologies or to specialisation in the production of lower-skill and lower technologically intensive sectors, but lies rather in innovations that are based on lower levels of research, size, technological experience and skills, and which achieve improvements by modernising older techniques, including outdated foreign technology. Without a strong reliance on R&D, a greater part of their innovatory capacity has been given to production engineering, learning by doing and using, and organisational capabilities although, as shown in Chapter 6, the future development of their technological capacities is largely dependent on their access to foreign technology. There is a steadily rising number of examples of Third World multinational enterprises (MNEs) that have become genuinely innovative even though their activities have tended to be less scientifically refined and have not generally involved frontier technologies.

The proposition adopted in this book therefore relates to the changing forms of innovation as opposed to their existence as a variable factor in stages of development. Hence, although the forms of innovation may be different in firms from the Third World from those in industrialised countries, the existence of innovation within the firm determines the growth of its international production, especially in manufacturing. Although the technological activities of Third World MNEs require an empirical investigation of technology creation beyond R&D and international patenting activities, the concept of rapidly emerging technological capabilities or technological competence is a useful means of analysing the international growth of manufacturing firms from quite different environments, and at different stages of technological development and capacity (Cantwell and Tolentino, 1990).

The process of independent technological accumulation on the part of Third World MNEs therefore makes it necessary to formulate theories on the growth of these MNEs that recognise that the role of innovation within firms has been increasing, and has become more sophisticated as international production has developed. The scope of innovation of Third World MNEs in the two main schools of thought differs between the framework of the product cycle model articulated by Wells (1983, 1986) and that of the model of localised technological change advanced by Lall (1983a). Some of these disagreements were clearly illustrated in Chapter 4, and only the essential points are discussed below.

In the framework of the product cycle model, the expansion of Third World MNEs need not be based on any technological or other oligopolistic advantages, but derives from the use of low-wage, labour-intensive production processes, and low salary payments to managers who are none the less adept at organising in developing country conditions. To the extent that these firms have unique technological advantages, these are limited to the imitation and adaptation of foreign technology in the production of standardised goods at the tail end of the product and industry life cycle. Nevertheless, their distinctive nature *vis-à-vis* First World MNEs is seen to derive from their capacity to adapt foreign technology in accordance with the requirements of Third

World markets and production conditions: for example, the adaptation of foreign technology to the circumstances of smaller plant and firm size and greater labour intensity. There is also a technological advantage in their ability efficiently to utilise locally available natural resources rather than imports (Wells, 1986). As a result, these firms are fundamentally engaged in industries characterised by low expenditure on R&D and low product differentiation. What is more, there is presumably no reason why the developing country affiliates of US or European MNEs cannot imitate and copy the technological improvements achieved by Third World firms, in the course of adapting an essentially foreign technology and product development.

Some evidence is presented in this chapter to support the model of localised technological change and the theory of technological competence. Unlike the predications of the product cycle model, the proposition is that indigenous firms in developing countries have the capability to follow an independent or localised technological trajectory in which imported technology is integrated with the development of indigenous technology in a manner consistent with the dynamic growth path of changing comparative advantage. Such integration is more successful when foreign-generated technology is complementary to the path taken by indigenous innovation, and can therefore be beneficially integrated and adapted within an unique programme of technology generation. Such a view of the technological advantages of Third World firms, which encompasses not only the simple imitation and adaptation of innovation carried out previously in a more advanced country by a competitive fringe of Third World firms but also a further set of related although differentiated innovation as a result of the unique technological experience of Third World firms, is also consistent with the views of Atkinson and Stiglitz (1969), Rosenberg (1976), Nelson and Winter (1977, 1982) and Stiglitz (1987). The innovative activities undertaken by firms are triggered and guided not only by evolving market signals as predicated by the product cycle model but also by evolving technologies (Atkinson and Stiglitz, 1969; Rosenberg, 1982; Pavitt, 1988).

Indeed, there are new multinationals operating in the Third World, and particularly those of the newly industrialised countries (NICs), that are genuinely and uniquely innovative. Indeed, in some cases discussed in this chapter these firms have progressed to early stages of R&D activities, in part through investments in research facilities in the developed countries, and although they are still at a relatively early stage of development their abilities in technological accumulation continue to expand. Some Third World MNEs now operate in sectors whose products are far from standardised, and indeed their activities are helping to generate a fresh pulse of innovation in some cases. Due to the localised and irreversible nature of technological change, their distinctive innovative capacities cannot be readily replicated by the affiliates of industrialised country MNEs in developing countries that have progressed to new technologies, since there are high costs in efficiently reproducing or transferring older technologies. For example, if MNEs from industrialised countries attempt to replicate the technological advantages of Third World MNEs in small-scale manufacture, they lose the advantages of accumulated experience in large-scale manufacture (Wells, 1978).

Evidence is presented here on the growing significance of Third World outward investment, and on how innovative activity among Third World MNEs, although currently at an early stage, has tended to become increasingly important and more significant as expansion towards international production develops and as the home countries advance through progressively higher stages of industrial development. The next section considers how the evolutionary pattern in Third World outward investment relates to other macroeconomic developmental approaches previously developed to describe outward investments from the industrialised countries. It describes how the general trend towards the internationalisation of business, owing in part to the rise of outward investment from the Third World and other new source countries, has affected the relationship between the investment position of countries and their national level of development. In addition, the framework adopted outlines how the sectoral and geographical structure of Third World outward investment might be expected to evolve. The trend in the industrial composition of Third World outward investment as development proceeds is analysed in this chapter, while the next chapter looks at the equivalent trend in the geographical distribution of outward investment. In each case, attention is drawn to the underlying accumulation of technological advantages as a means by which Third World MNEs have transformed the significance and the character of their international production activity.

THE EVOLUTION OF THIRD WORLD MULTINATIONALS AND THE MACROECONOMIC DEVELOPMENT THEORIES OF INTERNATIONAL PRODUCTION

Historically, the outward direct investment of countries has tended to follow a developmental course over time, in which emphasis gradually shifts from resource-based activity with fairly limited technological requirements towards more sophisticated types of manufacturing. For the mature multinationals of Europe and the USA such an evolution has been going on for the past hundred years or so. Before 1939 the bulk of direct investment was associated with primary commodity production, and only in the post-war period did outward investments in the manufacturing sector assume importance. Since the 1980s these investments have heralded an era of international technological competition in the industrialised countries.

The evolution of Japanese multinationals since the 1960s has been compressed into a much shorter time span. Investments in resource-related activity and import-substituting manufacturing in Southeast Asia led the way from the 1950s until the early 1970s, but after the late 1960s interest shifted to more sophisticated Japanese manufacturing investments in Europe and the USA (Ozawa, 1990). Although Third World MNEs are nowhere near as technologically advanced as the modern Japanese multinationals, the sectoral and geographical composition of their activity is also evolving much more rapidly than the investments of the traditional source countries. The main reason for this is that there has been a general trend towards the internationalisation of business which has been common to firms of all countries. The newer sources of outward direct investment, including the Third World, have in general embarked upon multinational expansion at an earlier stage of their development than industrialised country firms.

This chapter aims to expand the macroeconomic development theories of international production to explain the industrial development of Third World MNEs. As these theories were examined in considerable detail in Chapter 3, only their main highlights are set out below to provide a background to the formulation of the concept of stages of development in international production.

In the 1970s, on the basis of the theory of comparative advantage (or costs) as exemplified by the Hecksher-Ohlin model of trade between two countries, Kojima advanced an integrated theory of international trade and direct investment. He claimed that export-oriented Japanese direct investments in developing countries, in labour-intensive industries with simple technology, were trade oriented since these investments originated from the comparatively disadvantaged (or marginal) industry of the home country and led to lower-cost imports in Japan but an expanded volume of exports from the host country. On the other hand, import-substituting American FDI in the developed countries was anti-trade oriented since these investments originated from the comparatively advantaged industries of the home country and led to a misallocation of resources and a decreased volume of exports from the host country (Kojima, 1973, 1975, 1978, 1982).

Such a view now appears to be a reflection of the early stages of development of Japanese multinationals. The history of Japanese post-war trade and investment development does not lend support to this form of application of the static theory of comparative advantage which thirty years or so ago would have suggested that Japan should concentrate on the production of labour-intensive goods and import capital and technology-intensive goods. Kojima's theory is essentially industry based as opposed to country based as he claims, in as much as Japanese and American MNEs were concentrated in different industries, which helps to explain why their overseas production was aimed principally at serving either export or local markets. The extent to which the theory was country based derives essentially from the general way in which the stage of national development helps to explain the industrial structure of indigenous firms. Since the late 1960s, alongside the restructuring of Japanese industry, the characteristics of Japanese MNEs have increasingly evolved towards highly technologically intensive import-substituting manufacturing investments in the USA and Europe. This may be attributed in part to the efforts of the Japanese government systematically to upgrade the composition of the country's comparative advantage, aided in part by licensed foreign technology (Ozawa, 1974). The analysis of Japan's post-war development provides support for the development of 'future-oriented technologies', in that the sectors in which a country enjoys the greatest potential for innovation and in which investment may be most beneficial are not necessarily those in which that country currently has a comparative advantage (Blumenthal and Teubal, 1975; Pasinetti, 1981).

Although further support for Kojima's macroeconomic approach to FDI was provided by Ozawa (1971, 1977, 1979a, 1979b, 1985), he acknowledges in his later work that the distinctive characteristics of Japanese FDI from 1950 to the early 1970s in resource-based and labour-intensive or technologically standardised industries in developing countries belonged to the first two phases of Japanese overseas activity. Since the late 1960s, as a result of industrial restructuring in Japan, overseas investment has evolved to the third, import substituting or recycling, stage of international production. This is centred on the mass production of assembly-based consumer durables in high-income countries such as the USA and Europe, supported by a network of subcontractors. Moreover, since the early 1980s, a fourth phase of Japanese FDI can be discerned. This is characterised as the flexible manufacturing of highly differentiated goods, involving the application of computer-aided designing (CAD), computer-aided engineering (CAE) and computer-aided manufacturing (CAM) (Ozawa, 1990).

The growth of Japanese overseas investment has therefore followed an evolutionary course from resource-based and simple manufacturing towards more technologically sophisticated forms of outward investment. The difference from the evolutionary path taken by US overseas expansion lies essentially in the swiftness with which Japanese MNEs have made the transition. Hence, although US and Japanese firms and industries are at a different stage of evolution, Japanese firms have demonstrated their ability to catch up quickly and keep pace with advances in technological development in the West.

Although most Third World MNEs are unlikely to develop as fast as their Japanese counterparts, who can today draw on frontier technologies, the trend is towards a persistent upgrading in the types of activity for which these firms are responsible. The phases of development suggested by Ozawa for Japanese MNEs are thus relevant to the study of Third World multinationals. In the early stages, as domestic firms in the developing countries upgrade their domestic production activities, resource-based and less technologically sophisticated manufacturing activities are transferred to countries at an earlier stage of development that have lower levels of technological capacities and production costs. In the later stages, more sophisticated manufacturing activities are transferred abroad, even to the developed countries. The current overseas activities of the NICs in the developed countries, although based on a less research-intensive form of technological innovation than Japanese MNEs, can nevertheless be described as being in the early phases of the import-substituting or surplus-recycling form of international production.

The trend towards the internationalisation of business and the increasing complexity of their pattern and composition can be analysed within the more general context of an investment development cycle. The idea of an investment development cycle has been advanced by Dunning (1981a, 1981b, 1986a, c, 1988a) and is based on the proposition that the level of inward and outward investment of different countries, and the balance between the two, depend upon the national stage of development. However, the empirical evidence examined in Chapter 5 of the general trend towards internationalisation since the mid-1970s, with the increased significance of outward investments from newer source countries, both developed and developing, has shown that the formulation of the investment development cycle or path may require some qualification. The more rapid and earlier multinational expansion of firms from the newer source countries such as Germany, Japan, the smaller developed countries and the Third World has meant that the national stage of development is no longer a good predictor of a country's overall net outward investment position.

More recent versions of the investment development path have argued that apart from the level of net outward investment, the character and composition of outward direct investment of a country's firms vary with the national stage of development (Dunning, 1986a, c). The early forms of foreign investment are frequently resource based or sometimes import substituting, but in each case have a quite specific location associated with a particular type of activity. However, as firms mature their outward investments evolve from a single activity or product in a particular location towards a more international perspective on the location of the different types of production in which they are involved. Such MNEs are responsible for directly organising an international division of labour within the firm in place of markets.

This book examines the changing character and composition of outward direct investment as development proceeds. The general argument is that the outward direct investment of countries follows a developmental or evolutionary course over time and is predominant in resource-based production or in simple forms of industrial production that embody limited technological requirements in the earlier stages of development. At later stages, the sectoral path of these investments tends to evolve towards more technologically sophisticated forms of manufacturing investment. The developmental course of the most recent outward investors from the Third World has been faster and has a distinctive technological nature compared to the more mature multinationals from Europe, the USA and Japan, owing to the different stages of their national development. This chapter and the next focus on the role of newly emerging technological advantages in explaining the increasing complexity of the industrial composition and greater geographical diversification of Third World outward investment.

THE INDUSTRIAL DEVELOPMENT OF THIRD WORLD OUTWARD INVESTMENT: COMPARATIVE ASSESSMENT

This chapter seeks to address the central issue of the relationship between the pattern of domestic development, the emergence of competitive or ownership advantages on the part of indigenous firms and the growth of outward investments in particular sectors. The main argument is that the significance and complexity of the technological innovation of firms from developing countries are largely determined by the stage of industrial development of their home countries and the cumulative experience of firms in international investment.[1] Developing countries may therefore be divided into groups defined with reference to the stage of industrial development of their home countries, which is reflected in the industrial structure and technological complexity of their outward investment. Those belonging to the most advanced group have more complex forms of outward investment because of their greater capacity for technological accumulation, associated with the higher stages of industrial development of their home countries. The newly industrialised countries, defined in this book as including South Korea, Taiwan, Singapore and Hong Kong in Asia and Mexico, Brazil and Argentina in Latin America, are therefore assigned to these most advanced groups. The hypothesis is that the international production activities of these countries have already progressed beyond the earliest stages of development in which outward investment in the primary sector and simple manufacturing activities tend to lead the way. Since their capacity for localised technological change and accumulation has expanded and their international production experience has broadened, their outward direct investments are likely to be concentrated on more complex manufacturing and support service activities.

The sectoral evolution of outward FDI from developing countries generally follows the developmental path of the more traditional investor countries as well as the newer ones from the developed countries. Such outward investments are generally predominant in resource-based and simple manufacturing activities at an early stage of development and then in more complex manufacturing and associated service activities at a later stage as their capacity for technological accumulation expands and their multinational investment experience broadens.[2] A connection is therefore seen between the pattern of domestic development, the emergence of ownership advantages of local firms and the growth of outward investment.

However, although such an underlying pattern exists between the increasing importance and complexity of technological accumulation, as greater experience in international investment is acquired, and increasing sophistication in the industrial and geographical pattern of outward direct investment, the significance and form of the technological accumulation vary among the technological leaders of the US, the UK, Japan and Germany as opposed to the simpler and less research-intensive technology creation of Third World MNEs. Nevertheless, although the forms of innovation may be different owing to the different stages of development, a clear link was established in Chapter 6 between the diffusion of frontier advances in technological innovation in the advanced, industrialised countries and the growth of technological and organisational innovations by firms from developing countries, especially the NICs.

One such important link, especially for the Asian NICs, is provided by the establishment of the new international division of labour in the world economy which enabled the growth of intermediate product trade within industries on a global scale and led to a new type of productive organisation in the process of technological diffusion. In particular, the establishment of the new international division of labour has increased the scope of intermediate product trade within the firm by enabling labour-intensive production processes within industries to be increasingly carried out in the NICs as a result of the growth of large, disciplined and low-wage labour forces in these countries (Fröbel, Heinrichs and Kreye, 1980; Casson *et al*, 1986). The co-ordination of these separate production processes in different countries requires complex forms of organisational and managerial innovation in addition to the firm's technological innovation.

Such technological, organisational and managerial innovations undertaken by adaptive foreign multinational affiliates of MNEs from the more advanced countries engaged in export-platform investments in the NICs may have fulfilled a tutorial role for some technologically innovative local firms which engaged in international expansion through exports of labour-intensive manufactured products and outward investment at a later stage. Although the main thesis is that local firms in the developing countries follow an independent technological course and integrate foreign technology in the context of their own natural technology trajectory (Nelson and Winter, 1977, 1982), the organisational and managerial forms of innovation required in establishing labour-intensive production processes may have made a positive contribution to the enhancement of the organisational and managerial innovation of local firms (see also Aydin and Terpstra, 1981, for related ideas). In their turn, the local firms in the Asian NICs are at present taking advantage of a further international division of labour by transferring the more labour-intensive stage in the production process of their manufactured products to other lower-wage developing countries. Such a process has often been important in maintaining the international competitiveness of exporters since the host developing countries have lower labour costs and may either have preferential trading arrangements with or geographical proximity to their targeted export markets, or quantitative restrictions on exports may not yet have been imposed or may be less severe.

The significance of firms from different countries within each sector will be examined in this chapter within the context of their national stage of development as well as various country- and firm-specific characteristics. The method of analysis is to describe characteristics of firms and countries that are typically representative of the general trend.[3] However, where firms of a given country are either more or less heavily involved in international production in a particular sector than might be suggested by their national stage of industrial development, then reference is made to various country- and firm-specific characteristics that may help to explain such a departure. This survey of Third World multinationals only seeks to focus on locally based firms in developing countries that have invested abroad and not on third-country investments in a particular developing country that has in turn invested abroad.[4] Such a survey encompasses the primary, secondary and construction sectors whose growth may be explained by the application of the concept of technological competence.[5] The review begins with outward investment in sectors that are normally associated with lower levels of development and then progresses to a consideration of more sophisticated types of activities associated with higher stages of development.

RESOURCE-BASED INVESTMENTS

In general, the early forms of countries' outward direct investments are primarily resource based and are motivated by the need to integrate backwardly into the search for raw materials which either do not exist in the home country or are only available in amounts inadequate to support industrialisation. Several of the resource-based investments of developing countries seem to follow the earlier resource-based investments of the UK, Europe, the USA and, more recently, Japan, which are largely dependent on overseas natural resources for their domestic industrial requirements. The shortage of agricultural land, which is the main constraining factor on economic growth in the classical growth model of David Ricardo, can be interpreted in a broader sense to include shortage of natural resources. Hence, the Ricardian trap of industrialisation explains the major motive for resource-based investments of resource-poor industrialised countries (or Ricardian economies) in resource-rich countries (Ozawa, 1982). But despite the similarity in resource scarcity between Japan and the Asian NICs, the resource-based investments of the Asian NICs are considerably smaller than Japan's, due partly to the resource nationalism of host countries in recent years, which restricts resource-based investments, and partly to the large financial requirements of forestry and mining projects. In addition, the Asian city states of Hong Kong and Singapore have not emphasised resource-based industrialisation but instead imported finished goods rather than producing domestically using imported raw materials. In any case, the importance of resource-based investments declined even further in the 1980s as considerations of supply and security of natural resources weakened with the slump in world commodities and oil markets and as investments based on costs and markets became more crucial determinants of outward foreign investments (Chia, 1989).

The main hypothesis then is that the importance of outward direct investments in the primary sector declines as countries advance through higher stages of industrial development. Within the Third World, resource-rich countries at an earlier stage of development, such as the Philippines, Malaysia and Thailand in Asia, and Chile, Colombia and Costa Rica in Latin America, are postulated to have a higher proportion of their outward FDI in resource-based activities because the innovative activities of their firms are currently at an early stage and manifested in their experience in extraction of natural resources. By comparison, since the NICs are at a more advanced stage of development, the innovative activities of their firms in terms of localised technological change are generally increasingly more significant and therefore demonstrate greater capabilities for more sophisticated forms of outward investment.

Table 9.1 summarises some of the resource-based investments undertaken by different Third World MNEs. The particular nature of the resource-based investment differs according to the country in question and the different technological ownership advantages that its firms possess. Nevertheless, although the particular nature of resource-based investment varies between countries and firms, there is a logical connection between the pattern of domestic development, the emergence of ownership advantages by indigenous firms and the resulting growth of outward investment.

However, apart from technological advantages, which became more important with development, the abundance or scarcity of natural resources in the home country is an important country-specific determinant of the pattern of resource-based outward investments of Third World MNEs, which can explain the deviation of countries from the predicted trend given their stage of development. The evidence in Table 9.2 shows that countries at both higher and lower stages of national development, as measured by GNP per capita, can have significant shares of outward investment in the primary sector. Therefore, the expected relationship between the declining importance of outward investments in the primary sector as countries advance through higher stages of development as measured by GNP per capita and the share of FDI directed to the primary sector may not necessarily be true in all cases.

In particular, in the Ricardian economies such as South Korea, Hong Kong, Taiwan and Singapore, whose rapid industrial growth is constrained by limited domestic resource supplies, firms may undertake outward resource-based investments to gain control over foreign sources of supply of the raw materials vital for domestic industrial expansion. On the other hand, firms from resource-rich developing countries such as the Philippines, Malaysia, Argentina, Brazil and Peru that have accumulated expertise in the exploitation of the natural resources abundant in the home country may undertake overseas resource-based investments to exploit those advantages. Many of the resource-based firms from these countries sought and obtained extractive concessions in resource-rich host countries to exploit their technical and organisational skills in resource extraction which could not be further exploited in their home country (see also Wells, 1984). Technological advantages may have been developed in the first instance through colonial or other ties with foreign

Table 9.1 The nature of resource-based investments of Third World MNEs

Home country and firm	Nature of resource-based investment	Host country
India		
Firm unknown at source	To mine magnesite, zinc and lead ore for further processing	Nepal
Hydrocarbons India Pvt Ltd	To participate in oil concessions	Iran
Philippines		
Gonzalo Puyat & Sons Incorporated, P.T. Taliaban Timber, CDCP	To acquire timber for home and third country markets	Indonesia
Landoil Group	To acquire petroleum for home market	Middle Eastern countries
Peru		
Companhía Minera Buenaventura	To undertake small-scale mining projects	Venezuela, Ecuador and other Latin American countries
Mexico		
Cordemex	To acquire jute for home textile industry	Tanzania
Malaysia		
Malaysian Mining Corporation	To engage in the smelting, manufacture and marketing of tin, diamonds, coal and other base metals such as gold	Thailand and Australia
Sime Darby Holdings	To engage in the cultivation, manufacture and marketing of rubber and palm oil	Philippines
Brazil		
Petrobrás	To acquire petroleum resources for the home country	Several Middle Eastern and African countries, Norway, Colombia, Mexico, Guatemala
Siderbras	To acquire coal resources for the home country	USA, Colombia
Cia Vale do Rio Doce	To acquire coal resources for the home country	Canada

Table 9.1   (continued)

Home country and firm	Nature of resource-based investment	Host country
CPRM	To acquire mineral resources	Developing countries
Paranapanema	To acquire mineral resources	Developing countries
	To develop a gold mine	Guyana
Argentina		
Industrias Siderurgias Grassi SA 37 3	To acquire inexpensive manganese alloys for steel production for home and third country markets	Brazil
Cabsha	To acquire cacao for the home food processing industry	Costa Rica
A. Estrada	To obtain timber for the home paper industry	Brazil
Dunlit	To acquire copper resources for the home metal processing industry	Chile
Yacimientos Petroliferos Fiscales (YPF)	To acquire petroleum resources for the home country	Bolivia, Ecuador, Uruguay
South Korea		
Firms such as Pohang Iron and Steel Company	To acquire crude oil, wool, aluminium, raw sugar, cotton, rubber, iron ore and timber for the home market	Several resource-rich countries such as Australia, the USA, Canada and the West Indies
Korea Electric Power	To extract uranium ore	USA and Canada
Hyundai, Daesung, Lucky-GoldStar	To extract bituminous coal	Australia
Korea–Indonesia Resources Development Company	To extract bituminous coal and oil	Indonesia
Sun Eeel Shipping	To extract bituminous coal	USA
Yukong, Samwhan, Hyundai and PEDCO	To explore oil fields	North Yemen
Tae Sung Lumber	To extract wood	Indonesia

Table 9.1 (continued)

Home country and firm	Nature of resource-based investment	Host country
Halla Resources	To extract wood	Papua New Guinea
Dae Doo Battery	To extract sulphur and chrome	Costa Rica
Dong II Trading	To extract sulphur and chrome	Thailand
Korea Petroleum Development Corporation and KODECO	To develop oil fields	North Yemen and East Madura, Indonesia
Haitai Dairy Corporation Limited	To engage in livestock farming	Australia 3n 3
Taiwan		
Firms unknown at source	To acquire timber, bamboo, fishery and various agricultural resources for home and third country markets	Several neighbouring and resource-rich countries such as Thailand, Malaysia, Indonesia and Costa Rica
Chinese Petroleum Corporation	To acquire petroleum resources	Colombia
Singapore		
Intraco, Keck Seng, Guthrie, Sim Lim Group	To cultivate and mill palm oil in order to trade edible oil, the end-product	China
	To extract and process marble for trading	Malaysia
Wah Chang International Limited	To engage in petroleum exploration through the supply of vessels or the building of jack-up rigs and oil production platforms	China
Hong Kong		
Firms unknown at source	To utilise opportunities for exploitation and processing of agricultural and forestry products which are not available at home	Indonesia

Sources: Diaz-Alejandro, 1977; Wells, 1978, 1984; Jo, 1981; Kumar, 1981; Ting and Schive, 1981; White, 1981; Katz and Kosacoff, 1983; Koo, 1984, 1985; Lim, 1984; Pang and Komaran, 1984; Davis, 1987; Morris, 1987; Wells, 1988; Korea Institute for Foreign Investment, 1989; *Far Eastern Economic Review*, 18 January, 1990.

Table 9.2 The share of outward FDI in the primary sector and the GNP per capita of selected developing countries, 1988

Home country	GNP per capita, US $	Share of outward FDI in the primary sector
India	340	3.0
China	330	26.7 [1]
Philippines	630	19.2 [2]
Thailand	1,000	0.2
Colombia	1,180	3.7
Peru	1,300	1.2
Malaysia	1,940	7.8 [1]
Argentina	2,520	33.8 [2]
Korea (Republic of)	3,600	49.4
Taiwan	6,396	1.1
Hong Kong	9,220	18.8 [2]

Sources: Bank of Korea, *The Status of Outward Foreign Investment, 1988*; ESCAP/UNCTC, 1988 for Malaysia; Taiwan Ministry of Economic Affairs, Investment Commission, *Statistics on Overseas Chinese and Foreign Investment, Technical Cooperation, Outward Investment, Outward Technical Cooperation*, February 1989; World Bank, *World Development Report 1990*, New York: Oxford University Press, for data for GNP per capita; UNICMIO, *World Investment Directory*, 1992 and 1993, forthcoming, from primary data obtained from Lim, 1984; Banco de la República de Colombia and Oficina de Cambios Sección Análisis Económico de Colombia; Bank of Thailand, unpublished data; Central Bank of the Philippines; China Ministry of Foreign Economic Relations and Trade, unpublished data; Comisión Nacional de Inversiones Extranjeras y Tecnología Extranjeras (CONITE) de Perú and Ministerio de Economía y Finanzas de Perú; data taken from the Argentine Central Bank and the Registry of Foreign Investment at the Argentine Ministry for the Economy; Indian Investment Centre, unpublished data on Indian joint ventures abroad.

Notes: 1 Represents 1987.
2 Represents 1982.

MNEs, as a result of which local firms have acquired some technological capacities in methods of extraction as well as downstream processing.

As a result of their advanced stage of development and faster rate of industrial growth, firms from the NICs are expected to be more likely to engage in backward vertical integration into resource-based activity. Such an expectation is even stronger where countries are resource scarce rather than resource rich. On the other hand, firms from developing countries at a lower stage of development, particularly if they are resource rich, are more liable to generate outward investment in resource-based activities which amount to a purely horizontal geographical diversification of their domestic activities. This means that different kinds of Third World MNEs are likely to arise in more advanced resource-scarce countries compared with less developed resource-rich countries. In the first case, manufacturing firms with a broader range of technological advantages establish backwardly vertically integrated resource-based investments; while in the second case, resource-based enterprises with more narrowly specialised technological advantages undertake resource-based investments that simply extend their domestic activity.

Table 9.3 summarises the technological advantages of Third World MNEs engaged in resource-based investments according to their stage of national development as measured by GNP per capita. The analysis of the technological advantages of Third World MNEs engaged in resource-based investments according to their stage of national development shown in Table 9.2 provides evidence of the described pattern. The major technological advantages of Third World firms based in resource-rich countries, when beginning the early stages of outward investment in resource-based activities, derive principally from their accumulated experience in the exploitation of natural resources with which the home country is relatively well endowed. Firms from these re-source-rich and lower-income developing countries have simpler forms of technological innovation associated largely, if not solely, with resource extraction. By comparison, firms from the more resource-scarce NICs that undertake resource-based outward investments demonstrate a far greater technological capability for geographically diversified and vertically integrated activities in their foreign operations, which range from extraction and processing to marketing. For example, the city state of Singapore has substantial FDI in the marketing and distribution of resource-based products. In addition, the most striking example of geographical diversification of resource-based firms from developing countries pertains to the Mexican state-owned firm Cordemex, which established a subsidiary in Tanzania to manufacture jute on a large scale in an attempt to reinforce the firm's influence in the international hard fibre market.

By comparison with the city states of Hong Kong and Singapore, which have followed industrial development strategies of importing finished goods rather than producing domestically with imported raw materials, South Korea and Taiwan have attached greater importance to backward integrated outward investments to gain access to raw materials and natural resources.[6] South Korea has concentrated on mining and other natural resource-based industries to a much greater extent than other Asian NICs (World Bank, 1989). The country is most similar to Japan in that the shortage of indigenous natural resources and the lacklustre performance of the agricultural sector are the major factors behind the growth of resource-based overseas investments in resource-rich countries. In 1981, Korea had a total import dependence on such overseas natural resources as crude oil, aluminium, raw sugar, wool, cotton and rubber, and had 86.6 per cent and 84.8 per cent dependence, respectively, on overseas supplies of iron ore and timber for domestic industrial expansion. Korea's outward investment in resource-based industries is therefore geared largely to fulfilling the requirements of Korean industries for vital raw materials (Jo, 1981). Similarly, firms from Taiwan undertake resource-based investments in order to engage in the cultivation of various agricultural, forestry and fishery products which are vital raw materials for their paper, wood-processing and food product industries. These products are either marketed in Taiwan or

Table 9.3 The technological advantages of Third World MNEs in resource-based investments *vis-à-vis* their stage of development

Home country	GNP per capita, US $, 1988	Technological assets
India	340	Accumulated expertise in the exploitation and processing of magnesite, zinc and lead ores
Philippines	630	Accumulated expertise in the exploitation and processing of timber, in some cases supplemented by foreign technology; familiarity in working in virgin tropical forests and in utilising relatively light machinery
Peru	1,300	Process know-how in the operation of small-scale mining projects
Mexico	1,760	Accumulated expertise and long production experience in processing of hard fibres
Malaysia	1,940	Accumulated expertise and long production experience in the smelting, processing and marketing of tin, diamonds, coal and other base metals. Accumulated expertise and long production experience in the cultivation, manufacture and marketing of rubber and rubber products. Their expertise in rubber processing is supplemented by the acquisition of firms with advanced technology
Brazil	2,160	Process know-how and managerial skills required in the exploration for oil and other natural resources for the home country, supplemented by access to advanced technology from foreign firms. Outward investments in resource extraction have also been undertaken for the purpose of acquiring know-how that might be useful for resource extraction activities in the home country. Process know-how in the exploration for coal and other natural resources for the home country. Mineral exploration

Table 9.3 (continued)

Home country	GNP per capita, US $, 1988	Technological assets
		experience was acquired within Brazil's vast territorial expanse and is applied to resource-rich countries
Argentina	2,520	Process know-how in the extraction of cheap raw materials for steel production and of various agricultural, forestry and mining products for processing in the home country. Process know-how in the exploration for oil for the home country
South Korea	3,600	Process know-how in the extraction of many vital industrial raw materials for industrialisation in the home country
Taiwan	6,396	Accumulated expertise in the exploitation and processing of various primary products geared for the domestic and foreign markets
Singapore	9,070	Accumulated expertise in cultivation, processing and marketing of various resource-based products such as edible oil and marble Accumulated expertise in petroleum exploration including the provision of associated capital goods such as vessels, jack-up rigs (under foreign licence) and offshore oil production platforms
Hong Kong	9,220	Process know-how in the exploitation and processing of agricultural and forestry products which cannot be exploited in the home country

Source: As for Table 9.2.

exported to third countries (Ting and Schive, 1981). Firms from Argentina also engage in backwardly integrated resource-based investments to extract raw materials for industries such as steel, copper, food and paper in which they have become competitive in the home market. For example, Industrias Siderurgias Grassi SA required cheaper and more reliable sources of manganese alloys to maintain the competitiveness of its steel production plant. The increased output as a result of the process of backward integration enabled the firm to expand outside the home market and export to other Latin American countries, Europe and the USA.

Despite the lesser importance attached to Singapore's resource-based outward investments, there is evidence that its manufacturing firms engage in resource-based investments to integrate within the firm the processes of extraction, manufacture and packaging of natural resource-based products such as edible oil and marble, as well as trading to home and third countries. These firms have been able to utilise their firm-specific competitive advantages in product and process technology in combination with their sourcing and marketing skills in foreign markets (Lecraw, 1985). Similarly, the international operations of the Brazilian firm Cia Vale do Rio Doce, which is primarily engaged in iron ore mining and steel production, consist of forward integration into international production and marketing in order to be assured of steady buyers and to keep abreast of the latest technology in material use.

The greater technological capability of firms from the NICs is not only seen in the extent of their vertically integrated foreign operations but also, as demonstrated by firms from Singapore and Brazil engaged in petroleum exploration, there is evidence that resource-scarce NICs have gained the capacity to use the most advanced exploration and production technology. Advanced knowledge in the engineering aspects of petroleum exploration and production has enabled Braspetro, the overseas oil exploration arm of the Brazilian firm Petrobrás, to become one of the world pioneers in offshore production systems. Apart from having the expertise related to deep-sea oil exploration, with as much as 95 per cent of total equipment and materials used in overseas operations sourced from Brazil's local capital goods industry, the firm is also learning by doing in its operations outside Brazil.[7] In particular, the risk contracts undertaken by Petrobrás with foreign oil companies for oil research have resulted in invaluable technological interchange. The firm is now selling its technological expertise in the design, construction and installation of entire offshore platforms such as oil rigs, refineries, storage systems and pipelines and the handling of complex seismic research, deep-water drilling and production. Apart from having the know-how in the engineering aspects of oil exploration, the firm also has the organisational (managerial) skills required for handling hundreds of workers in the jungle, desert or off the coast of a foreign country. Similar technological advantages are demonstrated by the Singaporean firm Wah-Chang International Limited, which engages in petroleum exploration through the provision of vessels, jack-up rigs (under foreign licence) and offshore oil production platforms.

By contrast, the resource-based investments of Hong Kong firms in the exploitation and processing of agricultural and forestry products are unique to Third World NICs. Their resource-based foreign investments may represent aggressive forms of investment that are geared largely to exploiting profitable investment opportunities in the host country and not to exploiting technological ownership advantages generated in the home country. Although the end-products of their foreign resource-based activities may be exported to home and foreign markets, a number of these investments, especially those in timber extraction and processing in Indonesia, are usually made by investment companies which treat their investments simply as a form of portfolio choice in order to exploit the comparative advantage in natural resources of the host country.[8] Moreover, a majority of these investments are unregistered and consist of informal joint ventures with Indonesian military officials or with Indonesians of Chinese descent. They represent aggressive strategies in which new markets and lines of activities are developed for Hong Kong parent firms (Chen, 1981).

By contrast, as Table 9.1 shows, firms from countries such as the Philippines that engage in overseas resource-based investments are resource-based enterprises which have geographically diversified their domestic activities in foreign markets. The last chapter showed that several Philippine firms in the timber industry made fuller use of their access to large quantities of timber in Indonesia and, in addition, opened new markets outside the Philippines and exported to such third countries as Japan, Korea and Taiwan. Other Philippine firms have backwardly integrated into overseas petroleum exploration in order to overcome the shortage of petroleum resources needed for the expansion of the Philippine industrial sector.

The nature of Malaysian MNEs, which were once foreign owned and controlled but subsequently taken over and nationalised, has enabled them to exploit far more sophisticated forms of ownership advantages in vertically integrated resource-based activities than firms from other developing countries at a similar stage of development. Some of these firms are discussed further in the next section.

The empirical evidence on outward resource-based investments of Third World MNEs reviewed in this section lends credence to the product cycle model advanced by Wells in which the technological advantages of Third World MNEs derive from their ability to imitate and adapt foreign technology to suit Third World markets and production conditions. The major technological advantages of Third World firms engaged in the early stages of outward investment in resource-based activities, especially those that are resource rich, derive principally from their accumulated experience in the exploitation of natural resources with which the home country is relatively well endowed. The resource-based investments of firms from lower-income developing countries are seen to derive from simpler forms of technological advantages associated largely, if not solely, with resource extraction which conforms with the expectations of the product cycle model. By contrast, the greater technological capabilities of resource-based firms from the NICs lend credence to the concept of the cumulative process of technological accumulation and increasing technological competence as countries advance in their stage of development. Overall, the technological advantages of Third World MNEs in the resource-based sector are supplemented by the organisational and managerial skills acquired through accumulated experience in the implementation of various resource-based projects. The next section analyses the outward investments of Third World MNEs in the manufacturing sector.

MANUFACTURING SECTOR INVESTMENTS

Import-substituting investments

Tables 9.4 and 9.5 summarise the nature of import-substituting manufacturing activities of some Third World MNEs. As in the previous section on resource-based investments, the analysis of Third World MNEs' import-substituting investments will distinguish between the downstream vertical integration of resource-based firms into processing activities in the host country and the horizontal integration of manufacturing firms as a result of geographical diversification in foreign markets.

Forward vertical integration of resource-based firms

The sectoral pattern of the outward investment of resource-based firms is likely to develop from an initial concentration on extractive activity. For resource-based firms, a more advanced stage in the sectoral pattern of outward investment is their downstream processing activities in host countries, as a result *inter alia*, of the gradual development of localised technological capacities. There is thus a shift in the sectoral pattern of outward investment from purely resource-based investments towards forward vertically integrated manufacturing investments which represent the secondary processing of natural resources.

The development of firm-specific technological advantages by some resource-rich Third World MNEs in downstream processing activities that build upon the resource strength of the home country is analogous to the earlier development of firm-specific advantages by resource-rich developed countries such as Canada and Sweden. The outward investments of firms from these countries have since progressed into fully integrated operations, from extraction and production through to the marketing and trading of resource-based products. Most resource-based Canadian and Swedish MNEs, like firms from resource-rich Third World countries, built their initial technological advantages on the basis of favoured access to abundant resources in the home country. In the case of Canadian firms, favoured access to these resources was obtained either on the basis of ownership, as in the case of mineral deposits and energy resources, or long-term leases for rights in the case of timber resources. At a

Table 9.4   The nature of vertically integrated import-substituting manufacturing
investments of some Third World MNEs

Home country and firm	Nature of import-substituting investment	Host country
China		
China International Trust and Investment Corporation (CITIC), China National NonferrousMetal Industrial Corporation	Jointly invested to acquire 10 per cent equity interest in an aluminium smelting plant	Australia
China International Trust and Investment Corporation (CITIC)	Owns a) US $30 million timber extraction and processing plant b) Mining investments c) Pulp plant	Washington State, USA Bolivia Canada
China National Chemicals Import and Export Corporation	Purchased a stake in an oil refinery	USA
China National Metal and Mineral Import and Export Corporation	Invested in a relatively large iron works plant	Brazil
Indonesia		
Mantrust	Purchased Chicken of the Sea for the secondary processing and marketing of processed tuna fish to serve host country	USA
Thailand		
Unicord	Acquired Bumble Bee Tuna from Pillsbury for US $285million for the secondary processing and marketing of processed tuna fish to serve host country. The acquisition made Unicord the world's largest tuna processor	USA
Peru		
Cia Minera Buenaventura	Purchase of equity interest in mining projects geared to serve host country	Venezuela, Ecuador and other Latin American countries
Mexico		
Autlan	Secondary processing and marketing of Mexican minerals in host country	Alabama, USA, and Venezuela

Table 9.4 (continued)

Home country and firm	Nature of import-substituting investment	Host country
	and for export	
Malaysia Sime Darby	Secondary processing and marketing of various resource-based products such as rubber and palm oil for host country and export	Philippines, UK, Australia
Malaysian Mining Corporation	Secondary processing of tin, diamonds, coal and other base metals such as gold for host country and export	Thailand, Australia
Brazil Cia Vale do Rio Doce	Purchase of 25 per cent of the equity of a steel mill to process iron ore	USA
Copersucar	Purchased Hill Brothers Coffee in 1976 to vertically integrate its coffee production and to take advantage of the brand name	USA
Venezuela Petróleos de Venezuela (PDV)	Acquired refineries and petrol stations through joint ventures which enabled the firm to have a guaranteed market for over 500,000 barrels of oil per day and to sell 75 per cent of its OPEC quota as refined products which are outside the oil cartel's price control	West Germany, Sweden, USA
	Acquired Citgo Petroleum Corporation, the ninth largest marketer of petrol in the USA, in 1989	USA
Several Latin American countries	Paper and publishing, food products, petroleum, steel	Neighbouring Latin American countries
Korea Daewoo	Established a petroleum refinery	Belgium

Table 9.4 (continued)

Home country and firm	Nature of import-substituting investment	Host country
Daewoo and Pukyang Industries and Hasil of Indonesia	Construction of a $10 million tuna cannery	Indonesia
Taiwan 3n 3 Firm unknown at source	Established a petrochemical industry in the form of a joint venture	Malaysia
Hong Kong Firms unknown at source	Secondary processing and marketing of wood products for host country and export. Manufacture of food and chemical products for local market	Several resource-rich neighbouring countries such as Indonesia

Sources: Field research; Katz and Ablin, 1978; O'Brien *et al.*, 1979; Wells, 1980, 1983; Ting and Schive, 1981; Villela, 1983; Katz and Kosacoff, 1983; Lall, 1983b; Lim, 1984; Lecraw, 1985; Pang and Komaran, 1985; ESCAP/UNCTC, 1988; Wells, 1988; Forestier, 1989; Korea Institute for Foreign Investment, 1989; World Bank, 1989; Lee, 1989; Xiaoning, 1989; *The Economist*, 22 December, 1990; Lecraw, 1991; Industrial Development Authority, Ireland.

Table 9.5 The nature of horizontally integrated import-substituting manufacturing investments of some Third World MNEs

Home country and firm	Nature of import-substituting investment	Host country
India Birla, Tata, JK Group, Godrej, Kwality Textiles, Thapars, Mafatlal, Ranbaxy, ASC Enterprises, Hindustan Machine Tools, Kirloskar	Textiles, paper and pulp, food and palm oil processing, motor vehicles and engines, precision tools, metal products, fertilizers, pharmaceuticals, diesel engines, minicomputers	Southeast Asia, Africa, West Asia, South Asia, developed countries
Philippines San Miguel Corporation, Gokongwei Group	Food and beverages, textiles and leather, electrical and electronics	Indonesia, Thailand Hong Kong and Malaysia

Table 9.5   (continued)

Home country and firm	Nature of import-substituting investment	Host country
Thailand		
Saha Union	Opened a factory to produce sewing yarn under its Venus brand name	Georgia, USA
Siam Cement Group	Established in 1990, Tilecera Incorporated, a US $45 million joint venture with Italian-based Jiacobazzi Stilgres to produce ceramic tiles	Tennessee, USA
Mexico		
Cementos Mexicanos	Invested in cement companies in the USA, making the firm the largest cement manufacturer in the North American market and the fourth largest in the world	Texas and California, USA
Vitro	Purchased the second largest US glassmaker, Anchor Glass Container Corporation	USA
Brazil		
Embraer	Aircraft assembly operations	Egypt and Northern Ireland
Bardella	A leading Brazilian capital goods manufacturer, it has bought a 50 per cent stake in the US affiliate of Germany's Schuler GmbH. Both Embraer and Bardella needed a local presence to win contracts in the host countries	USA
Securit	Brazil's largest office furniture maker, it has opened an assembly 37 3 plant and showroom in the USA in order to be able to sell directly to customers	USA
Gradiente Electronica, Caloi, Inbrac, Villares and Eluma	Produce electrical appliances, bicycles, low-technology cables, elevators and motor vehicle parts respectively	Various developing countries
Several other Latin American countries		
Various firms	Invested in the mechanical engineering, transport equipment and electrical appliances industries	Neighbouring Latin American countries

Table 9.5 (continued)

Home country and firm	Nature of import-substituting investment	Host country
South Korea Samsung Electronics and GoldStar Company, Daewoo Electronics, Tong Yang Nylon, Rinnai Korea, Korea Marvel, Maxon Electronics and Tri Gem Computer	Invested in the production of consumer electronics such as televisions, video-recorders, computer-related products, microwave ovens and electronic components	USA, Canada, Japan, and the EC countries
Pohang Iron and Steel Company (steel), Hyundai Motor Company and Borneo International Furniture (BiF)	Similarly invested in production of steel, compact cars and furniture	USA, Canada, Japan and the EC countries, and a number of developing countries such as Turkey, Indonesia and Thailand
Others	Invested in the production of standardised products such as fountain pens, adhesives, construction materials and plastics	South Asian countries, Middle East and China
Dong-A Pharmaceuticals, Hosam Trading and Hanmi Brush Corporation	Set up joint ventures to produce glass bottles, processed corn and cosmetics in November 1989	China
Sinjoo Ind Company and Shinil Leather Garment Company	Launched construction of plant to make gloves, bags and leather garments in December 1989	China
Samsung Company and Daewoo Corporation	Opened branch offices in general trading in 1989 3n 3	Beijing, China
Hyundai Corporation	Seeking to construct a semi-finished tungsten production plant	Tienjin
Lucky-Goldstar International	Considering establishing a joint venture to produce ABS plastics and elevators	China
Young An Hats Company Limited, Baik Yang Company Limited, Kunja Industrial Company Limited and Sung Do Textile Company	Have invested in the textile and clothing industries	USA
Daewoo and Hanil Racket	Produce tennis balls	Indonesia

Table 9.5 (continued)

Home country and firm	Nature of import-substituting investment	Host country
Taiwan		
Tuntex Fiber Company and others	Food and beverages, textiles, plastic and plastic products, non-metallic industries	Various Asian countries
Firm(s) unknown at source	Food and beverages, pulp and paper, rubber and plastics, electronic and electrical appliances, non-metallic minerals, basic metals, mechanical engineering and chemicals	Developing Asian countries, USA
Tatung and others	Electronics and electrical appliances	Export markets such as the UK and Luxembourg
Formosa Plastics Group, China General Plastics	Established a large petrochemical complex ranging from the production of ethylene, the building block of plastic products, to the manufacture of plastic products	Texas and Louisiana, USA
Singapore		
Acma, Yeo Hiap Seng Keppel Shipyard and others	Refrigerators and home appliances, shipbuilding and food products	Developing Asian countries
Khong Guan Group	Established a joint venture with Hong Kong's Far East Consortium to manufacture biscuits	China
Hong Kong		
Bonaventure Textiles Limited, Fang Bros Knitting Limited and Shing Cheong Electronics Limited	Have invested in such sectors as clothing and watches	Ireland 3% 3

Sources: Field research; Katz and Ablin, 1978; O'Brien *et al.*, 1979; Wells, 1980, 1983; Ting and Schive, 1981; Katz and Kosacoff, 1983; Villela, 1983; Lall, 1983b; Koo, 1984, 1985; Lim, 1984; Lecraw, 1985; Pang and Komaran, 1985; Davis, 1987; Wells, 1988; ESCAP/UNCTC, 1988; Forestier, 1989; Korea Institute for Foreign Investment, 1989; Lee, 1989; Simon, 1989; World Bank, 1989; Xiaoning, 1989; *Korea Business World*, April 1990; *The Economist*, 22 December 1990; Industrial Development Authority, Ireland.

later stage, the technological advantages of these firms progressed from the initial ones, which emerged from country-specific advantages in the extraction of natural resources, towards more sophisticated forms of technological advantages in manufacturing and marketing *inter alia* as a result of the limited size of the Canadian market, which necessitated international production in order to maintain competitiveness. These vertically integrated international production activities enabled firms to have a cost advantage and erect new forms of barriers to entry (Rugman, 1981b, 1985b; Rugman and McIlveen, 1985).

Similarly, Sweden's endowment with iron ore and timber explains its comparative advantage in the metals (mainly steel), and the paper and pulp as well as wood products and furniture industries. As in Canada, the small size of the market compelled local firms to export and then produce abroad at a later stage in order to benefit from economies of large-scale production. The extent of foreign production by Swedish MNEs in these industries is concentrated in secondary processing of primary products such as fabricated metals and paper products (Swedenborg, 1979, 1985).

The increased level of sophistication in the sectoral pattern of outward investment of some Third World MNEs at an early stage of development is first reflected in the gradual concentration of investments in simple forms of manufacturing activities which are still primarily linked to country-specific advantages in natural resources. The emergence of an increasing capacity for localised technological capabilities is manifested at this early stage of vertically integrated investments. The underlying pattern is that developing countries at a more advanced stage of development, such as the NICs, exhibit a greater capacity for higher forms of technological innovation and their downstream vertically integrated activities may also be export oriented. Firms from developing countries at a lower stage of development tend to have less sophisticated forms of technological capabilities and their downstream vertically integrated investments may largely take the form of import substitution to serve host country markets.

Evidence of the greater sophistication and trade orientation of the outward investments of the NICs is shown in the case of Hong Kong manufacturing firms which have undertaken secondary processing and marketing of food and wood products for host country and export. Similarly, the Mexican firm Autlan established processing plant in Alabama, USA, and in Venezuela which also serve as marketing outlets for Mexican minerals. The outward investments of the Korean firm Daewoo in a petroleum refining firm in Belgium provide further evidence of the greater sophistication of the NICs' forward vertically integrated manufacturing investments. The firm took advantage of highly skilled technicians and scientists, and the ability to procure crude oil on the spot market in Belgium. On the other hand, firms from countries at a lower stage of development than the NICs are exemplified by the Peruvian firm Cia Minera Buenaventura which purchased equity interests in mining projects in

Venezuela, Ecuador and other Latin American countries primarily to serve domestic requirements.

Perhaps the most important exception that has to be made with regard to countries' stage of development and the complexity of firms' innovative activities is the case of Malaysian MNEs. These firms exhibit a higher degree of technological sophistication in their forward vertically integrated outward investment than firms from other countries at a similar stage of development, owing to their peculiar nature, having once been foreign owned then nationalised by the Malaysian government. For example, Malaysian Mining Corporation is considered to be the largest tin conglomerate in the world, owning at least seventeen tin mining companies and operating thirty-eight of the fifty-five dredges in Malaysia. It has fully integrated activities which include geological exploration, mine development and management, smelting and marketing. Seven affiliates and subsidiaries of the firm are involved in smelting, marketing and manufacture of tin, diamonds, coal and other base metals such as gold. The firm had six large mining ventures abroad in 1984, three involving tin mining in Thailand and three involving diamond and coal mining in Australia. The offshore mines of the firm in Thailand produced about 5 per cent of Thailand's total tin production in 1981 (Lim, 1984).

Sime Darby International Tire Company Incorporated represents another Malaysian conglomerate MNE which was once foreign owned but subsequently acquired by the Malaysian government. The basic activity of the firm is in rubber and palm oil cultivation. It owns and manages some 212,133 acres (86,000 hectares) of oil palm, rubber and cocoa land in West Malaysia and manages another 99,000 acres (40,000 hectares) of agricultural land belonging to the Sabah Land Development Board. In addition to these domestic activities, the firm has established and operates 49,400 acres (20,000 hectares) of rubber plantations in the Palawan Islands of the Philippines in collaboration with the Philippine government, the British Commonwealth Development Corporation and the World Bank. Evidence of the extent to which Sime Darby has successfully engaged in vertically integrated activities in its foreign operations is shown by its takeover of B. F. Goodrich Philippines, a local subsidiary of the American MNE Goodrich which manufactures rubber tyres. Other downstream processing activities of the firm include the acquisition of Carboxyl Chemical Limited in the UK which manufactures metallic stearates, drawn lubricants and defoaments and Surfactant Services Pte Limited in Australia which manufactures biodegradable detergent compounds from vegetable and palm oil.

Table 9.6 summarises the technological advantages of Third World resource-based firms which have engaged in forward vertical integration in their outward investment, according to their stage of national development. Five types of Third World MNEs' manufacturing technology can be distinguished following the categorisation of Lall (1983c), namely, the provision of capital goods; production execution functions; linkages establishment functions;

technical training and skill transfer functions; and the establishment of local research, development, engineering and similar R&D functions.

The evidence is consistent with the generalisation made about countries' stages of development and the sophistication of their firms' innovative activities. The more advanced stage of industrial development of the NICs such as Mexico, Brazil, Hong Kong and Singapore is shown in the greater complexity of innovative activities associated with their outward investments in forward vertically integrated manufacturing activities. Many of the firms from these countries have demonstrated their capacity to develop genuinely unique and localised technological innovations in the production of certain resource-based products, in part through the use of locally available natural resources. For example, the scarcity of coking coal has prompted a Mexican firm and a Brazilian firm to make use of locally available natural gas and charcoal respectively in the steel reduction process. Moreover, the scarcity of petroleum resources in Brazil has prompted indigenous firms to use alcohol and electricity instead of petroleum in the production of an alcohol-powered car, Gurgel, and an electric car. Further evidence of unique technological innovation is seen in Pilao, the Brazilian firm that has patented technology in processing the short fibres from eucalyptus trees. The firm has licensed the technology in Japan, the USA, Belgium and twenty-six other countries and, in addition, has formed a joint venture with a major competitor, the Finnish paper firm Paper Machine Group OY (Villela, 1983).

The evidence therefore shows that the technological skills of firms from the NICs are not limited to project execution and establishment of linkages but also include the provision of technical assistance and capital goods created by the home country and, in some cases, generate genuinely unique innovation. By comparison, MNEs from developing countries on the next tier, such as India, China, Indonesia, the Philippines, Thailand, Peru and Malaysia, have technological advantages generally limited to production execution and establishment of linkages.[9] For example, the purchase by Mantrust of Indonesia and Unicord of Thailand of the US tuna fish companies Chicken of the Sea and Bumble Bee Tuna respectively represents an attempt not only to enhance and support the export of tuna fish from the home country but also to expand their technical capability and skills in tuna fish processing and to gain access to superior managerial resources, established marketing and distribution networks, and the brand names of their acquired firms (Lecraw, 1991).[10] For these firms, the driving force is the fear that the United States will impose countervailing duties on foreign tuna producers in order to protect its dwindling fishing fleet. In addition, in the case of Unicord, the establishment of new markets for frozen Thai seafood and vegetables to be sold under the Bumble Bee brand name is also important (*The Economist*, 22 December 1990). Overall, Third World MNEs have been able to engage successfully in vertically integrated resource-based manufacturing investments in foreign countries through their technological advantages combined with good organisational capabilities in resource-based production.

Table 9.6 The technological advantages of Third World MNEs in vertically integrated import-substituting manufacturing *vis-à-vis* their national stage of development

Home country	GNP per capita, US $, 1988	Technological assets
China	330	Technological capacities for production execution and the establishment of linkages in the exploration for and processing of aluminium and iron
India	340	Process technology and long production experience in palm oil products and the capacity to provide capital equipment
Indonesia	440	Accumulated production execution and linkage capacities in the exploration for and secondary processing and marketing of tuna fish products
Thailand	1,000	Accumulated production execution and linkage capacities in the exploration for and secondary processing and marketing of tuna fish products
Peru	1,300	Long production experience and specialised process know-how in mineral processing in Peru; project execution capability; provision of technical assistance; development of linkages
Mexico	1,760	Process know-how in the extraction, processing and marketing of minerals, using in some cases locally available raw materials, based upon vertically integrated activities in Mexico; provision of technical assistance; project execution capability; development of linkages
Malaysia	1,940	Accumulated expertise in the extraction, processing and marketing of resource-based products such as vegetable and palm oil, tin, diamonds,

Table 9.6 (continued)

Home country	GNP per capita, US $, 1988	Technological assets
		coal and other base metals such as gold, supplemented by the provision of capital goods and technical assistance; project execution capability; development of linkages
Brazil	2,160	Process know-how in the extraction, processing and marketing of such resource-based products as petroleum, iron ore and coffee, in part through the establishment of linkages and the provision of capital goods. The scarcity of natural resources such as coal and petroleum has prompted a number of Brazilian MNEs with technological innovation capabilities to make use of locally available raw materials
Venezuela	3,250	Accumulated production execution and linkage capacities in the exploration for and secondary processing and marketing of petroleum and natural gas. Petróleos de Venezuela has upgraded its exploration and production technology through joint ventures with firms in Germany, Sweden and the USA
Hong Kong	9,220	Production execution capabilities in the extraction, processing and marketing of resource-based products such as wood and food products, helped by the development of linkages; provision of technical assistance and establishment of local R & D

Sources: O'Brien *et al.*, 1979; Wells, 1983; Lim, 1984; Korea Institute for Foreign Investment, 1989.

Horizontal integration of manufacturing firms

A different type of import-substituting investment by Third World MNEs is that undertaken by manufacturing firms which engage in horizontally integrated manufacturing activities in foreign countries in order to exploit their accumulated technological advantages in manufacturing. The importance of these investments has grown significantly since the mid-1980s, as the export-oriented industrialisation strategies of the NICs have been accompanied by the phenomenal growth of exports of manufactured products. This has resulted in balance of trade surpluses and the imposition of trade barriers in their export markets, especially in the developed countries. The NIC firms' pattern of import substituting-cum-surplus recycling form of international production in their export markets, geared to overcome trade barriers and to recycle trade surpluses, runs broadly parallel with the behaviour of Japanese MNEs in the period since the late 1960s.

These investments are also geared to support the industrial restructuring process at home towards more high value-added, technologically intensive activities and to maintain the international competitiveness of local exporters in the face of rising labour costs in the home country. This occurs when the lower technology, more labour-intensive mature industries are transferred to lower-income developing countries with lower levels of technological capacities and an abundant supply of low-cost labour. The pattern of these investments by manufacturing firms is predicted to follow the theory of localised technological change in which international production is associated with the increasing capacity of firms to follow an independent technological trajectory and to generate genuinely unique innovation. The general expectation is that developing countries at a lower stage of development than the NICs, such as the Philippines and the lower-income Latin American countries, have a greater proportion of their import-substituting outward investment in more mature and standardised industries and have less sophisticated forms of technological advantage. Firms from the higher-income NICs have advantages that are increasingly related to simpler kinds of research activity.

The more advanced level of domestic development in the NICs and the increasing sophistication and complexity of their indigenous capital goods sectors, largely as a result of determined government intervention, have enabled local firms to reduce their technological dependence on foreign technology and to generate higher forms of technological advantages. The ownership advantages of these firms, especially those that invest in the developed countries, are based on more sophisticated forms of technological advantages, unlike the ownership advantages of export-oriented investors in lower labour-cost developing countries that stem from the adaptation of foreign technology and/or standardised process to a relatively small scale of operations and some adaptation of product designs to developing countries conditions.[11] Evidence of their more sophisticated technological advantages is shown not only in their ability to transfer disembodied localised technology abroad in the form of technical know-how but also more complex embodied technological advantages in the form of capital equipment because of the well-established local capital goods industries in their home countries. Chapter 7 has shown that the establishment of local capital goods industries is instrumental in the promotion of local technological development, in the enhancement of the bargaining power of local firms in relation to the importation of foreign technology and in the improvement of localised technological change (Stewart, 1977, 1979; Mitra, 1979; Pack, 1981).[12]

Exports of machinery, especially of the second-hand kind, accounted for almost a third of the total investment capital of Taiwanese MNEs in the late 1970s (Ting and Schive, 1981). These exports of machinery are a reflection of Taiwan's rapidly growing indigenous capital goods industry, the products of which have become the fundamental transmitting agent of firms' innovative activities. Similarly, the technological advantages of Brazilian MNEs are reflected in terms of their capacity to provide machinery and equipment, which is a product of the well-developed, predominantly locally controlled capital goods sector in the home market. Argentina also has a fairly developed capital goods sector with sole local design capability which is evident in its low to medium technology turnkey activities abroad and in the overseas import-substituting investments of Argentine private sector enterprises in the pharmaceutical, food processing and light engineering industries (Katz and Ablin, 1978).

The Indian MNEs provide an excellent illustration of the strong link between the pattern of domestic development, the emergence of ownership advantages of indigenous firms and the growth of outward investment. The development of an indigenous capital goods sector in India, aided wholly or in part by government policies that emphasised the establishment of a strong and diverse industrial goods sector, has enabled Indian firms to generate highly sophisticated forms of technological and managerial advantages, competitive with those of firms from the more advanced NICs, which they exploited abroad in the form of outward direct investment. The establishment of a strong capital goods sector in India has enabled the MNEs to devote as much as 67 per cent of their total equity contribution to overseas joint ventures to capital equipment. Further evidence of the high technological assets of Indian firms is their ability to undertake indigenous basic design projects, as well as product and process adaptation, through their emphasis on the continual process of R&D (Lall, 1983b). As a result, Indian MNEs have not only been involved in a large number of projects in relatively simple, low technology areas such as textiles and allied products but also in large-scale, capital and technologically sophisticated industries such as motor vehicles, precision tools, steel tubes, mechanical seals, fertilizers, pharmaceuticals, synthetic fibres, carbon black and minicomputers (Bhatt and Dalal, n.d.)

An analysis of the import-substituting investments of Third World MNEs and the different nature of their technological advantages is undertaken on the basis of specific industries, which allows for more direct comparisons of outward investment between firms from different developing countries. The main industries to be analysed fall in resource-based sectors, such as food and beverages, textiles, metal products and paper, as well as non-resource-based sectors, such as mechanical engineering, motor vehicles, electrical appliances, pharmaceuticals and plastics. The *a priori* expectation is that although MNEs from both the NICs and the lower-income developing countries are engaged in the low and medium technology-intensive and resource-based sectors, those from the higher income countries are more actively involved in the capital and technology-intensive sectors such as chemicals and chemical products, mechanical engineering, electrical and electronic products, and transportation equipment, as a result of the more advanced stage of industrial development of their home countries. In these sectors the MNEs sometimes provide a complete package of embodied and disembodied technology in the form of technical machinery and know-how. Many of these Third World manufacturing MNEs based in the NICs have the technological and organisational capacity to produce high quality, research-intensive products and in some cases to engage in product differentiation as a technological and marketing advantage. This is seen in the case of South Korean enterprises such as Hyundai (motor vehicles), Samsung, Daewoo and GoldStar (consumer electronics), Leading Edge (computers); Taiwanese MNEs such as Formosa Plastics and China General Plastics Corporation (plastic products), Tatung (electrical products) and Acer and Mitac (computers); and also in Brazilian MNEs such as Gradiente Electronica (electrical equipment). Many of their successful overseas operations are largely attributed to their high level of technological competence and skills in product differentiation. These firms have established a good international reputation through the use of a well-designed brand name and logo and, as a result, have pursued a relatively sophisticated product modification strategy despite their limited capacity for any original R&D or new product innovation. The combination of a high level of technological innovation and marketing skills is also reflected in the overseas activities of Gradiente Electronica, Brazil's largest manufacturer of sound equipment, which has established an assembly plant in Mexico and purchased Garrard, a UK subsidiary of Plessey, as part of its long-term marketing strategy which aims to secure Garrard's distribution network in Europe and the USA.

Instances of Third World firms' product differentiation abilities in other less research-intensive sectors are provided by Caloi, Brazil's largest manufacturer of bicycles; Securit, a major Brazilian manufacturer of metallic office equipment, and Venus, the brand name of sewing yarn sold by Saha-Union, a Thai blue chip which produces textiles, garments and footwear. Brand names of soft drinks such as Incacola and F & N have also been promoted by firms from Peru and Singapore respectively, in Latin American and Asian markets. In addition, San Miguel beer from the Philippines is produced and actively promoted in Indonesia, Spain and Hong Kong.[13] The major strategy of these firms is to produce a differentiated product and to engage in price competition based on lower operating costs, owing in part to the use of more labour-intensive equipment. They have established a niche in the middle price range bracket between the lower priced, unbranded products of local firms and the higher priced products of MNEs from the developed countries (Wells, 1973).

Food products

The last chapter showed that the technological advantages of Philippine MNEs in the food and beverage sector such as the San Miguel Corporation and the Gokongwei Group, specifically those of Universal Robina Corporation and Consolidated Foods Corporation, derive principally from their ability to organise and implement the production of light consumer goods supplemented by management skills trained in labour-intensive and capital-intensive means of production, the development of linkages with local suppliers and the provision of continuous technical assistance. However, San Miguel has been successful in transferring production of the internationally known and product differentiated San Miguel beer to previous export markets. Similarly, the Peruvian soft drink firm Inca Cola expanded abroad in such neighbouring countries as Ecuador and Bolivia, where it captured a large share of the domestic market on the basis of patents and trademarks, and also in Los Angeles, California, where it caters to the immigrant Latin American population.

By contrast, an example of the higher capacity for localised technological advantages of food and beverage firms from the NICs is provided by a Taiwanese MNE which produces the food chemical monosodium glutamate (MSG) on the basis of long production experience combined with indigenously generated technology in this area. Such high forms of technological advantage are shared by Tata Oil Mills, a subsidiary of the Tata Group, an Indian oil refining and soap manufacturing firm which built the UNITATA palm oil complex in Malaysia, considered to be the largest, most integrated and diversified palm oil facility in the world. The major factor in explaining the high degree of technological advantages of Indian MNEs, despite their country's lower stage of development, lies in Indian government policies which placed emphasis on the development of an indigenous and predominantly locally controlled capital goods sector, aided in part by restrictive policies towards foreign investment and licensing in the period up to the mid-1980s (Lall, 1982a, 1983b, 1985a).

Textiles, leather and clothing

Another sector in which MNEs from both the NICs and lower-income developing countries have accumulated technological advantages is in textiles, leather and clothing. The Argentine MNE Alpargatas initiated the manufacture of shoes and textiles in Brazil and Uruguay in 1900 while Bunge y Born started operations in Brazil in 1929. Such long production experience is to some extent shared by Taiwanese MNEs in the textile industry, which are actively involved in more technologically intensive synthetic fibre processing as a result of the forward integrated activities of chemical firms into synthetic fibre processing. The Tuntex Fiber Company, which was established in the late 1960s, was the eleventh largest synthetic fibre producer in the industry by 1979, suppling up to 90 per cent of the local market demand. The firm initially acquired synthetic fibre technology through the purchase of equipment from a developed country MNE, which it then improved significantly through process changes.

Such high forms of technological accumulation are increasingly adopted by MNEs from the lower-income developing countries. For example, Coltejer, the largest Colombian textile firm, has planned or may have already established a textile subsidiary in the USA on the basis of outstanding process innovation in printing fabrics. Similarly, Indian MNEs in the textiles sector such as Birla have demonstrated a high degree of embodied and disembodied technological accumulation which is enhanced in some cases by the importation of new and sophisticated machinery from abroad. A notable example of Birla's capacity for indigenous technological innovation is seen in its development of a distinct and unique form of rayon technology (Lall, 1983b). In addition, mention has already been made of Saha-Union, a major Thai manufacturer of textiles, garments and footwear, which established a factory in Tallapoosa, Georgia, to produce sewing yarn under its Venus brand name. The motive for this investment was the need to expand production capacity in the United States in the face of quantitative restrictions on exports of yarn from Thailand (*The Economist*, 22 December 1990).[14]

Metal products

A number of MNEs from the NICs have also been able to develop higher forms of technological advantage in the metal products sector on the basis of long production experience and localised technological innovation. For example, the Argentine firm Siam di Tella established manufacturing plant for metallurgical products in Brazil, Uruguay and Chile and commercial agencies in New York and London as early as 1928. In addition, Industrias Siderometalurgicas Grassi, an Argentine firm engaged in the production of metallic alloys, has undertaken successful overseas joint venture production facilities in Brazil through the development of proprietary technology which is embodied in machinery and equipment designed for small output capacities. As was shown in the previous section, indigenously generated technology in the metal goods sector can also be manifested in the substitution of more locally available natural resources for scarce natural resource inputs. The pioneering technology of the Brazilian steel firm in using charcoal instead of coking coal in the steel reduction process has been successfully transferred to other developing countries which lack adequate local supplies of coking coal. Other firms from lower-income developing countries, such as those from the Philippines, are also rapidly accumulating technological advantages in the metal products sector.

Paper products

Further evidence of the capacity for Third World MNEs for localised technological change is reflected in the paper sector. Apart from Pilao, the Brazilian manufacturer of paper equipment mentioned in the previous section, which has patented technology in paper production from eucalyptus trees, a few firms from the lower-income developing countries have developed from small ventures in the domestic market into internationally oriented organisations, with a regional focus which enables them to pursue a strategy of complementation. For example, Carvajal, a family-owned Colombian firm which manufactures paper products and supplies and which has become one of the biggest firms in the sector, initiated a chain of joint venture investments in other Latin American countries such as Costa Rica, Panama, Nicaragua, Ecuador, Chile and Venezuela, according to a regional division of labour between these countries.

The Indian MNE Birla mentioned in the previous section also owns and operates the most profitable paper plant in Kenya, which is considered to be the largest in the African continent and is equipped with sophisticated machinery and technology.

Mechanical engineering and transport equipment

The greater degree of sophistication of localised technological advantages accumulated by firms from the NICs, allied to the development of an indigenous capital goods sector and greater production experience, is reflected in their increasing importance in higher capital- and technology-intensive sectors such as mechanical engineering, transportation equipment, electrical equipment and chemicals. For example, an Argentine firm which created small-scale machinery to produce generic-type exhaust pipes which fit specific models of automobiles has successfully transferred the technology to Uruguay in the form of a joint venture. In addition, Wobron, Argentina's main producer of automobile clutches and other motor vehicle parts, has developed a competitive asset through the development of less automated discontinuous production technology as a substitute for the highly capital-intensive in-line technology employed by a foreign licenser.

Mention should also be made again of the Brazilian automobile firm that innovated the production of an alcohol-powered car, Gurgel, and an electric car built of steel-reinforced plastic and fibreglass in response to the high cost of petroleum in Brazil. While 25 per cent of the firm's total vehicle production is currently directed to export markets, it is contemplating establishing plant in Panama and Ecuador. This strong capital goods manufacturing capability is shared by a subsidiary of the Indian conglomerate MNE Tata Engineering and Locomotive Company (TELCO), which has generated a high degree of localised technological innovation aided initially by the importation of advanced foreign technology from Daimler Benz. Such high technological capability is demonstrated by the firm's ability to design special machine tools for truck manufacture as well as all the equipment and fixtures required for the establishment of an assembly plant in Malaysia. TELCO is one of the first Third World MNEs in the motor vehicles sector to export equipment, components and know-how.

Electrical equipment

Chapter 6 showed that the production of electrical equipment is an area in which firms from the NICs such as South Korea, Hong Kong, Taiwan, Singapore, Argentina and Brazil have accumulated a respectable level of technological expertise as a result in part of the dynamic role of foreign technology in the electrical and electronic sectors in these countries. Such foreign technology has encouraged the growth of production by indigenous firms through strengthened capacities for localised technological innovation. While a majority of outward investments in the production of electrical and electronic equipment undertaken by firms from the NICs are export oriented, there are a few import-substituting investments, especially in developed countries. These firms have developed a unique and competitive technological advantage *vis-à-vis* firms from the developed countries which enables them to fulfil primarily, but not exclusively, the demand of Third World markets. For example, an Argentine firm has developed the technology for re-designing radios in order to conform to smaller-scale operations. On the other hand, the largest Brazilian manufacturer of sound equipment, Gradiente Electronica, has been able to establish an internationally known trademark and a distribution network on the basis of its experience in production in Brazil's large domestic market, which enabled it to cater to a wide range of consumer and industrial needs.

Chemical products

Perhaps the highest degree of indigenous technological innovation of MNEs from the NICs is reflected in their outward investments in more technologically intensive sectors such as chemical products. For example, Bagó SA, the largest Argentine pharmaceutical firm, developed the capacity to produce chemical raw materials and active components, and in the early 1980s established a modern R&D laboratory which synthesises known chemicals and undertakes both experimental and clinical pharmacology. The firm's technological sophistication is also seen in the area of process engineering and new product design. A combination of technological and professional expertise enabled the firm to design, construct and operate complete turnkey pharmaceutical plant abroad and to engage in FDI. The empirical evidence provided by Bagó SA enables us to detect the early stages of technological accumulation which is expected to become increasingly important and complex as countries advance through higher stages of development and as firms acquire greater experience in international investment.

Chapter 7 showed that Philippine firms have also acquired some technological accumulation in the pharmaceutical sector. However, in contrast to Argentine pharmaceutical firms, the technological advantages of the Philippine ones are limited to process technology in re-packing and dosage formulation. These firms have not developed the technological capacity to undertake the synthesis of raw materials, owing *inter alia* to the high capital and technological intensity of their production and the less than optimal size of the Philippine market which does not allow for profitable economies of scale in drug production at the present time.

Other chemically based sectors in which MNEs from the NICs have accumulated high forms of technological advantage are plastic products. One Taiwanese company, Formosa Plastics, operates six foreign subsidiaries and is a major producer of polyvinyl chloride (PVC) plastic products. This firm acquired proprietary technology and production experience in the establishment and operation of the fifth largest PVC processing plant in Taiwan in 1975.

Table 9.7 illustrates in tabular form the link between the stage of development of Third World MNEs, as measured by GNP per capita, and the nature of the technological advantages of some Third World MNEs in horizontally integrated import-substituting industries. The evidence shows that the degree and form of innovative activities of firms in a particular industry are generally associated with the stage of development of their home countries. Even as countries advance through higher stages of development, the important link between stage of development and degree of firms' innovation remains (see Patel and Pavitt, 1991), although it may gradually weaken as firm-specific factors assume importance as determinants of technological capacity. Such a view of the association between national stage of development and the degree of innovation of local firms is consistent both with the model of the product cycle and that of localised technological change of Third World multinationals.

The point at which these theories differ is in the scope of such innovation. The dynamic theory of Third World outward investment adopted in this book recognises the importance of Wells' product cycle model in explaining the limited innovativeness of Third World MNEs in the early stages of outward investment. The technological development of these enterprises engaged in outward direct investment in mature, standardised and often resource-based industries such as textiles, metals and non-metallic mineral products is strongly linked to the technological course of firms from the developed countries, which is assimilated and adapted to suit Third World markets and production conditions. The adaptations pertain mainly to the scale of operations, factor intensity, lower production costs including factor costs, use of locally available raw materials and equipment, and other product and process design, which represent accumulated production experience and learning by doing and using.

Table 9.7 The technological advantages of Third World MNEs in horizontally integrated import-substituting manufacturing *vis-à-vis* their stage of development

Home country and firm	GNP per capita, US $, 1988	Technological advantages
India	340	
Birla		High degree of embodied and disembodied localised technological change both indigenously generated and enhanced by the importation of new and sophisticated machinery from abroad. The main technological advantages of Birla and other Indian MNEs engaged in the production of textiles in developing countries are based on providing an adapted and older process, product or technical service that is more appropriate to the needs of host developing countries because of its smaller scale and greater ease of operation, better adaptation to the local environment or better match between product attributes and local requirements. However, the Indian–Indonesian joint venture that operates the largest rayon (synthetic fibre) producing plant in Indonesia uses state-of-the-art technology comparable to any Japanese or other rayon-producing plant. Such access to state-of-the art technology and capital resources was made possible through a tie-up with an Austrian company which is prominent in the manufacture of rayon producing equipment
Tata Engineering and Locomotive Company		Design of special machine tools for truck manufacture, aided initially by the importation of advanced

Table 9.7 (continued)

Home country and firm	GNP per capita, US $, 1988	Technological advantages
		foreign technology from Daimler-Benz
Tata Oil Mills		Process technology and long production experience in products based on palm oil; provision of capital equipment
Philippines	630	
San Miguel Corporation, The Gokongwei Group		Organisation and implementation of production of light consumer goods and management skills combined with the development of linkages with local suppliers and the provision of continuous technical assistance; ability to engage in product differentiation
Colombia	1,180	
Coltejer		Capability for outstanding process innovation in printing fabrics
Carvajal		Sophisticated technological innovation in paper processing that resulted in rationalised investments in several Latin American countries
Peru	1,300	
Inca Cola		Ability to engage in product differentiation
Mexico 37 3	1,760	
Firm unknown at source		Technological innovation to make use of locally available raw materials such as gas instead of coking coal in the steel reduction process
Cementos Mexicanos, Vitro		Advanced technological capabilities to become world-class players in the cement and glass sectors respectively. Vitro is a

Table 9.7 (continued)

Home country and firm	GNP per capita, US $, 1988	Technological advantages
		technologically dynamic, vertically integrated glass group that has had extensive experience in the export of glass products, glassmaking equipment and process technology. The firm initially acquired the process technology and production design through licensing and technical assistance contracts with a US firm and then adapted them to suit a smaller market. The adapted process technology combined with the glassmaking equipment provided by Fama, a local capital goods producer with design and production capabilities, has enabled Vitro to expand its product line and to considerably improve the productivity of the originally imported process technology. Fama's design and production of glassmaking equipment was prompted by the unavailability of glassmaking equipment from the USA, Germany and the UK during the Second World War
Brazil	2,160	
Firm unknown at source		Technological innovation to make use of locally available raw materials such as charcoal instead of coking coal in the steel reduction process and alcohol and electricity in the production of the alcohol-powered car, *Gurgel*, and the electric car
Gradiente Electronica and other Brazilian electrical appliance manufacturers		Technological adaptation that results in the production of tropicalised electrical appliances suited for Africa

Table 9.7 (continued)

Home country and firm	GNP per capita, US $, 1988	Technological advantages
Caloi, Securit		Ability to engage in product differentiation as a technological and marketing advantage in the production of bicycles and metallic office equipment. The advantages of these firms are based mainly on the management skills for manufacturing in small volume which is more appropriate to the needs of host developing countries
Pilao		Unique technological innovation in processing the short fibres from eucalyptus trees which the firm has patented and licensed in Japan, the USA, Belgium and 26 other countries
Cotia		Localised technology related to bottling a soft drink derived from local fruits
Marcopolo		Localised technology related to building buses designed for warm climates
Inbrac		Low technology for cable making machinery
Villares		Technology for production of sophisticated elevators and escalators
Eluma		Leading Brazilian maker of copper products and auto parts, Eluma established foreign operations in Latin America and South Africa through technology developed through its joint ventures with Bundy of the US
Argentina	2,520	
Firms unknown at source		Design capability for small-scale machinery which produces generic-type exhaust pipes to suit specific vehicles

Table 9.7 (continued)

Home country and firm	GNP per capita, US $, 1988	Technological advantages
		Process technology in redesigning radios to conform to smaller-scale operations
Wobron		Development of less automated discontinuous production technology as a substitute for the highly capital intensive in-line technology employed by a foreign licenser
Industrias Siderometalurgicas Grassi		Proprietary technology embodied in machinery and equipment designed for small-scale processing of metallic alloys
Bagó SA		Product and process innovation to produce raw materials and active components of drugs, to synthesise known chemicals and to undertake both clinical and experimental pharmacology
Angel Estrada, Bunge y Born, Cabsha, Arcor, Yelmo SA, Siam di Tella		Ability to engage in product differentiation as a technological and marketing advantage
South Korea		
Various firms	3,600	Possesses idiosyncratic or specific product and/or production know-how and some marketing development abilities in specific products that cater to a segmented market in the developed host countries, or alternatively possesses the intermediate technology to produce standardised products for the entire market of developing countries
Hyundai Motor Company		Successfully engaged in the export and overseas production of motor

Table 9.7 (continued)

Home country and firm	GNP per capita, US $, 1988	Technological advantages
		vehicles, through technological, management and marketing expertise in the production of small and medium-sized, inexpensive motor vehicles, areas which the Japanese motor vehicle firms have left behind in their move to the production of larger, more expensive vehicles. Hyundai was able to upgrade its technology on the basis of active adaptation and assimilation of foreign technology in relation to motor vehicle production for which the firm had obtained licence until 1982. South Korea was estimated to have a share of nearly 4 per cent of the US automobile market in 1987, up from zero in 1985
Gold Star, Samsung, Leading Edge, Hyundai		Technological and marketing skills to engage in product differentiation in consumer electronics, computers and subcompact automobiles. Within two years of its introduction in 1984, the subcompact Korean car, Pony, had become the best-selling foreign car in Canada
Samsung, Daewoo and Pohang Iron and Steel Company		Advanced technological know-how and marketing skills allied to their sophisticated function in the establishment of linkages in the production and marketing of consumer electronic products and steel
Borneo International Furniture (BiF) Korea		One of Asia's largest furniture manufacturers, it has a global perspective on the design and manufacture of contemporary home and office furniture. With the use of furniture designs from Italy, the firm manufactures furniture using the most

Table 9.7 (continued)

Home country and firm	GNP per capita, US $, 1988	Technological advantages
		sophisticated German technology and US assembly line techniques
Taiwan	6,396	
Firm unknown at source		Long production experience and technology in the production of monosodium glutamate
Tuntex Fiber Company		Technological expertise and long production experience in synthetic fibre processing, developed initially through the purchase of equipment from a developed country MNE but which the firm adapted to suit local conditions; provision of capital equipment
Formosa Plastics		Proprietary technology and production experience in polyvinyl chloride (PVC) plastic products. The firm acquired a chemical plant of Imperial Chemical Industries (UK) located in Baton Rouge, Louisiana, USA, in 1980
Tatung		One of the leading electrical and electronic manufacturers in Taiwan, the firm has technological expertise in the production of consumer household appliances that can meet European demands. The firm also has the marketing ability to engage in product differentiation using American advertising
Other Taiwanese MNEs		Competitive assets in developing countries are their capability to produce a product more compatible with the requirements of host developing countries through the use of intermediate technology and superior marketing skills

Table 9.7 (continued)

Home country and firm	GNP per capita, US $, 1988	Technological advantages
Singapore	9,070	Ownership-specific advantages in manufacturing and international production in areas related to their business
Wearne Bros, Fraser & Neave Limited		Some of the more advanced technological advantages of their MNEs in the distribution of motor vehicles and equipment and a soft drink company are explained by their status as former British firms which were bought out by local firms
Acma		Production experience in the manufacture of refrigerators and other home appliances, nurtured during the period of tariff protection in the 1960s and 1970s
Hong Kong	9,220	Technological advantages in the manufacture of low-technology, labour-intensive and mature industries have been developed in some cases through joint ventures with partners from Europe, Japan and the USA while others have built on technology and expertise brought to Hong Kong from mainland China after 1949

Sources: Primary data obtained from field research; Diaz-Alejandro, 1977; Katz and Ablin, 1978; Rhee and Westphal, 1978; Wells, 1980; Ting and Schive, 1981; White, 1981; Lall, 1982b, 1983b, 1985a; Katz and Kosacoff, 1983; Villela, 1983; Dahlman and Sercovich, 1984; Pang and Komaran, 1985; Ford and Clifford, 1987; Kraar, 1987; Kwon and Ryans, 1987; ESCAP, 1988; Kucway, 1988; Wells, 1988; Baker, 1989; Korea Institute for Foreign Investment, 1989.

However, as countries advance through higher stages of development and as firms acquire greater experience in international production, Lall's model of localised technological change becomes increasingly significant. Such a model, which is implicit in the theories of technological accumulation and technological competence, suggests that firms from developing countries have the ability to follow a technological course independent of that of the more developed countries, through either indigenous technological creation, adaptation of foreign technology or the revival of techniques which were used at an earlier stage in developed countries. Whereas the growth of external market demand constitutes the primary determinant of relocation of production by local firms in the product cycle model in the early stages of Third World outward investment, the existence and continuation of technological accumulation become the primary determinant of international production by local firms in the model of localised technological change and technological accumulation in the later stages of Third World outward investment.

Allied to the technological advantages and size of a firm is production experience. Such experience has been interpreted in two different ways by Wells and Lall. Production experience in Lall's sense is part and parcel of the process of localised technological accumulation by which Third World firms, in particular, acquire experience through learning by doing and learning by using. By contrast, the distinctive nature of production experience for Third World MNEs in Wells' sense derives not so much from the acquisition of unique technological advantages as from managerial experience in serving the middle and lower ends of the market and cost structures, which are enhanced by the lower costs of managerial and technical staff and their greater flexibility and adaptability (Wells, 1978).

Related to the nature of technological advantages of Third World MNEs is the size of the firm. Most of the MNEs that originate from developing countries, like those from the developed countries, are large firms, especially those that invest in the more advanced, industrialised countries to establish access to markets as well as new technologies and skills. Among the NICs in Asia and Latin America, South Korea's outward investments are generally more associated with large conglomerate firms or *chaebol*. Table 9.8 gives a descriptive cross-section of the size of import-substituting firms emerging from a number of developing countries. The table suggests that a majority of Third World MNEs in import-substituting investments are large firms, able to devote more financial resources to R&D both relatively and absolutely, and able to generate technological ownership advantages that help to reinforce the firm's large size and ability to exercise increased market power, in the Hymer sense, which is necessary for world competitiveness. The large size of a firm and its conglomerate nature also enable it to be an efficiency-seeker in the Coasian sense by engaging in economies of scale and scope. The conglomerate nature of some Third World MNEs has enabled them to gain access to larger financial resources, technologies, management and other skills and market information, as well as to benefit from various forms of political influence and

Table 9.8 The nature of MNEs from developing countries

Home country and firm	Nature of the size of the firm
China	
Shenzhen Electronics Group	The largest electronics conglomerate in China
India	
Birla, Tata	The largest industrial groups in the country which together control about 20 per cent of the assets of the entire corporate sector of India
JK Group, Shahibag Enterprises, Thapars Group	Large, diversified and long-established firms which have a broader range of manufacturing activity in the home compared to overseas markets. These groups along with Birla and Tata account for 67.8 per cent of Indian overseas investment in ventures in operation
Philippines	
San Miguel Corporation, Gokongwei Group	The largest manufacturers of consumer and producer goods in the Philippines
Mexico	
Cementos Mexicanos, Vitro	The largest cement and glass companies in Mexico
Malaysia	
Sime Darby Group	Conglomerate firm whose total assets and sales amounted to US $0.9 billion and US $1.2 billion in 1988. Listed as the 387th largest company in Asia by ELC International ranked by sales in 1988
Malaysia Mining Corporation	Second largest Malaysian MNE with vertically integrated activities in the exploitation, processing and marketing of various primary products. Had total assets and sales of US $324 million and US $238 million in 1988
Multi-Purpose Holding BhD	Essentially a holding firm with investments in banking, hotels, insurance, property, trading, shipping, manufacturing and plantations. The firm has expanded phenomenally through the acquisition of large foreign firms abroad
Malaysian United Industries	Initially a small private limited firm that manufactured light consumer goods such as toothbrushes and plastic products. The expansion into international production was initiated by the purchase of 20 per

Table 9.8 (continued)

Home country and firm	Nature of the size of the firm
	cent equity interest in Peters (WA) Limited, the largest firm in Western Australia engaged in food products
Brazil	Two-thirds of the 33 Brazilian multinationals are leaders in their respective sectors
Petrobrás	One of the world's largest corporations with 1985 sales of US $17 billion. The firm is a multi-faceted industrial conglomerate with diversified domestic and international activities
Argentina	
Bunge y Born	Has annual sales of over US $10 billion, owns 89 Argentine companies and more than 1,730,000 acres (700,000 hectares) of agricultural land; employs tens of thousands and would be the world's third largest grain trader with offices in 80 countries. Moreover, it owns Argentina's largest flour producer, Molinos Rio de Plata, two of Brazil's top food manufacturers, Sanbra and Samrig, Spain's second largest edible oils company, Arlesa, as well as operations in textiles, finance, chemicals, packaging, computers and paints. Political reasons prompted the firm to move its international headquarters from Argentina to Brazil in 1974 and by 1981 it had 28,000 employees there
Singapore	
Acma, Prima, Yeo Hiap Seng	Large manufacturing firms in the electrical appliance and food sectors which have expanded abroad in activities related to their main area of business
Jack Chia MPH, Wah Chang, Haw Par Brothers, Intraco, Joo Seng Group, Keck Seng Group	Conglomerate firms involved in highly diversified activities in the primary, secondary and tertiary sectors; have become established international investors
Latin America	A majority of Latin American MNEs are medium-sized firms
South Korea	
Hyundai Motor Company	The largest automobile manufacturer in South Korea

Table 9.8 (continued)

Home country and firm	Nature of the size of the firm
Samsung Daewoo	The largest conglomerate firms in South Korea
Taiwan	MNEs from Taiwan are large relative to other firms in their specific industries

Sources: Field research; Diaz-Alejandro, 1977; Lall, 1983b; Lim, 1984; Schive and Hsueh, 1985; Wells, 1988; Forestier, 1989; Staubus, 1989.

from a greater ability to bear risks and initial loss. However, in recent years, as a result of the general trend towards internationalisation of business with more recent outward investor firms engaging in international production at an earlier stage, compared to more traditional investors, and the growing importance of outward investments in export-oriented, labour-intensive manufacturing by small- and medium-sized firms in the Asian NICs that have experienced the greatest difficulty in maintaining price competitiveness for their exports, investments by such firms are becoming increasingly significant. For example, in the case of South Korea, the share of outward foreign direct capital stock accounted for by small- and medium-sized firms increased from less than 2 per cent before 1985 to over 4 per cent in 1988 and 31.5 per cent by the first quarter of 1989 (Jun, 1989).

Export-oriented manufacturing investments

Outward direct investments in the export-oriented manufacturing sector are more significant for the NICs than for the lower-income developing countries because of the rapid growth, increased outward orientation and enhanced competitiveness of firms from the NICs. By 1989, Hong Kong, Taiwan, South Korea, China, Singapore, Mexico and Brazil belonged to the twenty-five largest leading exporters in world merchandise trade (GATT, 1991). Table 9.9 summarises the nature of the export-oriented investments of some Third World MNEs.

Evidence suggests that firms from the NICs, in particular South Korea, Hong Kong, Taiwan, Singapore and, to a much lesser extent, Brazil, are most actively engaged in export-oriented manufacturing investments, particularly because of their wider policy strategies to shift the industrialisation process from one based on import substitution to one of export promotion.[15] The larger volume of export-oriented foreign investment from Asian countries has been generally attributed *inter alia* to several institutional factors, such as the absence of constraints on foreign exchange, the presence of overseas ethnic Chinese and Indian communities and the political geography of Asian countries. However, one of the most important factors in the faster growth of

Table 9.9   The nature of export-oriented manufacturing investments of some Third World MNEs

Home country	Nature of export-oriented investments	Host country
Philippines Construction & Development Corporation of the Philippines	Secondary processing of timber for export	Sabah, Malaysia
Brazil Cia Vale do Rio Doce	Secondary processing and marketing of iron ore and steel products for export	Egypt
Firms unknown at source	Processing of Brazilian timber for export to European and Arab markets	Malta
	Electrical and electronic equipment	Mexico, UK
Labra	Claims to be the world's second largest manufacturer of pencils; invested US $130 million in pencil factory in Portugal to increase its exports to the EC	Portugal
Nansen	Exports between 10% and 20% of electrical products to Latin America. Opened a plant in Colombia to serve as a springboard for Third World country sales	Colombia
Korea Various firms[1]	Artificial chemicals, cement, garments, shoes, electric cables, motors and diesel engines, paper and plywood, steel, ship building, electrical and electronic equipment	Neighbouring Asian countries, Latin America, developed countries
GoldStar	Established an overseas television factory in Mexico to produce colour television chassis for export to its US plant	Mexico
Taiwan Firm unknown at source	Garments, toys, electronic and electrical equipment	Lower-income developing countries
Singapore Intraco, Keck Seng, Guthrie, Sim Lim Group	Secondary processing and packaging of various agricultural, forestry and natural resource products for export	Neighbouring Asian countries, Australia, Canada

Table 9.9 (continued)

Home country	Nature of export-oriented investments	Host country
Prima Flour Other firms unknown at source	Textiles and garments, food and beverages, leather products, wood products, rubber products, glass and non-metallic mineral products such as bricks, tiles and clay products, jewellery, plastics, paper and printing, fabricated metal products, machinery	Lower-income developing countries in Asia, Africa and the Middle East
Hong Kong Semi-Tech	Produces low-priced, high-quality consumer durables such as tape decks, telephones, sewing machines and electronic toys under the Singer brand name. These products are shipped to a retail network spanning 100 countries	Asia
Other firms unknown at source	Textiles, toys, garments, footwear, watches, chemicals, consumer electronic equipment and electrical appliances	Lower-income developing countries such as China, Indonesia, Thailand and Sri Lanka

Sources: Field research; Jo, 1981; Ting and Schive, 1981; Chen, 1981, 1983a, 1983b; Villela, 1983; Kumar and Kim, 1984; Lim, 1984; Lecraw, 1985; Wells, 1988; Korea Institute for Foreign Investment, 1989; Friedland, 1990.

Note: 1 Some of these Korean firms such as the clothing manufacturers Dong Bang & Company Limited, Kolon International Corporation, Shinsung Corporation, electrical and electronic manufacturers such as Daewoo Electronics (refrigerators and compressors), Rinnai Korea (gas ranges), GoldStar (colour TVs, audio equipment, washing machines), Samsung Electronics (TV and TV components, refrigerators and video-recorders), Korea Marvel (electronic components), Dai Shin Denken Company (electronic components), Sammi Sound Tech, Korea Toptone (speakers), Korea Development (telecommunication equipment), GoldStar Telecommunication (telephone sets), Maxon Electronics (consumer electronic products) and footwear manufacturers such as Dong Hae Chemical, Starwin, Korindo, H.S. Corporation, Sung Hwa Corporation, Tong Yang Rubber, Tae Kwang Industries., Kukje Corporation, Buhan Chemical and Sung Hwa Corporation have invested in neighbouring Asian countries, Latin America and Caribbean countries for export. In order to exploit Mexico's favoured status as a duty-free country, GoldStar established a plant in Mexico to produce colour television chassis for export to its US plant. MNEs from Taiwan have invested in electronic and electrical equipment sectors in lower-income developing countries for export.

export-oriented Asian MNEs is government policies that emphasise export-oriented industrialisation with growth dependent on export earnings, which has resulted in large amounts of exports of manufactured goods from Asia. Regardless of the motives for foreign investment, developing country firms in general and Asian firms in particular exploit an asset built from export experience (Wells, 1980).[16]

The greater importance of export-oriented industrialisation strategies and rapid growth in Asia and the slower pace of growth of Latin American countries, due to macroeconomic difficulties, debt service and competitiveness problems, helps to explain why the greater maturity of Latin American MNEs, owing to the earlier industrial development of their home countries, has nevertheless been surpassed by the rapid development of Asian MNEs in recent years (World Bank, 1989).[17] By 1982, the value of outward direct investments of Asian MNEs was estimated to be US $7,300 million compared to US $5,205 million for Latin American MNEs.

The importance of exports of manufactured products has induced Third World firms to overcome protectionist pressures in traditional export markets as well as rising labour costs and/or diminishing labour supplies in their home countries in order to preserve their international competitiveness and open new export markets based on export experience.[18] In the case of Taiwan and South Korea, the effects of currency appreciation combined with large balance of payments and trade surpluses since the mid-1980s provided an even greater incentive to rationalise and maintain the competitiveness of the export-oriented industries that are most vulnerable to cost fluctuations (World Bank, 1989; Chia, 1989). This behaviour parallels the development of Japanese MNEs from 1950 to the mid-1960s and has been encouraged by home governments.[19] More importantly, export-oriented foreign investments can support the industrial restructuring process at home by the transfer of lower value-added, labour- and resource-based industries to more competitive foreign locations and the upgrading of domestic activities towards high value-added, technologically intensive industrial activities. Evidence in favour of this view of the role of outward direct investment in domestic industrial restructuring is found in almost all the Asian NICs that have invested in other developing Asian countries and have only transferred or relocated the production of standardised, lower value-added products in such industrial sectors as apparel, toys and watches or electronics in the host countries, while maintaining the higher value-added, more technologically complex and skill-intensive production in the home country (World Bank, 1989).

As a consequence of their high dependence on export earnings for growth, firms from the Asian NICs in particular have established export platform production facilities in standardised, low-technology and labour-intensive products in lower-income developing countries.[20] Although these investments were initially final assembly or packaging operations, the domestic content regulations of the Multi-Fibre Arrangement forced some firms to relocate their textile and clothing manufacturing operations (Chia, 1989).

During the early stages, manufacturing firms from the NICs engage in horizontally integrated investments principally for the local assembly of a product identical to the parent company's product. At a later stage, as home country operations become more sophisticated, firms may engage in vertically integrated investments by transferring the more labour-intensive stage of their multi-stage production processes to lower-income developing countries. Four factors have proved conducive in attracting investments to these new sites. The first is an abundant supply of lower-cost labour enabling firms from the NICs to lower their costs of production further. The second factor is that quantitative restrictions may not yet have been imposed or may be less severe in these countries, or the host country in question may have preferential trading arrangements with targeted export markets. The third factor is that production in these countries saves transportation costs and transit time in shipping because of their geographical proximity to targeted export markets. Examples of this are found in the outward direct investments of South Korean MNEs in such industries as textiles, clothing, toys, hats and electrical and electronic components, not only in South Asia but also in the Caribbean region, which lower their production costs in terms of labour, transportation and transit time as well as overcoming import restrictions imposed by the North American market on their export products. In addition, direct investments in the Caribbean Basin countries are particularly attractive because of their free trade status with the USA, accorded through the Caribbean Basin Initiative, which makes most exports from these countries exempt from payment of US customs duties for twelve years.[21] In the case of Taiwan MNEs, an additional incentive for investment in the Caribbean Basin area is that twelve of the twenty-five countries that still maintain formal diplomatic relationships with Taiwan belong to this region (Chen, 1986; Korea Institute for Foreign Investment, 1989). The fourth factor particularly pertains to firms from Hong Kong and Taiwan, which seek production sites in politically stable countries in response to pressure from their customers, many representing large retail establishments in the USA and Europe. Some of these customers are reluctant to maintain long-term supply contracts with firms that produce in politically unstable locations such as Hong Kong and China.

These export-oriented outward investments enable NICs firms to maintain a greater competitive edge when exporting to established markets in the developed countries, to overcome the quantitative import restrictions on their manufactured exports, to take advantage in some cases of preferential trading arrangements accorded their host countries and to diversify their investment risks by production in several locations. At a later stage, the rationalised production and product segmentation associated with the transfer of production that embodies higher technological intensity is found, especially in Taiwanese firms. For example, one Taiwan firm simultaneously assembles electronic filters and magnetic devices in Thailand, technology- and skill-intensive power supply units in Mexico, and final PC units in Taiwan (World Bank, 1989). Similarly, the international operations of another Brazilian firm

Table 9.10 The technological advantages of some Third World MNEs in
export-oriented investments

Home country	Nature of technological assets
Philippines	Process know-how in the exploration for, secondary processing and marketing of timber which includes design, layout and continuous technical assistance based upon vertically integrated manufacturing facilities in the Philippines; project execution capability
Brazil	Process know-how in the extraction, processing and marketing of iron and steel products; design, layout and engineering know-how; project execution capability; provision of technical assistance; development of linkages; provision of capital goods. Labra is able to sell pencils competitively to the EC from its production base in Portugal partly through its access to licensed technology from Europe.
Korea	Ability to initiate and operate overseas projects at relatively low costs. Original technology sourced mainly from developed countries through licensing, outright purchase or joint ventures and then adapted through experience to relatively small-scale operations and product designs to suit Korean conditions in the first instance and then transferred to lower-income developing countries
Taiwan	Technological and marketing advantages reflected in the ability to engage in product differentiation in the production of electrical and electronic equipment
Singapore	Accumulated expertise in the extraction, processing and marketing of various resource-based products such as edible oil, through the provision of capital goods, development of linkages and the provision of technical assistance; product and process technology leading to products and processes suitable to conditions in host countries; ability to keep abreast of latest technology and to engage in product differentiation. Initial technology largely obtained through licensing agreements with firms from developed countries but subsequently established own R&D which generated

Table 9.10 (continued)

Home country	Nature of technological assets
	the capabilities for designing components for electrical appliances and/or product and process innovation. Unique product innovation in the development of a strong plastic film
Hong Kong	Extensive experience in the production of labour-intensive products. Some elements of manufacturing know-how are embodied in their access to, or design of, machinery generally designed for shorter production runs and lower volume manufacture. Know-how is also embodied in the minds of managers and engineers who have long problem-solving experience. Hence there are elements of more sophisticated innovation in the form of new product designs, new machinery and equipment, new production processes, in addition to their skills in adaptation and modification of foreign technologies on the basis of long production experience. Initial technology is largely acquired through assistance from a foreign firm, Hong Kong manufacturing associations, in-house innovations and technology marketing agencies. Machinery, equipment and raw materials largely sourced from abroad. Technological advantages are supplemented by marketing advantages achieved in some cases through product differentiation, e.g. Singer by Semi-Tech. Similarly, Hong Kong watch manufacturers such as Asia Commercial, National Electronics and Stelux Holdings claim the use of prestigious brand names through the acquisition of Swiss and French watchmakers

Sources: Field research; Wells, 1973, 1978; Ting, 1980a, 1980b; Jo, 1981; Ting and Schive, 1981; Villela, 1983; Chen, 1983a, 1983b; Euh and Min, 1986; Kumar and Kim, 1984; Lim, 1984; Lecraw, 1985; Wells, 1988; Goldstein, 1990.

consist of a manufacturing plant to process timber in Malta, which is neither the source of raw materials nor the principal market for the finished products. The wood-based products of this plant, established by a Brazilian enterprise in collaboration with Maltese and Libyan interests, are directed to European and Arab markets (O'Brien *et al.*, 1979).

Table 9.10 summarises in tabular form the major technological advantages of export-oriented Third World MNEs. The major ownership advantages of these firms from the NICs emerge from their abilities to organise and implement the production of various resource-based industrial products or light consumer goods such as textiles, clothing and footwear, toys and sophisticated electrical goods. The technological innovation of these firms lies not so much in their ability to innovate new product or process technology or in providing machinery and equipment sourced from the home country but in integrating foreign product and process technology with their own localised technological innovation in order to suit production conditions in host countries. These modifications pertain mainly to the scale of operations and labour intensity and productivity, and represent long production experience and learning by doing and using in their home countries.

The distinctive nature of firms from the different NICs derives from their different capacities to engage in product and process innovation. The particular nature of the industrial development and the small market size of the Asian city states has contributed to the more limited technological competence of their firms and the greater need for foreign technology to strengthen local efforts in technological change and research and development (Chia, 1989). For example, Hong Kong's limited industrial development has meant that its MNEs are not as capable of transfering the sophisticated technology necessary to conduct more technologically intensive export-oriented foreign operations. Technological advantages in Hong Kong are largely based on the mastery of foreign technologies in the assembly and manufacture of light consumer products for export that are suited to production conditions in lower-cost and labour-abundant developing countries, and on efforts in improving the efficiency of such production processes through managerial and export marketing advantages (Chen, 1983b; Lall, 1991). However, in recent years Hong Kong firms have begun to produce more technologically sophisticated electronics and other products in developing countries where conducive locational advantages exist, such as technical infrastructure in Malaysia (World Bank, 1989).

Similarly, firms from Singapore have limited capacities to innovate in product and process technology, as shown by their lower level of scientific, engineering, technical and management skills. The major constraint on the development of an indigenous R&D capacity by Singaporean firms has been largely attributed to their continued reliance on foreign MNEs to provide a package of technology, capital, management and access to markets (Ting, 1980a). Nevertheless, the future course of their technological advantages will largely be influenced by their greater dynamism and flexibility owing to their capacity to keep abreast of the latest technology. Singaporean conglomerate firms such as Jack Chia-MPH, Wah Chang International, Haw Par Brothers, Intraco, Joo Seng Group and Keck Seng Group have long engaged in international operations, in the initial stages as trading firms but over time they have accumulated enough capital and localised technological advantages to diversify into different sectors which may or may not have been related to their original activity. As a result, these conglomerate MNEs are able to undermine competitors through the practice of cross-subsidisation, where profits can be shifted between product lines.

However, apart from the Asian city states, there is increasing evidence that a number of these MNEs, particularly those based in the more industrially diversified South Korea and Taiwan, has graduated to using more advanced technologies requiring greater use of capital goods and engineering and technological expertise. As a result of massive promotion on the part of the home government and private sector, as shown in Chapter 6, MNEs from South Korea and Taiwan have figured prominently in the high-technology electronics and semiconductor industries in recent years.

Although firms from the NICs feature prominently in export-oriented manufacturing investments, there is growing evidence that firms from lower-income developing countries are increasing their international production in export-oriented manufacturing activities, often in less technologically sophisticated and resource-based sectors. For example, the Philippine MNE Construction & Development Corporation of the Philippines (CDCP) has vertically integrated into the secondary processing of timber in Malaysia in a joint venture arrangement with Malaysian firms whose products are primarily geared for the export markets in Japan, Korea and Taiwan. Hence, CDCP has been able to supplement its capacity for localised technological innovation in timber extraction, gained through long production experience, with the more sophisticated forms of technological advantages of Malaysian firms, which enabled it to engage in more competitive forms of trade-oriented outward investment.

The technological advantages of export-oriented MNEs from the Third World and especially those from the Asian NICs are enhanced by superior but lower-cost management and marketing skills, which are products of extensive export experience and the establishment of contacts with customers in developed countries. Many of these firms have established the reputation of being reliable suppliers for their customers in export markets. Their marketing advantages have been developed through the establishment of international distributive networks as well as through their ability to engage in price competitiveness rather than product differentiation. In addition to these advantages, the well-developed capital markets in many of the NICs such as Hong Kong, Singapore and Brazil have enabled these firms to gain access to finance for their outward investments. The extent to which they capitalise on their accumulated technological expertise is reflected not only in their direct investments abroad but also in other forms of foreign economic involvement, such as plant exports, licensing, consulting and technical services (Dahlman and Sercovich, 1984).

RESEARCH-INTENSIVE INVESTMENTS

Another illustration of the more sophisticated types of overseas activities undertaken by firms from the most advanced developing countries, or NICs, is the access gained to the most sophisticated and advanced forms of manufacturing technology in developed countries. These overseas investments in R&D are expected to increase since the NICs and some lower-income developing countries are rapidly upgrading their domestic industrial structure away from the production of unskilled, labour-intensive products towards higher skill and more research-intensive activities.[22] A number of these investments are encouraged by the governments of the relevant countries in order to promote industrial development in the direction of high technology. An example of this is provided by the Economic Development Board, the Singaporean government's investment promotion agency, and Singapore Technology Corporation which not only actively assist local firms to undertake research linked investments but are themselves investing in US technology companies. Firms from Brazil, South Korea, Taiwan, Singapore, Hong Kong and some other developing countries, such as Venezuela and India, are already producers and, in some cases, significant exporters of products in such research-intensive sectors as biotechnology, software, computers and peripherals, semiconductors and telecommunications.[23]

A few examples of these research-linked investments by Third World MNEs in developed countries is given in Table 9.11. The table shows that they have been undertaken both by the NICs, mainly Taiwan, Singapore and South Korea, and to a lesser extent by the lower-income developing countries such as the Philippines and Malaysia.[24] An estimated thirty or more companies in Silicon Valley – or some 50 per cent more than in 1989 – received Taiwan financing in 1990.[25] Unlike the other Asian NICs, the investments of Hong Kong firms in the developed countries have concentrated on real estate and retailing ventures (*Asian Wall Street Journal Weekly*, 14 January 1991).

The more complex nature of activity funded by these investments is related to the more advanced industrial development of these NICs and their desire to obtain more complex technological capabilities in the production of research-intensive and profitable manufactured products. These have included chemicals, semiconductors and computers in the case of Taiwanese MNEs; semiconductors in the case of the South Korean and Hong Kong MNEs; computers, microelectronics, biotechnology and genetic engineering in the case of Singaporean MNEs; and motor vehicle components in the case of Brazilian MNEs. In the late 1980s and early 1990s, Asian investors in particular provided crucial early-stage funding in product sectors such as semiconductors and personal computers which American venture capitalists now frequently ignore. In addition, these investments have opened doors for young

Table 9.11 Some examples of research-intensive investments by Third World MNEs

Home country and firm	Nature of research-intensive investment
India	
Hindustan Machine Tools	Established a small electronics operation in the USA to gain access to advanced electronics technology necessary in the production of electronically controlled machine tools
Philippines	
Landoil Resources Corporation	Obtained an equity interest in Charriot Resource, a Canadian energy company, in order to supplement the firm's limited technological expertise in petroleum exploration
	Obtained an equity interest in Siframe SA, a French construction and engineering company, in order to gain access to more sophisticated construction technology
	Obtained an equity interest in Consulan Group, a Swiss MNE specialising in hotel management, in order to upgrade the firm's hotel management capabilities
P.T. Taliaban Timber	Entered into a joint venture with Mitsubishi Shoji Koisha Limited of Japan in order to gain access to machinery, technology and finance needed in timber exploitation
Malaysia	
Sime Darby	Invested in agro-genetic engineering through a joint venture with the US firm International Plant Research Institute which specialises in the research application of genetic technology to tropical crops. These two firms agreed to establish the ASEAN Biotechnology Corporation in Malaysia which is involved in research on new types of perennial plants and training Malaysian scientists; and the ASEAN Agri–Industrial Corporation which is involved in agro–industrial projects in the ASEAN region based on advanced technology
Brazil	
Firm unknown at source	Direct investment in Portugal has enabled a Brazilian firm to gain access to abundant and relatively inexpensive

Table 9.11   (continued)

Home country and firm	Nature of research-intensive investment
	computer equipment, enabling it to use computer technology more intensively than the parent firm. Portugal has therefore become an invaluable testing ground for the development of new technological and organisational skills of Brazilian firms
Metal Leve	Established an R&D centre in Ann Arbor, Michigan, in order to acquire the latest technology in the manufacture of auto engine pistons
Petrobrás	Joined a consortium of non-Latin American oil independents to bid on North Sea concessions off Norway, not only to gain reliable access to sources of oil but also to acquire know-how that could be exploited in oil exploration activities in other countries
Venezuela	
Petróleos de Venezuela	Acquired refineries and petrol stations in Germany, Sweden and the USA to tap the expertise of foreign operators
South Korea	
Various firms[1]	Established manufacturing affiliates in Silicon Valley, California, USA, for the purpose of acquiring the latest technology in semiconductor production. The technology acquired is then translated into the fabrication, assembly and testing of semiconductors in Korea
Samsung	Established a subsidiary in Tristan Semiconductors, located in Silicon Valley, to acquire design capabilities in semiconductor production
Hyundai	Established a subsidiary in Modern Electrosystems located in Silicon Valley
Seoudu Logic	This small, technology-based Korean firm has developed computer-aided design software for the Korean market based on design techniques absorbed from Silicon Valley where the firm pursued the idea of industrialising techniques developed in research laboratories
The Electronics and Telecommunications Research Institute (ETRI)	Attempted to develop a 64-bit super mini-machine in conjunction with AIT, a small US company based in California.

Table 9.11 (continued)

Home country and firm	Nature of research-intensive investment
	Although this venture was unsuccessful, scientists from ETRI were given an excellent opportunity to work on problems of computer architecture in Silicon Valley
Taiwan	
Formosa Plastics	Established a joint venture with the US firm Atlantic Richfield Company to gain access to the manufacturing technology of sophisticated forms of manufactured chemicals, such as polyvinyl chloride, vinyl chloride monomer and ethyl dichloride. In 1980, the firm acquired a plant of Imperial Chemical Industries (UK) in Baton Rouge, Louisiana, USA
Channel International	This Taiwan-based consortium, funded in part by the government, acquired Wyse Technology, the ailing US computer maker, in December 1989, for US $268 million, largely to acquire foreign technology. Wyse Technology, founded in 1981, has become successful through the sales of inexpensive computer terminals for mainframe and minicomputers. The deal marks Taiwan's largest overseas acquisition; it has been actively pursuing the acquisition of foreign technology
Sampo, Advanced Devices Technology, China Development Corporation, United Microelectronic Corporation (UMC), Tatung and the state-owned Electronics Unicorn Research and Service Organisation (ERSO)	Have invested in design facilities in semiconductor production in Silicon Valley. For example, UMC established Unicorn in Silicon Valley while maintaining close relations with three other R&D-based overseas Chinese-operated firms, Mosel, Qusel and Vitelic
Acer Group	Completed a US $94 million purchase in 1990 of Altos Computer Systems, a maker of multi-user computer networks in San Jose, California
Tatung	Invested US $1 million in GraphOn Corporation, an 8-year old maker of computer window displays, based in San Jose, in order to use GraphOn's products in the systems it makes
Hualon Microelectronics Corporation	Purchased 10 per cent of Seeq Technology Incorporated, a San Jose computer-chip maker, for $5.3 million. As part of the deal, Seeq chips will be

Table 9.11 (continued)

Home country and firm	Nature of research-intensive investment
	made in Taiwan, helping Hualon to use more of the capacity as a foundry
Kung Ying Enterprises	Took over Mouse Systems Corporation, a maker of computer input devices in Fremont, California
Pacific Wire and Cable Company	Established a high-technology joint venture, Mosel, in Silicon Valley
Singapore	
National Iron & Steel, Wearne Brothers and other firms unknown at source	Established manufacturing affiliates in the developed countries for the purpose of acquiring the latest technology in computers, microelectronics, robotics biotechnology and genetic engineering
National Iron & Steel	Invested US $3 million in a few small venture capital companies in California, to gain access to new technologies and to persuade the more successful US firms to relocate their manufacturing operations in Singapore
Wearne Brothers	Purchased an 88.75 per cent equity interest in a leading US agro-technology firm in order to diversify the activities of the firm and to persuade the US firm to relocate some R&D activities in Southeast Asia
Ssangyong Cement Limited	Participating in a $5.8 million financing for Edsun Laboratories Incorporated, Waltham, Mass. Edsun is using the money to make integrated circuits for graphics enhancement in personal computers
Firm unknown at source	This large publishing conglomerate established 18 of its 27 overseas subsidiaries in the industrialised countries to keep abreast of rapidly changing printing and publishing technology. This was the primary motive behind its $100 million takeover in the early 1980s of a large British publishing group
Hong Kong	
Firm unknown at source	Aggressive foreign investments have been undertaken in such industries as chemicals, wood and wood products, food processing and machinery in order to acquire skills for the production of these products which can be subsequently undertaken in the home country. The

Table 9.11 (continued)

Home country and firm	Nature of research-intensive investment
	purchase of the US firm Bulova by Hong Kong's Stelux Manufacturing Company in 1976 is an excellent case in point in which Stelux gained access to a company with an established reputation without incurring the higher costs of acquiring the skills needed to build a reputation for itself
Hua Ko, Elcap, RCL	Acquired equity participation in semiconductor design houses in California. For example, Elcap owns 20 per cent of Universal Semiconductor Incorporated in San Jose, California, which provides training and design to Elcap. Although Hua Ko has in addition its own design capability, Chipex Incorporated, its subsidiary in San Jose has been a site for training engineers from Hua Ko. All these firms design and produce their masks in California. The fabrication of wafers, assembly and testing is undertaken in Hong Kong

Sources: Wells, 1978; Ting and Schive, 1981; Villela, 1983; Lall, 1983b; Lim, 1984; Pang and Komaran, 1984; UNCTC, 1986; Schive, 1988; Wells, 1988; Evans and Tigre, 1989; Henderson, 1989; Korea Institute for Foreign Investment, 1989; Kang, 1990; Pang and Hill, 1991; Valeriano, 1991.

Note: 1 Among these firms are the Samsung Group, Daewoo Group, Hyundai Group, Zymos Corporation, Lucky-Biotech Corporation, Maxon System and GoldStar Technology Incorporated.

US companies seeking lower-cost manufacturing in Asia and entry to Asian markets (Valeriano, 1991). Nevertheless, these investments in R&D by MNEs from developing countries are largely geared to gaining access to the most sophisticated and advanced forms of manufacturing technology.[26]

By contrast, research-linked investments by lower-income developing countries are still largely concentrated on the acquisition of advanced technology either in resource-based sectors such as food, petroleum or timber exploitation and processing or in services sectors such as construction and hotel management. For example, the research-linked investments of the Malaysian MNE Sime Darby are a reflection of the extent to which the firm has engaged in complementary activities allied to its main area of activity in rubber exploitation and processing. The firm has invested in R&D in the USA in agriculture and agro-related industrial projects. By contrast, the research-intensive investments of Indian MNEs in the USA, geared to acquiring advanced electronics technology in the production of machine tools, reflect the advanced development of the indigenous capital goods sector in India despite its lower stage of development, compared to the NICs, in terms of GNP per capita. Although the advanced industrial development of the capital goods sector may be attributed to earlier government policies that encouraged domestic technology creation by restricting the availability and use of foreign technologies (Lall, 1985a), since the mid-1980s the industrial potential of India in the production of more highly technology-intensive goods such as electronics, food-processing equipment and pharmaceuticals has been enhanced by inward foreign investment, particularly from the USA (Debes, Holstein and Javetski, 1987).

In a number of cases, the acquisition of firm-specific assets such as product and process technology, managerial resources, brand names and access to marketing and distribution channels is made through the purchase of established firms or through other forms of strategic alliances with firms in the developed countries. A number of these alliances consists of joint ventures or licensing agreements with US firms.[27] These outward investments are also important ways in which Third World firms can enhance their ownership-specific advantages.

These research-linked outward investments in centres of innovation in the developed countries, as well as investments to acquire firm-specific assets through strategic alliances, have been an important avenue for rapid technology transfer (diffusion) and development for firms from developing countries and especially those from Asia. For example, since the Taiwan computer industry is strong in the basic type of personal-computer hardware, its interest in research-intensive investments emerges from its aim of supporting the development of the next level of technology, such as electronic workstations. International production or licensing as a means of engaging in backward technology transfer by firms from both developed and developing countries is further discussed by Dunning and Cantwell (1990).

These sophisticated forms of research-related investments undertaken by Third World MNEs are part and parcel of the cumulative process of technological accumulation through which firms gain access to complementary but more advanced technology which could be combined and localised with their indigenous technological innovation. The high degree of indigenously generated technological capability of some Third World firms has enabled them to participate in more sophisticated forms of research-intensive investment and to integrate the products of the R&D activities of firms from developed countries with their own independent technological trajectory. Research investments offer an avenue by which developing countries can successfully penetrate sectors with greater technological opportunities in order to lessen their technological dependency and effectively catch up in the process of technological advancement. In this respect, the existence and accumulation of technological capabilities both affects, and is affected by, the pattern and growth of international production.

CONSTRUCTION SECTOR INVESTMENTS

The major Third World investors in the construction sector have emerged from South Korea, Brazil, Argentina, the Philippines and India. The major competitive technological advantages of these firms derive from their experience in construction activities in their respective home countries which was exploited in foreign markets at a later stage. The major competitive edge of these firms *vis-à-vis* both indigenous and foreign construction firms in host countries derives from their access to a large supply of cheaper skilled production factors such as construction labour and engineering talent in the home country; their longer experience in construction in developing countries leading to a labour-intensive, though modern, localised technology in construction and engineering-related operations; and their established reputations.

As with Third World investments in other sectors, the technological advantages of Third World construction firms derive from their ability to integrate advanced construction technology from the developed countries with their own indigenous technological tradition in accordance with the requirements of Third World markets and production conditions. Such localised technology is either re-exported to other developing countries which have similar factor endowments and resource constraints or re-adapted to suit the peculiar needs of the host country.

The main difference between overseas manufacturing and construction activities derives from the greater importance of the cyclical nature of the construction business in the home market as a major factor in the overseas investments of construction firms. A case in point is the civil engineering projects of Argentine MNEs in activities related to engineering design, site preparation and construction supervision which are products of indigenously generated technology that cannot be exploited in the home market due to recessionary pressures resulting in over-capacity and high fixed costs (White, 1981). The nature of the investment, however, means that little direct investment is involved.

As with overseas construction investments by firms based in the developed countries, the competitiveness of construction investments by NICs such as South Korea, Singapore, Brazil and Argentina is sustained in part through long experience in construction. For example, Korean construction firms were the second largest in the world construction market in the 1980s, largely due to their accumulation of technological and organisational capabilities and experience in large-scale civil construction projects undertaken for the USA in Guam and Vietnam during the Vietnam War. The long construction experience of Korean firms is exemplified by Hyundai Engineering and Construction Company, the largest Korean construction firm, which started international operations as early as 1965 in Southeast Asia and by 1970 had broadened its international markets to include the Middle East, the Pacific Region and Africa. Korean contractors have comparative advantages over other foreign competitors based in developing countries because of their supply of skilled and semiskilled manpower, modern labour-intensive technology for construction and engineering-related operations, the availability of sophisticated and specialised modern equipment, and flexible attitudes in dealing with local authorities (Jo, 1981).

The accumulation of advanced construction technology by Korean firms was also enhanced by the emphasis laid by the Korean government on the establishment of heavy industries such as cement, steel, shipbuilding and other basic metals, the promotion of exports of goods, capital and services and, finally, the acquisition of sophisticated and specialised modern equipment and capital through joint venture arrangements with foreign firms that had more sophisticated technological expertise in engineering and construction know-how (Jo, 1981; Westphal *et al.*, 1984).

Long construction experience is also the source of Brazilian construction firms' technological advantages. The successful overseas construction contracts of the Brazilian firm Constructora Mendes Junior are the result of experience gained through operations in self-sufficient work camps in isolated places, building roads and railways over very rough terrain and working with an unskilled low-cost labour force. The successful construction ventures of the firm are reflected in the fact that by 1980 it already ranked thirteenth among the 250 largest world construction firms classified by their foreign sales (Wells, 1983).

Other Brazilian engineering firms such as Andrade Gutierrez, Hidroservice Engenharia de Projetos Limitada, Promon Engenharia SA, IESA and Themag Engenharia Limitada have spearheaded other construction firms in Third World countries, largely through provision of technical know-how supported by machinery and equipment exported from Brazil. Although their overseas construction activities mostly represent sales operations, the technical knowledge of Brazilian construction and design engineering firms has been nurtured over decades by learning by doing in their construction activities in a developing country environment and the mastery of some specific transferable and modern civil engineering technologies, in addition to their organisational and managerial advantages. Such a process of technological accumulation has been facilitated by large investment programmes undertaken by the Brazilian government, particularly during the 1970s, and the oligopolistic structure of the industry, associated with the emergence of few construction and engineering firms (Guimaraes, 1986).

Similarly, the advanced urban planning and housing development of Singapore have enabled some firms to develop ownership advantages in architecture and town planning that are of special interest to neighbouring countries. Although their overseas operations do not encompass cutting edge technology, and for the most part represent successful architectural modifications for middle- to low-income tropical cities, one Singaporean firm is among the most advanced in the region in terms of employing computer assisted design in Southeast Asia and in the management of complex design work such as that involved in the construction of large hotels and shopping complexes (Pang and Hill, 1991).

Notwithstanding their more advanced forms of technology, harnessed through long construction experience and combined in some cases with the export of machinery and equipment, construction firms from the NICs have voiced concern about their less than bright overall business prospects, owing in part to their weakened comparative superiority and the fierce competition offered by construction firms from both developed and developing countries. The latter are increasingly becoming competitive in the provision of low-cost but highly skilled construction and engineering labour and expertise.

The previous chapter showed that the indigenously generated technological advantages of Philippine construction firms, partly sustained by the government's 1974 labour policy to promote the export of Filipino manpower vigorously, combined with investments in backward technology transfer to gain access to more sophisticated and advanced forms of construction technology, are a clear sign that although the innovative activities of these firms are still at an early stage and based largely on construction experience such activities are expected to become increasingly sophisticated as the firms acquire greater experience in international construction activities. Hence, the greater role of localised technological accumulation and increasing international competitiveness in terms of lower costs should enable them to compete more effectively with construction firms from the NICs in the future.

CONCLUSION

This chapter has extended the analysis of Philippine MNEs towards a broader perspective which includes other significant investors from developing countries, particularly those from Asia and Latin America, which are at the same or a different stage of industrial development. It sought to examine the relation between the pattern of domestic industrial development, the gradual process of expanding technological capabilities of selected Third World firms and the cumulative growth in their international direct investment. The analysis of Third World MNEs was undertaken within the framework of dynamic theories of international production as well as the Third World MNE theories, namely the product cycle model and the theories of localised technological change and technological competence.

The cross-sectional analysis of Third World outward investment undertaken in this chapter suggests that its sectoral composition is such that resource-based and simpler manufacturing investments embodying lower levels of technological accumulation are generally predominant forms of outward investment for firms from countries at lower stages of development. Firms from countries at more advanced stages of industrial development that have accumulated a greater amount of technological competence have the increasing capability to sustain more complex forms of manufacturing and research-linked outward investments. With some exceptions, notably in Hong

Kong, Singapore and India, this finding is further collaborated by time-series data that allow for a dynamic analysis of the sectoral evolution of the outward investments of some Third World MNEs. The outward direct investment of developing countries was seen generally to follow the developmental path of the more traditional investor countries as well as the newer ones from the developed countries, which is initially predominant in the primary and simpler forms of secondary sectors. At a later stage, the sectoral structure of investments tends to evolve towards more complex forms of manufacturing and service activities as the capacity for technological innovation expands and as experience in international investment broadens.

However, although the sectoral development of Third World outward investment bears a resemblance to that of the more traditional sources of outward investment from the advanced, industrialised countries, the innovative capacities of Third World firms are distinguished from those of firms from the developed countries as a result of the cumulative process of technological change, based on unique learning experience. The earlier stage of national development of developing countries and their less technologically sophisticated domestic industrial structure, compared to that of the developed countries, have meant that the technological advantages of Third World firms are of a less sophisticated kind. By comparison, firms from the more developed countries have accumulated greater production experience and have more significant investments in R&D in research-intensive and technologically intensive industries.

Owing to their earlier stage of technological development, the distinct innovative capacities of Third World MNEs are characterised by lower levels of research intensity, size, technological experience and skills. They undertake technological improvements in products, processes or equipment which may represent a revival or modernisation of older technologies, including technology considered outdated in the developed countries. The generation of unique innovative capacities of Third World firms, especially those at a lower stage of development, represents in most cases an accumulation of production experience and learning by doing and using rather than a product of deliberate R&D activities. However, with the establishment of R&D laboratories, science parks and technological research institutions by governments in collaboration with the private sector, combined with overseas investments in research and efforts to seek foreign technology through various modalities, the future technological course taken by these firms may be expected to be increasingly influenced by efforts made in R&D.

The two schools of thought on Third World MNEs can therefore be applied to the outward direct investment of countries at different stages of technological development or alternatively to the outward direct investment of firms within countries over time. In the earlier stage, the product cycle model provides a useful framework within which to analyse the internationalisation of firms whose technological capacities are limited to the assimilation and adaptation of the innovative activities of the technologically leading firms in accordance with the requirements of Third World markets and production conditions (Wells, 1986). In the later stages, as home countries advance through progressively higher stages of development and as their firms acquire greater experience in international investment, the notion of localised techno-logical change and eventually technological accumulation assumes greater significance. This allows for an independent or differentiated technological path on the part of Third World firms, which contrasts with the more limited scope of technological innovation embodied in the product cycle model. The latter model cannot explain the different patterns of innovative activities apart from those determined by market demand factors. In fact, innovative activities of firms are triggered and guided both by evolving market signals and evolving technologies (Atkinson and Stiglitz,1969; Rosenberg, 1982; Pavitt, 1988).

Evidence of dynamic changes in the area of Third World outward invest-ment is provided by the newly industrialising countries whose outward investment is increasingly associated with early stages of R&D activity. Some of these countries have advanced beyond the lower-income developing coun-tries whose outward investment still embodies imitation and adaptation of foreign technology to suit Third World production conditions as well as localised technological change. For the NIC firms, the theory of technological accumulation is increasingly becoming a more useful framework within which to explain the expansion of their production networks across national boun-daries. The further outward expansion of these firms in more technologically sophisticated activities may be expected to increase as these countries advance through higher stages of development and as their firms acquire greater experience in international investment. The emergence of these firms' capacity for technological accumulation explains their greater ability to catch up quick-ly with the older and more established MNEs from the more developed countries.

The analysis of the technological advantages of Third World MNEs across a wide range of industries suggests a rapid pace of industrial development in developing countries accompanied by the existence and accumulation of indigenous technological capability, especially in those sectors in which in-digenous firms have accumulated production experience. The existence and accumulation of this indigenous technological capability are an important cause and effect of the growth of international production networks and the process of domestic industrial restructuring. The existence of indigenous technological capability influences firms' ability to engage in international production, which in turn influences the accumulation of further technological capabilities that enable firms to upgrade domestic industry towards more technologically intensive activities and to engage in more sophisticated forms of outward investment.[28]

Exceptions to this finding are the city states of Hong Kong and Singapore whose outward investments are not prompted by rapid industrialisation or the need for industrial restructuring in their home countries but by the necessity to expand productive capacity outside the limited domestic market and to overcome protectionist pressures and maintain export competitiveness. In the case of Hong Kong, for example, the growth of exports from outward investments in labour-intensive assembly in lower labour cost developing countries has expanded at an annual rate of 25–30 per cent in recent years compared to the growth of exports from the home country at the much slower pace of 2–3 per cent (Lall, 1991, from data taken from the *Survey of Hong Kong Re-exports, 1988*). As a result, the share of Hong Kong's domestic exports fell from 70 per cent in 1979 to 40 per cent in 1989 (GATT, 1991). Apart from circumventing the limited size of the domestic market and the shortage of land and labour, there is an additional reason to diversify investment risks – the economic and political uncertainty in Hong Kong with the impending return of the British territory to the People's Republic of China in 1997.

Outward direct investment as an instrument of domestic industrial restructuring from industries described as labour intensive to those that embody greater capital and technology intensity is not the case in Hong Kong, whose outward investments are not part of a government-directed industrial policy, perhaps because of the uncertainties over 1997. Whatever the reason for the lack of active government support for outward investments in Hong Kong, characteristic of the other Asian NICs, it has resulted in Hong Kong firms investing outside their national boundaries, motivated by the generation of quick returns rather than long-run growth based on planning and R&D (ESCAP, 1986) or fulfilling any macroeconomic development objectives for their home country. This explains why the development of their technological capabilities is confined along fairly narrow and defined areas of activity which limit the growth of their future comparative advantage in higher technologically intensive industries (Fransman, 1984; Chen, 1988; Lall, 1991). Nevertheless, their rather limited competitive advantages in the area of technology are somehow sustained through their large reservoir of entrepreneurial and marketing know-how and access to global distribution channels. Although Singapore is in some respects different from Hong Kong because of the more interventionist and paternalistic role of the state and its lower value-added manufacturing, the country shares many common features with Hong Kong in that investments in education and R&D and the breadth of its industrial and technological base are inferior to those of the other Asian NICs such as South Korea and Taiwan (see Pang and Lim, 1989).

Hong Kong's narrow industrial base is seen by Lall (1991) as an added pressure to internationalise, as opposed to countries with a broader scope for industrial diversification. His argument is based on the premise that countries that specialise in a wider range of industries can retain a strong production and export base through a continual process of domestic industrial restructuring in the home country. However, although the pressure to internationalise at a much earlier stage may be greater for Hong Kong firms, owing to weak local technological capabilities and the narrow scope for diversification into other industrial sectors in the home country, there may be difficulty in sustaining competitive advantages in labour-intensive manufacturing in the longer term,

Table 9.12 The industrial distribution of outward direct investment of developed and developing countries, 1988, US $ million

Industry	Developed countries[1]	Developing countries[2]
Primary	129,279.70	706.14
Agriculture	5,668.48	171.42
Mining and quarrying	51,046.88	507.65
Petroleum	72,514.51	NA
Secondary	334,347.90	1,249.13
Food, beverages and tobacco	35,753.25	87.69
Textiles, leather and clothing	5,808.90	71.31
Paper products	19,593.93	45.60
Chemical products	68,164.50	116.84
Coal and petroleum products	4,111.63	0.83
Rubber products	7,717.43	43.89
Non-metallic mineral products	4,934.02	28.88
Metal products	44,925.59	79.99
Mechanical equipment	49,007.16	10.56
Electrical equipment	31,764.64	233.10
Motor vehicles	27,048.90	1.60
Other transportation equipment	8,013.20	5.46
Other manufacturing	27,462.77	16.94
Tertiary	427,501.3	2,128.60
Construction	4,885.39	42.29
Distributive trade	98,616.93	231.77
Transport and storage	19,767.99	20.71
Communication	716.17	7.65
Finance and insurance	207,190.10	1,452.78
Real estate	36,577.71	34.86
Other services	59,747.04	313.59
Total[3]	891,128.90	4,086.674

Sources: Author's calculations based on primary data obtained from Australian Bureau of Statistics; Statistics Canada; Danmarks Nationalbank; Suomen Pankki Finland Bank; Banque de France; Deutsche Bundesbank; Banca d'Italia; Japan Ministry of Finance; De Nederlandsche Bank; Central Bank of Norway; Bank of Portugal; Bank of Sweden; UK Department of Trade and Industry; US Department of Commerce; China Ministry of Foreign Economic Relations and Trade; Banco de la República de Colombia and Oficina de Cambios Sección Análisis Económico de Colombia; Reserve Bank of India; Korea Institute for Foreign Investment; ESCAP/UNCTC, 1988; Central Bank of the Philippines; Bank of Thailand; Taiwan Investment Commission; Ministry of Economic Affairs.

Notes: 1 The developed countries include Australia, Canada, Denmark, Finland, France, Germany, Italy, the Netherlands, Norway, Portugal, Sweden, the United Kingdom and the USA. Because of the unavailability of data for 1988, the total outward direct investment stock used in the calculations refers to 1987 for Canada and Italy.

2 The developing countries include the People's Republic of China, Colombia, India, the Republic of Korea, Malaysia, the Philippines, Taiwan and Thailand. The data on China and Malaysia refer to 1987.

3 The sub-totals and total may not add up because of discrepancy in the original data.

as predicated by the product cycle model, owing *inter alia* to the growth of firms from lower-income developing countries gaining greater comparative advantages in simpler forms of manufacturing activities over time. In an era of technological competition in international industries, comparative advantage must be directed and sustained in activities associated with more sophisticated forms of innovation, including R&D. Outward direct investment activities may constitute an important modality in achieving future comparative advantages for countries that follow both narrow and diversified industrial development policies. The relocation of their sunset industries, described by declining international competitiveness in production in the home country, to more competitive foreign locations may be an alternative to their managed decline in the home country. Foreign production may prolong the ultimate decline of these sunset industries and at the same time meet the need to sustain the process of domestic industrial restructuring in the home country which will be crucial in the development of future comparative advantages. These can be undertaken both directly through specialisation and concentration in the home country on more highly value-added, technologically intensive industries, with the relocation of sunset industries abroad, and indirectly through the infusion of new technology through access to complementary foreign technologies abroad. Such a strategy enables dualistic development to be made on an international scale.

The case of Hong Kong and other lower-income developing countries shows that, as well as technological advantages, managerial experience, training and international orientation are also significant determinants of the international production and competitiveness of Third World MNEs (see Lecraw, 1991, for a case study on Indonesia). The strengths of Hong Kong MNEs in marketing and distribution as well as management and entrepreneurship, as opposed to technological accumulation considerations, render the product cycle model relevant to explain their growth.

The changing technological capabilities of the current Third World MNEs compared to MNEs from the developed countries are reflected in the comparative analysis of the industrial distribution of their outward investments as shown in Table 9.12. An analysis of the data for 1988 shows that the most important industrial sectors for MNEs based in the developed countries, in declining order of importance, are chemicals, mechanical engineering, metals, food, beverages and tobacco, electrical and electronic equipment, and motor vehicles. By contrast, the most important industrial sectors for Third World MNEs are electrical and electronic equipment, chemicals, food, beverages and tobacco, metals, textiles, leather and clothing, and rubber products. The increasing convergence of the industrial sectors in which both First World and Third World MNEs invest suggests dynamic changes in comparative advantage taking place in the world economy with firms from the Third World increasingly gaining the capacity to follow and catch up with technological developments in the West and to achieve international competitiveness. However, significant differences are still prevalent between the operations of First and Third World MNEs despite their increasing concentration in the same industries. MNEs based in the developed countries are engaged in international technological competition in high research- and skill-intensive sectors such as chemicals, mechanical engineering, electrical and electronic equipment and motor vehicles, where the world technological frontier is moving at a rapid pace. By contrast, the outward investments of Third World MNEs which are at an earlier stage are still primarily linked to lower research-intensive, resource-based and mature industrial sectors such as metals, food, beverages and tobacco, textiles, leather and clothing, and rubber products, where the world technological frontier has not been changing very rapidly (Dahlman and Sercovich, 1984). Because of the concentration and specialisation of Third World MNEs in these activities, these firms have often become the main source of new technology in these industries. Apart from local firms in South Korea, which have become the third largest competitive producers of semiconductors in the world after Japan and the USA, and their advanced technological capability to keep pace with the latest developments in semiconductor production generated in the West, by and large the Third World firms based in the NICs that have achieved the capacity to invest in more research-intensive sectors have more limited technological capacities. These firms remain competitive on the basis of using an older technology developed in an earlier period by firms from the more advanced countries.[29]

The findings of this chapter suggest that the process of technological accumulation characteristic of the growth of international production of firms from the industrialised countries also finds application as an important determinant and effect of the pattern and growth of the international production activities of Third World MNEs. The growth of outward direct investment is a cumulative process, where the form of technological accumulation rather than its existence varies with the stage of development. The more limited capacities for technological innovation in the product cycle model provide a useful framework for an analysis of the early stages of outward investment of Third World firms, associated with their accumulated production experience in developing countries. However, for the more advanced stage of development of the NICs, with their broader experience in international investment and their increased capacity to generate more complex and more significant forms of localised technological innovation in their outward investments, the model of localised technological change and technological competence provides a more useful framework for analysis. NIC firms are showing early signs of capacity for R&D, intrinsic to the concept of technological accumulation, which is expected to become increasingly important and complex as their countries advance through progressively higher stages of industrial development and as their firms acquire greater experience in international production.

10 The geographical development of Third World outward investment

INTRODUCTION

In Chapters 8 and 9, the shift in the sectoral pattern of Philippine outward direct investment as a direct result of the process of firms' technological accumulation from purely resource-based investments towards manufacturing investments was seen to have developmental implications for the geographical pattern of their outward investments. For example, resource-based investments in the early stages of outward direct investment are initially particularly important in neighbouring developing countries with the closest geographical distance. The increased sectoral sophistication of international production activities, from purely extractive resource-based investments directed largely to the export of primary products to the home country, towards the downstream processing of these primary products in the host country, in the first instance, increased the geographical scope of these investments. At this stage, investments directed towards more sophisticated markets in other countries with increasingly further psychic distance may become increasingly more important. In addition, the geographical distribution of more sophisticated manufacturing investments and the allied forms of service sector investments such as trade and banking, and finance, may be directed increasingly to the more developed countries.

The term 'psychic distance' adopted in this book refers to the concept adopted by Hornell and Vahlne (1972) which refers to circumstances which prevent or restrain the flow of goods and/or payments between businesses and markets, such as differences in the level of development between the home country and foreign market in terms of their education, language, culture, customs and legal and commercial systems. However, Luostarinen (1978) also refers to economic distance, which is defined as the inverse of the economic pull of a particular market as measured by GNP or growth of GNP.

This chapter extends the analysis presented in Chapter 8 of the developmental repercussions of the sectoral distribution of Philippine outward investments on their geographical distribution to those of other Third World MNEs in Asia and Latin America at the same or a different stage of development as the Philippines. The stages of development approach in the analysis of the geographical distribution of Third World outward direct investment adopted in this book can be thought of in the context of Johanson's and Vahlne's (1977) *model of knowledge development and increasing foreign commitments.* In their work, the internationalisation process of the firm is regarded as following a sequential pattern in which initial foreign investments are directed to those markets with the greatest proximity in terms of psychic distance. The psychic distance at which foreign ventures are located is likely to increase directly with the complexity of the sectoral pattern of outward direct investments.

In the context of Philippine MNEs in particular, and Third World MNEs in general, the increasing diversification in the geographical pattern of Third World outward direct investment, as a result of the increasing complexity of the sectoral pattern, which in turn is determined by the stage of development of home countries and the cumulative process of technological accumulation, is described as following three sequential stages.

The first stage, which is considered the most important for new international investors involved largely in resource-based and simpler manufacturing and service investments, is directed towards investments in neighbouring and/or ethnically related territories. Evidence for this is shown in the relatively high share of intra-regional direct investment of developing country firms from Asia and Latin America, most of which, as predicated by the product cycle hypothesis, is located in other developing countries that are further down in the pecking order than the home country. However, the evidence suggests that the importance of investments in these neighbouring and/or ethnically related territories declines as investing firms gain greater experience in international production and as their home countries progress to higher stages of industrial development.

The second stage entails production in several neighbouring and/or ethnically related territories or in other non-ethnically related developing countries of simple forms of manufacturing and associated service investments. Lower production and transport costs, and close psychic distance, as well as the presence of favourable investment opportunities and the desire to make fuller use of regional economic integration seem likely to explain significant direct investments within distinct regional groupings at this stage of development. The technological advantages of these Third World MNEs in the first and second stages are based on an adapted product or process innovation obtained through experience in developing countries which can be applied to other developing countries at a lower stage of development. In the third stage, firms with more complex technological capabilities become better able to conduct more sophisticated manufacturing and service industries in countries further away from their home base and in countries further away in terms of psychic distance, possibly including investments in developed countries.

The general hypothesis is that the geographical development of MNEs from the more advanced developing countries such as South Korea, Taiwan, Singapore and Hong Kong in Asia and Mexico, Brazil and Argentina in Latin

America is steadily moving into the third stage. By comparison, the geographical development of MNEs from lower-income developing countries, whose international investments are at a relatively early stage, is still largely in the first and second stages, within neighbouring ethnically related developing countries and other non-ethnically related developing countries.

STAGES 1 AND 2: INVESTMENTS IN NEIGHBOURING AND/OR ETHNICALLY RELATED DEVELOPING COUNTRIES AND OTHER DEVELOPING COUNTRIES

Chapter 8, on Philippine MNEs, showed that ethnic ties and geographical proximity have been essential factors in decisions involving location of foreign production for a number of Philippine firms. The cost of acquiring reliable information about foreign markets is believed to be considerably lower when ethnic ties exist in a potential market. Moreover, the costs associated with transport and the movement of personnel as well as the psychic distance involved in investment in ethnically related areas are considerably reduced as cultures, customs, traditions and language are similar. However, ethnic ties may also place constraints on the location of production and create difficulties by requiring firms to manage geographically scattered units.

Apart from the ethnic factor, the general trend towards internationalisation of business, with recent outward investor firms engaging in international production at an earlier stage than traditional investors, has made investment in neighbouring developing countries further down in the pecking order and with lower labour costs particularly important. These investments are typically undertaken by small- and medium-sized firms in the Asian NICs engaged in export-oriented, labour-intensive manufacturing which have experienced the greatest difficulty in maintaining the price competitiveness of their export products in the home country because of increasing labour costs and are forced to relocate production to countries where labour costs are lower. Some of the other factors favouring relocation to lower-cost developing countries, such as the need to overcome protectionist pressures in traditional export markets, the presence of preferential trading arrangements, geographical proximity, political stability and the need for domestic industrial restructuring, were discussed in the last chapter. In the case of Taiwan, as in other Asian NICs, the appreciation of the currency, increasingly strict pollution controls and shortages of raw materials, in addition to the need to overcome higher wages and land costs, are factors that threaten the further growth of exports from the home country. Taiwan's outward investments in Asia also fulfil the important function of promoting trade within the region as its overseas factories source their supplies and machinery from the home country, thereby diversifying trade and decreasing export dependence on the USA. More importantly, with the on-going prohibition of direct trade between China and Taiwan, affiliates in Southeast Asia gain a base from which to export to China (Moore, 1990).

The main hypothesis suggests that the importance of ethnic factors and

Table 10.1 The stock of Indian direct investment abroad, 1988, US $ million

Country	Amount	% distribution
Developed countries	5.87	7.72
Western Europe	4.62	6.08
European Community	4.61	6.07
France	0.01	0.01
Germany	0.42	0.55
Greece	0.17	0.22
United Kingdom	4.01	5.29
Other Western Europe	0.01	0.01
Switzerland	0.01	0.01
North America	1.06	1.39
USA	1.06	1.39
Other developed countries	0.19	0.25
Australia	0.19	0.25
Developing countries	69.74	91.78
Africa	27.88	36.68
Egypt	0.90	1.18
Kenya	8.90	11.71
Mauritius	0.07	0.09
Nigeria	6.37	8.38
Senegal	9.41	12.39
Seychelles	2.04	2.68
Uganda	0.19	0.25
East, South and Southeast Asia	38.25	50.34
Bangladesh	1.61	2.12
Hong Kong	0.03	0.04
Indonesia	10.42	13.71
Malaysia	6.62	8.71
Philippines	0.24	0.32
Singapore	3.40	4.47
Sri Lanka	4.17	5.49
Taiwan	8.16	10.74
Other East, South and Southeast Asia	3.60	4.74
Western Asia	1.82	2.40
Jordan	0.31	0.41
Saudi Arabia	0.42	0.55
Bahrain, Oman, United Arab Emirates	1.09	1.44
Europe	1.59	2.09
Malta	0.01	0.01
Yugoslavia	1.58	2.08
Latin America and the Caribbean	0.05	0.07
Panama	0.05	0.07
Oceania	0.15	0.20
Fiji	0.10	0.13
Tonga	0.05	0.07

Table 10.1   (continued)

Country	Amount	% distribution
Eastern Europe and the USSR	0.38	0.50
Hungary	0.03	0.04
Union of Soviet Socialist Republics	0.35	0.46
Total	75.99	100.00

Source: UNCTC (1992) from unpublished data on Indian joint ventures abroad provided by
Ministry of Commerce, Government of India.

Note: 1   Data represents the value of Indian investments in joint ventures abroad in
production or under implementation as at the end of calendar year.

neighbouring territories has declined for firms from the NICs that have gained multinational investment experience in particular countries and whose innovative activities have grown wider and more complex. However, for the smaller and more recent Third World investor firms from the lower-income developing countries, whose innovative activities are at an early stage, ethnic contacts continue to have a fundamental impact on the extent and form of their initial foreign investments.

Evidence presented in Tables 10.1 to 10.11 on the geographical distribution of outward foreign direct investment stock by firms from a selected group of developing countries for which statistics for 1988 were available from home country sources offers some support for the theory concerning the inter-relationship between the increasingly complex industrial composition of outward FDI, as technological learning develops, and the geographical distribution of outward FDI.[1] These tables suggest that countries such as India, Pakistan, the Philippines, Thailand, Colombia and Peru have a greater proportion of their outward FDI in developing countries – between 76 per cent and 94 per cent. The geographical concentration of Indian outward investment in developing countries suggests that the technological sophistication of Indian firms is more limited than that of firms from the NICs, despite the sectoral evidence examined in Chapter 9. By comparison, the outward FDI in other developing countries of the higher-income and more advanced NICs such as South Korea, Taiwan and Brazil is significantly lower – between 29 per cent and 45 per cent.[2] The marked exception to the correlation between increasing national stage of development and the declining geographical concentration of international production in developing countries is seen once again in Singapore and Hong Kong. Despite their advanced status as NICs, their firms' international production activities are still largely concentrated in developing countries – 77 per cent and 88 per cent respectively. Some of the reasons behind this phenomenon, such as the more limited technological advantages of firms from these city states despite their higher stage of national development, as measured by GNP per capita, and the greater importance of export-oriented, labour-intensive

Table 10.2 The stock of Pakistan's direct investment abroad, 1988, US $ million[1]

Country	Amount	% distribution
Developed countries	12.95	5.67
Western Europe	12.26	5.37
European Community	11.60	5.08
Germany	0.57	0.25
UK	11.03	4.83
Other Western Europe	0.65	0.28
Gibraltar	0.65	0.28
North America	0.13	0.06
USA	0.13	0.06
Other developed countries	0.57	0.25
Developing countries	215.42	94.33
Africa	88.33	38.68
Libya	10.00	4.38
Nigeria	0.56	0.25
Sudan	77.77	34.05
East, South and Southeast Asia	11.06	4.84
Hong Kong	1.00	0.44
Malaysia	4.51	1.97
Singapore	5.20	2.28
Sri Lanka	0.35	0.15
Middle East	106.47	46.62
Iran	1.99	0.87
Lebanon	2.34	1.02
Saudi Arabia	88.78	38.88
Oman and United Arab Emirates	13.36	5.85
Latin America and the Caribbean	0.17	0.07
Unallocated developing countries	9.39	4.11
Total	228.37	100.0

Source: UNCTC (1992) from primary data obtained from State Bank of Pakistan, *Foreign Liabilities and Assets and Foreign Investment in Pakistan*, 1988.
Note: 1 Data obtained on the basis of cumulative outflows since 1972.

outward investments in neighbouring countries with lower labour costs and close ethnic ties, have already been analysed in the last chapter.[3]

With the exception of the Asian city states, the evidence in these tables, together with the sectoral evidence examined in the last chapter, confirms the initial hypothesis made about the interrelationship between the increasingly complex sectoral distribution of outward FDI as countries advance through higher stages of development and firms advance through the cumulative process of technological accumulation, on the one hand, and geographical

Table 10.3 The stock of Philippine direct investment abroad, 1988, US $ million[1]

Country	Amount	% distribution
Developed areas	5.37	13.60
North America	5.37	13.60
USA	5.37	13.60
Developing areas	34.10	86.40
Africa	15.02	38.05
Libya	15.02	38.05
East, South and Southeast Asia	10.46	26.50
Brunei Darussalam	5.63	14.26
Hong Kong	4.83	12.24
Latin America and Caribbean	8.60	21.80
Bermuda	0.01	0.03
Panama	7.81	19.79
Other	0.78	1.98
Unallocated developing countries	0.02	0.05
Total	39.47	100.0

Source: UNCTC (1992) from primary data obtained from Central Bank of the Philippines, Foreign Exchange Operations and Investments Department, unpublished data.

Note: 1 Data represent cumulative outflows of registered investments with the Central Bank of the Philippines since 1980.

distribution of outward FDI on the other. Firms from the advanced developing countries, such as the NICs, which have more complex forms of technological advantages are able to engage in more sectorally sophisticated forms of international production activities which often necessitate locating in countries with a further psychic distance, including non-ethnically related developing countries and developed countries.

By contrast, the innovative activities of firms from the lower-income developing countries such as China, India, Pakistan, the Philippines, Thailand, Colombia and Peru are at an early stage and are based on production experience in developing countries.[4] Their outward investments tend to be concentrated more in resource-based, import-substituting manufacturing and less complex forms of service investments in neighbouring developing countries with a close psychic distance.

Such an internationalisation process conforms to Johanson's and Vahlne's (1977) *model of knowledge development and increasing foreign commitments*. This model focuses on the impact of gradual acquisition, integration and use of knowledge about foreign markets and operations on the incrementally increasing commitments to foreign markets, ranging from exporting to the establishment of a sales subsidiary and eventually to the establishment of

Table 10.4 The stock of Thailand's direct investment abroad, 1988, US $ million[1]

Country	Amount	% distribution
Developed countries	50.93	24.0
Western Europe	0.55	0.26
European Community	0.55	0.26
Denmark	0.10	0.05
France	0.01	0.005
Germany	0.02	0.009
Netherlands	0.15	0.07
United Kingdom	0.27	0.13
North America	48.96	23.07
USA	48.96	23.07
Other developed countries	1.42	0.67
Australia	0.19	0.09
Japan	1.08	0.51
Turkey	0.15	0.07
Developing countries	161.23	76.0
Africa	0.43	0.20
Liberia	0.43	0.20
East, South and Southeast Asia	159.48	75.17
Bangladesh	0.04	0.02
Brunei Darussalam	0.21	0.10
China, People's Republic of	1.48	0.70
Hong Kong	105.04	49.50
Indonesia	20.29	9.56
Malaysia	0.18	0.08
Philippines	0.01	0.01
Singapore	31.97	15.07
Sri Lanka	0.06	0.03
Taiwan	0.16	0.08
Other East, South and Southeast Asia	0.04	0.02
Western Asia	0.31	0.15
Iraq	0.04	0.02
Saudi Arabia	0.19	0.09
Other Western Asia	0.08	0.04
Latin America and the Caribbean	1.01	0.48
Panama	0.80	0.38
Other Latin America and the Caribbean	0.21	0.10
Total	212.16	100.00

Source: UNCTC (1992) from primary data obtained from Bank of Thailand, unpublished data.

Note: 1 Data represent cumulative flows of investment abroad since 1978.

Table 10.5   The stock of South Korean direct investment abroad, 1988, US $ million[1]

Country	Amount	% distribution
Developed countries	620.84	55.47
Western Europe	41.07	3.67
European Community	41.07	3.67
Belgium and Luxembourg	2.17	0.19
France	3.25	0.29
Germany	12.20	1.10
Ireland	4.50	0.40
Italy	0.23	0.02
Netherlands	0.34	0.03
Portugal	0.27	0.02
Spain	0.25	0.02
United Kingdom	17.86	1.60
North America	490.56	43.83
USA	398.26	35.58
Canada	92.30	8.25
Other developed countries	89.21	7.97
Australia	58.84	5.26
Japan	29.71	2.65
New Zealand	0.06	0.01
Turkey	0.60	0.05
Developing countries	498.32	44.53
Africa	12.79	1.14
Cameroon	1.28	0.11
Ivory Coast	0.40	0.03
Egypt	0.26	0.02
Liberia	0.01	0.01
Morocco	0.21	0.02
Nigeria	1.36	0.12
Senegal	0.43	0.04
Other Africa and Sudan	8.84	0.79
East, South and Southeast Asia	254.41	22.74
Bangladesh	0.57	0.05
Brunei Darussalam	0.01	0.01
Hong Kong	15.45	1.38
India	0.14	0.01
Indonesia	188.74	16.87
Malaysia	29.02	2.59
Philippines	3.05	0.27
Singapore	4.74	0.42
Sri Lanka	2.51	0.22
Thailand	10.15	0.91
Macau and other East, South and Southeast Asia	0.03	0.01
Western Asia	175.74	15.70
Islamic Republic of Iran	0.26	0.02
Qatar	0.20	0.02

Table 10.5 (continued)

Country	Amount	% distribution
Saudi Arabia	39.29	3.51
United Arab Emirates, Yemen	135.99	12.15
Latin America and the Caribbean	25.98	2.32
Argentina	4.31	0.40
Bermuda	1.00	0.09
Brazil	1.61	0.14
Cayman Islands	0.50	0.04
Chile	3.95	0.35
Costa Rica	3.02	0.27
Dominican Republic	0.82	0.07
Jamaica	2.15	0.19
Panama	7.20	0.64
Other Latin America and Caribbean	1.42	0.13
Oceania	29.40	2.63
Fiji	0.75	0.07
Papua New Guinea	19.50	1.74
Other Oceania	9.15	0.82
Total	1,119.16	100.00

Source: UNCTC (1992) from primary data obtained from the Bank of Korea, *The Status of Outward Foreign Investment*, various issues.
Note: 1 Data represent cumulative flows of realised investment abroad since 1968, i.e. authorised investments less capital withdrawn.

Table 10.6 The stock of Taiwan's direct investment abroad, 1988, US $ million[1]

Country	Amount	% distribution
Developed countries	501.31	71.25
North America	425.65	60.50
USA	425.65	60.50
Other developed countries	75.66	10.75
Developing countries	202.27	28.75
East, South and Southeast Asia	171.75	24.41
Indonesia	40.20	5.71
Malaysia	19.27	2.74
Philippines	52.86	7.52
Singapore	20.72	2.94
Thailand	38.70	5.50
Unallocated	30.52	4.34
Total	703.58	100.00

Source: UNCTC (1992) from primary data obtained from Ministry of Economic Affairs, Investment Commission, *Statistics on Overseas Chinese and Foreign Investment, Technical Co-operation, Outward Investment, Outward Technical Co-operation, the Republic of China*, February 1989.
Note: 1 Data represent cumulative approved outflows since 1959.

Table 10.7   The stock of Singapore's direct investment abroad, 1987, US $ million[1]

Country	Amount	% distribution
Developed countries	322.39	22.91
Western Europe	175.33	12.46
European Community	168.17	11.95
Belgium and Luxembourg	58.64	4.17
Germany	0.30	0.02
Italy	2.15	0.15
Netherlands	82.76	5.88
United Kingdom	24.32	1.73
Other Western Europe	7.16	0.51
Switzerland	7.16	0.51
North America	40.13	2.85
USA	31.17	2.21
Canada	8.96	0.64
Other developed countries	106.93	7.60
Australia	92.18	6.56
Israel	4.25	0.30
Japan	5.55	0.39
New Zealand	4.95	0.35
Developing countries	1,084.71	77.09
Africa	74.66	5.31
Liberia	73.91	5.26
Mauritius	0.15	0.01
Other Africa	0.60	0.04
East, South and Southeast Asia	899.32	63.91
Bangladesh	0.05	0.004
Brunei Darussalam	27.37	1.95
China, People's Republic of	45.63	3.24
Hong Kong	268.95	19.11
India	0.50	0.04
Indonesia	33.78	2.40
Malaysia	477.41	33.93
Philippines	6.40	0.45
Republic of Korea	1.10	0.08
Sri Lanka	10.31	0.73
Taiwan	11.56	0.82
Thailand	16.26	1.16
Western Asia	3.50	0.25
Saudi Arabia	3.50	0.25
Latin America and Caribbean	81.76	5.81
Mexico	4.55	0.32
Netherlands Antilles	5.20	0.37
Panama	0.05	0.004
Other Latin America and Caribbean	71.96	5.11

Table 10.7 (continued)

Country	Amount	% distribution
Oceania	6.25	0.44
Papua New Guinea	6.25	0.44
Unallocated developing countries	19.22	1.37
Total	1,407.10	100.0

Source: UNCTC (1992) from primary data obtained from Singapore Ministry of Trade and Industry, unpublished data.

Note: 1 Data represent domestic investors' ordinary paid-up shares in overseas subsidiaries and associated enterprises as at end of year and the net amounts due to overseas branches.

Table 10.8 The stock of Hong Kong's direct investment abroad, 1987, US $ million[1]

Country	Amount	% distribution
Developed countries	1,704.9	12.2
North America	1,389.3	9.9
USA	939.7	6.7
Canada	449.6	3.2
Other developed countries	315.6	2.3
Japan	315.6	2.3
Developing countries	12,246.9	87.8
East, South and Southeast Asia	12,246.9	87.8
China, People's Republic of	8,236.8	59.0
Indonesia	1,867.5	13.4
Malaysia	421.7	3.0
Pakistan	14.4	0.1
Philippines	175.6	1.3
Korea, Republic of	100.9	0.7
Sri Lanka	161.9	1.2
Taiwan	990.6	7.1
Thailand	277.5	2.0
Total	13,951.8	100.0

Source: UNCTC (1992) from primary data obtained from Hong Kong Government Industry Department, *Report on the Survey of Overseas Investment in Hong Kong's Manufacturing Industries, 1989*, and unpublished data from Hong Kong Government Industry Department.

Note: 1 Data have been obtained on the basis of inward stock attributed to Hong Kong in the various host countries shown above.

Table 10.9   The stock of Colombia's direct investment abroad, 1988, US $ million

Country	Amount	% distribution
Developed countries	73.07	19.72
Western Europe	7.17	1.93
European Community	7.17	1.93
Belgium and Luxembourg	0.97	0.26
France	0.05	0.01
Germany	2.41	0.65
Italy	0.07	0.02
Spain	0.95	0.26
United Kingdom	2.72	0.73
Other Western Europe	0.003	0.0008
Switzerland	0.003	0.0008
North America	65.90	17.79
USA	65.90	17.79
Developing countries	297.50	80.28
West Asia	13.84	3.73
Bahrain	13.84	3.73
Latin America and the Caribbean	283.66	76.55
Antigua and Barbuda	0.30	0.08
Argentina	1.88	0.51
Bahamas	9.54	2.57
Brazil	0.003	0.0008
Cayman Islands	1.60	0.43
Chile	8.59	2.32
Costa Rica	0.22	0.06
Dominican Republic	0.01	0.003
Ecuador	5.64	1.52
Guatemala	0.04	0.01
Mexico	0.01	0.003
Nicaragua	0.02	0.005
Netherlands Antilles	0.56	0.15
Panama	187.31	50.56
Peru	14.42	3.89
Puerto Rico	0.47	0.13
Uruguay	0.09	0.02
Venezuela	26.20	7.07
Other Latin America and Caribbean	26.76	7.22
Total	370.57	100.0

Source: UNCTC (1993) from primary data obtained from Banco de la República, Oficina de Cambios, Sección Análisis Económico, Inversiones.

Table 10.10   The stock of Peru's direct investment abroad, 1988, US $ million[1]

Country	Amount	% distribution
Developed countries	12.51	21.12
North America	12.51	21.12
USA	12.51	21.12
Developing countries	46.72	78.88
West Asia	3.77	6.36
Bahrain	3.77	6.36
Latin America and the Caribbean	42.95	72.52
Bolivia	21.43	36.17
Brazil	0.33	0.56
Cayman Islands	15.15	25.58
Colombia	1.23	2.08
Mexico	0.30	0.51
Panama	0.04	0.07
Uruguay	0.16	0.27
Venezuela	4.31	7.28
Total	59.23	100.0

Source: UNCTC (1993) from primary data obtained from Comisíon Nacional de Inversiones y Tecnología Extranjera (CONITE), Ministerio de Economía y Finanzas, unpublished data.
Note: 1 Data represent investments approved by CONITE.

Table 10.11   The stock of Brazil's direct investment abroad, 1988, US $ million[1]

Country	Amount	% distribution
Developed countries	1,057.00	60.37
Western Europe	193.27	11.04
European Community	179.13	10.23
Belgium and Luxembourg	18.12	1.03
France	22.30	1.27
Germany	17.92	1.02
Italy	0.12	0.007
Netherlands	8.14	0.46
Spain	14.83	0.85
Portugal	7.15	0.41
United Kingdom	90.55	5.18
Other Western Europe	14.14	0.81
Austria	10.00	0.57
Sweden	0.011	0.0006
Switzerland	2.25	0.13
Other Western Europe	1.88	0.11
North America	853.67	48.76
USA	853.39	48.74
Canada	0.28	0.02
Other developed countries	10.06	0.57

Table 10.11   (continued)

Country	Amount	% distribution
Japan	10.06	0.57
Developing countries	693.77	39.63
Africa	16.14	0.93
Algeria	0.58	0.03
Gabon	0.80	0.05
Ghana	0.63	0.04
Ivory Coast	11.64	0.66
Liberia	0.001	0.00006
Nigeria	2.37	0.14
Togo	0.01	0.0006
Other Africa	0.11	0.006
East, South and Southeast Asia	16.39	0.94
Singapore	11.39	0.65
Other East, South and Southeast Asia	5.00	0.29
West Asia	13.85	0.79
Saudi Arabia	0.18	0.01
Other Western Asia	13.67	0.78
Europe	0.42	0.02
Malta	0.18	0.01
Other developing countries in Europe	0.24	0.01
Latin America and the Caribbean	646.97	36.95
Argentina	74.56	4.26
Bahamas	39.05	2.23
Bolivia	21.54	1.23
Cayman Islands	171.59	9.80
Chile	56.85	3.25
Colombia	6.74	0.38
Costa Rica	0.17	0.0097
Dominican Republic	0.14	0.008
Ecuador	0.97	0.03
Guyana	0.002	0.0001
Mexico	2.24	0.13
Netherlands Antilles	78.93	4.51
Panama	25.56	1.46
Paraguay	55.97	3.20
Peru	5.00	0.29
Trinidad and Tobago	1.94	0.11
Uruguay	36.16	2.07
Venezuela	15.34	0.88
Other Latin America and Caribbean	54.22	3.10
Total	1,750.77	100.0

Source: UNCTC (1993) from primary data obtained from Banco Central do Brasil.
Note: 1 Data represent historical value of Brazilian capital abroad registered with the Banco
Central do Brasil. These include outward investments in the banking sector but
exclude oil exploration by Petrobrás, the national oil corporation, and other capital
invested overseas but registered as expenses. As of April 1986, Petrobrás had
outward investments amounting to US $832.2 million (Wells, 1988).

production facilities in the host country. The time order of such activities seems to be related to the psychic distance between the home and the host countries. The lack of market knowledge due to differences between countries with regard to language and culture is an important obstacle for decision making. This phenomenon of the process of incremental adjustments to the changing conditions of the firm and the environment as applied to Third World MNEs necessitates further investigation.

Table 10.1 shows that despite the relatively advanced forms of technological advantages of Indian MNEs, almost 92 per cent of Indian outward direct investment in 1988 in joint ventures was directed towards developing countries, particularly East, South and Southeast Asian and African countries, especially those with a substantial ethnic Indian community such as Malaysia, Indonesia, Thailand, Sri Lanka, Nigeria, Kenya and Senegal.[5] Proximity and linguistic, historical and ethnic ties have determined the concentration of Indian direct investment in these countries (Lall, 1983b). Similar reasons may also explain the significantly large outward investments of Pakistan, the Philippines and Thailand. About 94 per cent of Pakistan direct investment abroad is directed to developing countries, particularly in the Middle East (47 per cent) and Africa (39 per cent); and in the case of Philippine outward investments, 86 per cent is directed towards developing countries, in particular Africa (38 per cent), East, South and Southeast Asia (27 per cent) and Latin America (22 per cent). Similarly, more than 75 per cent of Thailand's direct investments abroad are directed towards developing countries, mainly in East, South and Southeast Asia.

The lack of data on the outward direct investment of Malaysian firms does not enable an accurate quantitative comparison to be made of the geographical distribution of their outward investment *vis-à-vis* that of other countries. Nevertheless, the geographical distribution of some Malaysian overseas ventures reveals a high proportion of investments in neighbouring and/or ethnically related territories. For example, an analysis of the 153 overseas ventures of twenty-four of the largest Malaysian MNEs in the early 1980s suggests that more than two-thirds were located in Asia, particularly in Singapore and Hong Kong, which accounted for 38 per cent and 21 per cent respectively of the number of their overseas ventures (Lim, 1984). Investments by small and medium-sized Malaysian firms also predominate in other Asian countries such as Indonesia, Sri Lanka, Thailand and the Philippines. Even the Malaysian conglomerate MNEs such as Malaysian Mining Corporation and Sime Darby have resource-based investments in neighbouring Asian countries such as Thailand and the Philippines. These countries support predominantly Chinese business communities and are also the home of Malaysia's largest foreign investors. The close ethnic and economic ties between these countries are a major reason for the presence of such a large number of Malaysian companies.

Similarly, despite their higher stages of national development, MNEs from Singapore and Hong Kong concentrate their overseas activities around the

Table 10.12   Singapore's direct investments in Asia

Host country	Sector
Malaysia	Manufacturing, in particular food and beverages; trading and distribution of goods produced by Singaporean affiliates in Malaysia
Hong Kong	Trading, finance and investment. A few manufacturing activities by Yeo Hiap Seng, Haw Par Brothers and Jack Chia-MPH
Indonesia	Electrical equipment and food industries
	Agricultural plantation; services
	Textile and chemical industries
China	Manufacturing; petroleum exploration; oil palm cultivation; hotels and construction
Sri Lanka, Pakistan	Export-oriented manufacturing investments in labour-intensive, low value-added activities such as flour milling, garments, carbon film, resistors, jewellery and printing. Outside free trade zones, Singaporean MNEs have investments in civil engineering, construction and in the manufacture of sewing threads, rubber bands, sewing machines and commercial refrigerator assembly plant
Bangladesh	Export-oriented manufacturing investments such as garments; hotels and property development

Sources: Pang and Komaran, 1984; Lim, 1984.

Asian region. In the case of firms from Singapore, the explanation lies primarily in the close ethnic, personal and family ties that the Chinese community enjoy throughout Asia. The management of practically every Singaporean firm with overseas operations in China and the ASEAN countries shows that linguistic and cultural familiarity greatly enhanced their competitiveness (Pang and Hill, 1991). Of the 126 Singaporean overseas affiliates known to exist in the early 1980s, 110 were located in Asia (Pang and Komaran, 1984), particularly in Hong Kong, China, Brunei, the Philippines, Indonesia and Malaysia.[6] Table 10.12 summarises the activities of foreign affiliates of Singaporean MNEs in these countries. Although a substantial proportion of Singaporean direct investment in the Asian countries is in the resource-based, trading and finance sectors, a significant amount is in export platform, labour-intensive manufacturing sectors such as electrical appliances and is made in response to the rising labour costs in Singapore as well as trade protection in export markets, chiefly those in developed countries. Singapore MNEs have also been investing in capital-intensive sectors such as chemicals, rubber and petroleum-related products and also in non-Asian developing countries such as Papua New Guinea, Tahiti, Mauritius, Oman, Liberia and Zambia (Aggarwal, 1987). Nevertheless, direct investments in developing countries by Asian NICs in general and Singaporean MNEs in particular have been directed almost exclusively towards Asian countries, primarily because of familiarity with the economic, social and political conditions of these countries. Malaysia has always been a favoured location for Singapore direct investments because, as well as the availability of lower land and labour costs and the encouragement provided by the host country government, the close geographical proximity and economic ties reduce information and transaction costs, and facilitate management control of investments aimed at increasing production capacity and diversifying product range. Moreover, the favourable exchange rate between the Singapore and Malaysian currencies has also contributed to the cost competitiveness of production in Malaysia (Chia, 1989). Similarly, Singaporean investments in China have benefited from the encouragement given by the Chinese government while those in Sri Lanka have benefited from the ten-year tax holiday given to export-oriented investors. These firms have specified that where there are no locational advantages arising from geographical proximity or historical or ethnic ties, international production is largely undertaken in the form of minority-owned joint ventures or other forms of international economic links. For example, Singaporean firms that have expanded into other developing countries outside Asia have largely preferred the technological licensing or export route because the direct investment risks associated with unfamiliarity with the social, economic and political conditions of these countries are considered too high (Lim, 1984).

For similar reasons, a considerable proportion of Hong Kong direct investments abroad is geared towards countries with substantial ethnic Chinese communities such as China, Indonesia, Thailand, Malaysia and the other Asian NICs.[7] In Indonesia and Thailand, Hong Kong firms produce a wide range of higher value-added and more technologically intensive products such as electronics and electronic appliances, in addition to standardised products (World Bank, 1989). In Malaysia, Hong Kong direct investments are concentrated both in export-oriented industries such as textiles, garments and electronics, to take advantage of lower production costs, and in industries in which Malaysia has comparative advantages, such as food, chemicals and wood, that have not been important in the Hong Kong economy itself. These industries offer Hong Kong parent firms new markets and lines of activities (Chen, 1981). The ownership advantages of Hong Kong firms is seen to derive partly from their closer language and cultural affinity with Chinese businesses. However, although Hong Kong direct investments predominate in China and in Asian countries with large Chinese communities, some are also made in countries with further psychic distance, such as Latin American, Caribbean and Australasian countries – in particular, Brazil, Argentina, Mexico, Venezuela, Jamaica, Papua New Guinea and even Mauritius where their success has been attributed to the presence of ethnic Chinese entrepreneurs.[8] These investments in developing countries have largely been in export-oriented light manufacturing such as seasonal novelty items, toys, apparel and lower-end consumer electronic products that are mostly re-exported to Hong Kong, although some of their Asian investments such as in food manufacturing, wood processing and chemicals have been import substituting or resource based. Service sector investments have also been significant in hotels, real estate, construction, banking, trading and recreational services, as have some resource-based investments in resource-rich countries.

In recent years, Hong Kong investments have expanded in Southeast Asia because of concerns about the impact of political upheaval in China, the traditional area of overseas investment expansion, on Sino–USA trade relations, with the possible withdrawal of China's most favoured nation status. Diversification into other production bases in Asia, helped by a network of overseas Chinese communities in such countries as Indonesia, Thailand and Malaysia that enjoy the generalised system of preferences trade status with the USA, has often been necessary and encouraged by long-term customers in the USA as protection against political and economic risks. These US importers often have elaborate and expensive marketing plans and therefore do not want them threatened by interrupted deliveries.[9]

Table 10.5 shows that only 23 per cent of Korean FDI in 1988 was directed to Asian and Pacific countries, compared to a much higher proportion in 1978, when direct investments by Korean MNEs in timber extraction were exclusively confined to Southeast Asia as was 48.6 per cent of the value of direct investments in manufacturing, 21 per cent of direct investments in construction, 12.3 per cent of direct investments in trading and 87.8 per cent of direct investments in real estate. These manufacturing investments in neighbouring countries were primarily concentrated in food chemicals, cement, fountain pens, agricultural implements, shoes, garments, electric cables, plywood and jewellery. By contrast, only 38.5 per cent of the value of Korean direct investment in manufacturing in 1978 was directed to Africa and 7.4 per cent to Oceania. Most of Korea's direct investment in developed countries in 1978 was in construction and on-site services (Jo, 1981; Kumar and Kim, 1984).

The main determinants of location of export-oriented Korean manufacturing subsidiaries are the availability of relatively cheap inputs, in particular, labour costs; availability of necessary raw materials; general business environment and incentives provided by the host government; and the prospects for mutually advantageous economic activities. However, geographical proximity and political ties, as opposed to ethnic ties, have greatly influenced the location decisions of Korean manufacturing firms in neighbouring Asian countries at an early stage.

At a later stage, as the localised technologically innovative activities of Korean firms became more sophisticated and as their experience in international production broadened, they graduated into the more advanced stages of our model by establishing manufacturing investments in other non-ethnically related developing countries in the Middle East, Africa and Latin America, as well as in the developed countries. The stock of outward FDI by Korean firms by 1988, as shown in Table 10.5, illustrates that the geographical distribution of Korean MNEs in 1978 had been reversed and that they had increasingly graduated into the third and highest stage of the geographical development model. Although developing countries accounted for a substantial proportion of outward FDI in 1978, they accounted for 45 per cent of the total stock of Korean FDI by 1988, mainly in Asia, particularly Indonesia and Bangladesh, in the Middle East and, to a lesser extent, in the Caribbean Basin countries to take advantage of their duty-free trade status with the USA. Around 64 per cent of approved outward investments in developing countries are in the extractive sector, particularly mining, and around 20 per cent are in manufacturing. These manufacturing investments are mostly undertaken by smaller firms or *chaebol* subsidiaries producing labour-intensive or GSP-dependent manufactured products such as textiles and apparel. On the other hand, investment in the developed countries, which was initially concentrated in construction and on-site services such as trading in 1978, had become much more important by the late 1980s. The table suggests that 55 per cent of Korean FDI in 1988 was directed to the developed countries, mainly by the *chaebol*, and concentrated in North America and Australia. The last chapter showed that increasing amounts of import-substituting manufacturing investments as well as research-oriented investments, to gain access to the most sophisticated and advanced forms of manufacturing technology, and export-oriented service investments are undertaken by South Korean MNEs in the developed countries. This is a manifestation of their higher forms of technological advantages and the early signs of capabilities for technological accumulation.

Similarly, Table 10.6 shows the geographical distribution of Taiwanese direct investment in 1988, of which 29 per cent was directed to developing countries. In the 1960s and 1970s, more than 50 per cent of Taiwan's outward investment was directed to Thailand, Malaysia, Indonesia, the Philippines and Singapore (Ting and Schive, 1981). Even by 1982, as much as 45 per cent of these investments were directed to developing countries, mainly neighbouring Southeast Asian countries such as Indonesia, Thailand, Malaysia, the Philippines and, directly and indirectly, China (World Bank, 1989). Approximately 98 per cent of investments in these countries in 1982 were geared to the manufacturing sector and largely in import-substitution sectors such as chemicals (30.4 per cent), paper and paper products (12.9 per cent) and food and beverages (6.9 per cent), and some in export-oriented sectors such as electronics and electrical appliances (7.3 per cent) and textiles (19.7 per cent). Only

1.0 per cent of investments in these countries were in the extractive industries such as agriculture and forestry (Taiwan Research Institute, 1983).

These investments by Taiwan businesses in Southeast Asia benefit, particularly in Malaysia, from the presence of Chinese communities that share the common Hokkien dialect. These communities also form powerful local business allies with an increasingly dominant role in the commerce of Thailand, Indonesia and the Philippines (Moore, 1990). As pointed out in the preceding chapter, the limited geographical diversity of Taiwanese outward FDI may be explained by the absence of reciprocal trade and investment agreements with all but a few dozen countries, mostly small economies in the Caribbean Basin area that enjoy free-trade status with the USA under the Caribbean Basin Initiative. Although export-oriented, labour-intensive and natural resource-based manufacturing and resource-based investments in developing countries in Asia and Latin America continued to be the more significant sectors of Taiwan's outward investment in developing countries in the late 1980s, some import-substituting and research-intensive investments in the more developed countries, particularly in the USA, increasingly influenced the pattern of Taiwan's direct investment abroad.

The Latin American countries provide a useful basis of comparison. An analysis of the data on the geographical distribution of outward direct investment from Latin America shows that ethnic ties and geographical proximity remain generally important. However, the comparative analysis of the data on geographical distribution of outward direct investment by eight Asian countries and three Latin American countries presented in Tables 10.1 to 10.11 shows that the importance of ethnic ties and neighbouring territories differed significantly between Asia and Latin America in 1988. Investments in neighbouring countries or ethnically related territories accounted for 78 per cent and 45 per cent respectively of the total outward direct investment stock in 1988 of these eight Asian and three Latin American countries.[10] This conforms to the earlier findings of Wells (1983) who suggested that ethnic ties play a much more significant role for investments originating from Asian countries than from Latin American countries. Indeed, an analysis of recent foreign direct investment inflows in Asia shows an intra-Asian investment boom by the four Asian NICs, especially in such countries as Malaysia, Thailand, the Philippines and Indonesia.[11]

By contrast, the common language and national origin of the Latin American countries have not led to flows of reliable market information across their borders, and the resultant high cost of information has led to less trade and foreign direct investment among them. Moreover, the lack of trust within most extended families of Spanish origin in Latin America coupled with the difficulty of keeping family members in the business have imposed a significant constraint on the growth of Latin American firms in foreign markets (Wells, 1980, 1983).

Apart from the reasons suggested above, one of the more important reasons for the greater significance of investments in neighbouring countries or ethnically related territories for MNEs from Asia emerges from the shift in their industrialisation policies from import substitution in the early 1960s towards export promotion in the late 1960s, which led to the growth of trade in manufactured products and then to overseas expansion of Asian firms increasingly engaged in export-oriented manufacturing investments. Many of these locally based firms in Asia, particularly in Hong Kong, shifted from domestic production towards exports in the late 1960s, then shifted to direct investment in neighbouring lower labour-cost developing countries when exports were threatened by tariff barriers imposed in the export markets in developed countries and when high labour costs in their home countries threatened the competitiveness of their export products.

The role of ethnic ties can also be assessed against total outward direct investment stock in developing countries. Investments in neighbouring countries or ethnically related territories accounted for between 94 per cent and 95 per cent of the total outward stock of direct investments in developing countries by MNEs from both Asia and Latin America in 1988. This suggests that within developing countries, firms from both Asia and Latin America concentrate their investments intra-regionally or in ethnically related territories.[12] However, mention must be made here of Hong Kong, the largest Third World outward investor, which significantly influences the share of neighbouring countries or ethnically related territories in the total outward stock of Asian MNEs' direct investments in developing countries. If Hong Kong is excluded from the analysis, the importance of outward investments in neighbouring countries or ethnically related territories declines significantly from 95 per cent to 68 per cent.

Ethnic factors have played an important role among Latin American countries of non-Spanish origin, in particular those from Argentina. For example, ethnic ties have been important in the growth of foreign investment of Alpargatas, an Argentine firm of British origin, and SIAM di Tella, a firm founded by immigrant Italians (Wells, 1984). Between 1965 and 1981, 88 per cent of the total authorised investment abroad by Argentine firms was located within Latin America and Caribbean countries, compared to the 99 per cent share in 1978. This declining share of investments directed to neighbouring and/or ethnically related territories within Latin America is indicative that Argentine firms are moving from the first stage towards the second stage in the model of the geographical development of Third World outward investment.

A significant proportion of these investments in neighbouring Latin American countries represents resource-based as well as manufacturing investments, in particular, oils and fats, sweets and jams, wine, packing materials, publishing, chemicals, pharmaceuticals, toiletries, plastics, cement and clay, steel, metallic products, agricultural machinery, construction machinery, other machinery and equipment, electrical equipment and motor vehicles (Katz and Kosacoff, 1983). The more sophisticated export-oriented manufacturing investments by such Argentine MNEs as Peñaflor, Giol, Cabsha, Wobron and Yelmo favour a location in which subregional or bilateral agreements or specific cultural or ethnic relationships exist. In fact, regional complementarity agreements have acted as an incentive for a significant amount of Argentine FDI by granting preferential treatment for foreign firms originating from countries within the region. In addition, the import-substituting Argentine MNEs such as IME, FABI, Bagó and Alpargatas have located their investments throughout the Latin American region primarily to overcome the deleterious impact on Argentine exports of the import-substitution strategies adopted by many of these countries. Finally, the importance of resource-based investments by Argentine firms seems likely to be concentrated in neighbouring Latin American countries.

The importance of neighbouring Latin American countries in the geographical distribution of outward FDI is also witnessed in the case of Colombian MNEs. The stock of Colombian FDI as estimated in Table 10.9 amounted to US $371 million by 1988, of which 77 per cent was directed to neighbouring Latin American countries. This share has not changed significantly from that of 74 per cent in 1978, which suggests that although the value of Colombian direct investment abroad has almost quadrupled in the period 1978–88, neighbouring Latin American countries remain important areas for investment. By contrast, evidence for a declining share of intra-regional investments is shown in the case of Peruvian outward investments, where 73 per cent of approved investments were directed to Latin American countries in 1988 compared to 100 per cent in 1978.

Table 10.11 shows the estimated stock of Brazilian FDI, valued at US $1,751 million for 1988, of which investments in Latin American countries in particular accounted for only 37 per cent. Such a share directed to neighbouring countries represents a marked increase from that of 16 per cent in 1978.[13] For Brazilian MNEs in the construction and engineering sectors, and some in the manufacturing sector, considerable investments in neighbouring Latin America countries may be explained largely by their competitive advantages in technical and management skills adapted to environments of other developing countries. For these Brazilian MNEs, as well as those in the non-construction sectors, investments in other Latin American countries may also be explained in terms of constraint on the growth of firms in the home country, owing to lacklustre demand brought about by the debt crisis and political uncertainty. Almost half of thirty-three Brazilian multinationals undertook their first international production activity between 1984 and 1988, linked to the sale of such items as aircraft, electricity meters, textiles and iron ore.[14] These MNEs invested abroad not just for markets, but also for second bases (Wells, 1988) and investments in neighbouring Latin American countries may fulfil that objective, especially for the new international investors.[15]

Table 10.13 summarises the share of Third World direct investment directed to developed and developing areas according to different source countries in Asia and Latin America for which comparative data on outward FDI obtained from home country sources were available, with the stage of development of countries as measured by GNP per capita. The data support the view that with the exception of Hong Kong and Singapore, the stage of national development is a good indicator of the relative geographical distribution of Third World outward investment. Generally, countries at higher (lower) stages of development have a greater (lesser) proportion of outward direct investments directed to developed countries. The geographical distribution of the outward investment of firms from the more advanced developing countries, or the NICs, and the newer investors from other less advanced developing countries confirms the hypothesis made about the declining significance of investments in developing countries as home countries advance in their stage of development. Hong Kong and Singapore remain the marked exceptions to this generalisation with 88 per cent and 77 per cent respectively of their total outward stock by 1988 directed to developing countries. With the exception of these Asian city states, which still have considerable investments in neighbouring or ethnically related territories, by and large outward direct investments based on ethnic and cultural ties tend to become less important at later stages as these investments become more significant and as firms acquire greater experience in international production. The cross-sectional and time-series data examined in this chapter show that the more advanced of the developing countries easily conform to the most advanced stage of our geographical development model.

Outward investments by Third World MNEs in neighbouring and/or ethnically related territories, especially those whose home countries are at earlier stages of industrial development and whose firms are at an early stage in their international production activities, lend support to the product life cycle hypothesis which predicates that Third World MNEs can best exploit

Table 10.13 The geographical distribution of the outward direct investment of Third World MNEs *vis-à-vis* their stage of development, 1988

Country	GNP per capita, US $	Developed countries, %	Developing countries, %
Asia			
India	340	7.7	91.8
Pakistan	350	5.7	94.3
Philippines	630	13.6	86.4
Thailand	1,000	24.0	76.0
South Korea	3,600	55.5	44.5
Taiwan	6,396	71.2	28.8
Singapore[1]	9,070	22.9	77.1
Hong Kong[1]	9,220	12.2	87.8
Latin America			
Colombia	1,180	19.7	80.3
Peru	1,300	21.1	78.9
Brazil	2,160	60.4	39.6

Source: Tables 10.1 to 10.11.
Note: 1 The figure for geographical distribution represents 1987.

their competitive advantages in other developing countries, at least as far as their initial investments are concerned. However, in the second stage, as countries advance through higher stages of development and as the localised technologically innovative activities of firms become more significant, there are increased possibilities for economies of scale arising out of operating subsidiaries in several neighbouring and/or ethnically related territories or in other, non-ethnically related developing countries. This enables their firms to benefit more from internalisation advantages in terms of capturing economies of externalities and interdependent activities.

Such a dynamic view of the increasing sectoral and geographical complexity of Third World outward investment as countries advance through higher stages of development can be reinterpreted within the framework adopted by Dunning (1981a, 1988a) of firms possessing asset and transaction advantages. In the second stage, the asset advantages of Third World MNEs are supplemented by the more important transactional advantages because the hierarchies of the firm benefit more in terms of the common governance of separate but interrelated activities located in different countries.

Such a distinction between asset and transaction advantages relates closely to the distinction made in the preceding chapter between new, technologically active resource-based firms and manufacturing firms which engage in backward integrated activities in the resource-based sector. In the dynamic context adopted in this book, which is in line with that of Teece (1977, 1990), asset and transactional advantages are analysed in terms of the extent to which firms engage in vertically integrated activities as a feature in the growth of production, within which each firm has distinct technological advantages which are difficult to transfer to and implement by other firms with different technological advantages. The last chapter showed that the extent of vertically integrated investments is greater for firms from the NICs, whose innovative activities are more complex and sophisticated because of the more advanced stage of their countries' development and the firms' broader experience of international production.

In sum, the dynamic view adopted in this book refers to the increasing scope for vertical integration of firms as countries advance through higher stages of development and as the localised technologically innovative activities of firms become more complex and important. The presence of more sophisticated forms of innovation enables firms from the NICs to reduce their reliance on neighbouring and/or ethnically related territories as they are able to exploit their advantages in places further away from their home countries and possibly in developed countries. The next section looks at investments of Third World MNEs in developed countries.

STAGE 3: INVESTMENTS IN DEVELOPED COUNTRIES

The last section showed that outward direct investments by Third World MNEs in neighbouring and/or ethnically related territories are generally more important for MNEs from the lower-income developing countries than those from the NICs, owing largely to the more sophisticated forms of manufacturing and service investments undertaken by the latter, which necessitate location in countries with a further psychic distance, including the more developed countries. Indeed, such NICs as South Korea, Taiwan and Brazil have progressed into the third stage of our model of the geographical development of Third World outward investment and had respectively 56 per cent, 71 per cent and 60 per cent of the total stock of their outward investments in the developed countries in 1988. By comparison, the share of developed countries in the total outward investments of India, Pakistan, the Philippines, Thailand, Colombia and Peru were respectively 8 per cent, 6 per cent, 14 per cent, 24 per cent, 20 per cent and 21 per cent. The last chapter showed that some of these investments were linked to local research facilities, and were undertaken by firms, especially from the NICs, to gain access to the most advanced and sophisticated forms of manufacturing technology. These investments are geared to gain or increase competitive advantage through access to more advanced technology, particularly in electronics and electrical appliances, and higher value-added or luxury products where competitiveness is based more on research and development.

Apart from research-intensive investments undertaken in the developed countries, by firms typically from the higher income and more advanced of the developing countries but also by some of the lower-income developing countries, firms from the NICs have also undertaken a significant number of import-substituting investments in the developed countries. This is especially true of direct investments originating from South Korea, Taiwan, Brazil, Singapore and, more recently, Mexico.[16] Table 10.14 summarises some of these investments. An indication of the relative importance of these manufacturing investments is shown in the case of firms from Taiwan where at least 55 per cent of their approved outward investments in Japan, the USA and European countries in 1988 were in the manufacturing sector and, in particular, electronics and electrical appliances (37 per cent), chemicals (9 per cent), and rubber and plastics (3 per cent). Similarly, almost 50 per cent of Korean approved outward investments in 1988 in North America and Europe were in the manufacturing sector.[17] Some of these investments represent local assembly operations to circumvent trade barriers as well as to upgrade technology through location of production in centres of innovation.

In addition, in anticipation of the increased trade protectionism to be brought about by unification of the European Community by the end of 1992, a number of Third World companies, notably those from the NICs such as Brazil, South Korea, Taiwan and Hong Kong, have been trying to establish a niche through direct investments in these countries. Statistics from the Euro–Asia Trade Organisation show that about 200 Taiwanese companies have established branches, subsidiaries or distribution centres in Europe, mostly in Germany or the Netherlands, since 1986. One of the more recent investments is the 200-acre (80-hectare) within Far East Industrial Park in Cork, Ireland,

Table 10.14   The nature of direct investments by Third World MNEs in the developed countries[1]

Home country and firm	Nature of investment
China	
Shenzhen Electronics Group (SEG)	This, the largest Chinese electronics conglomerate, purchased 46 per cent of a Canadian company that has 95 electrical shops in the USA to be used as sales outlets for SEG products. In addition, the firm has plans to set up a TV plant in Canada and a base in the EC
The China International Trust and Investment Corporation and China National Nonferrous Metal Industrial Corporation	Jointly invested US $120 million to acquire 10 per cent equity interest in an aluminium smelting plant in Australia
India	
Kiloskar	Established diesel assembly plant and milling machinery operations in Germany
Philippines	
12 Philippine conglomerate firms	Established 12 trading companies in 1979 in USA and Japan in connection with the export promotion drive of the Philippine government to boost exports, especially of non-traditional products and to reduce marketing costs through consolidation
San Miguel Corporation	Acquired George Muehleback Brewery Company in Kansas City in 1937 and another brewery in San Antonio, Texas, in 1939. These ventures were divested in the late 1950s owing to the severe competition faced through the emergence of large national breweries
Philippine National Bank, Philippine Commercial International Bank, Metropolitan Bank & Trust Company, Allied Banking Corporation, Bank of the Philippine Islands, Manila Banking Corporation	Established overseas branches in the major financial centres and in the Middle East to diversify activities, acquire the latest banking technologies and practices, to raise capital, to capture the remittances of Filipino workers and to manage portfolio investments of their clients
Thailand	
Siam Cement Group	In 1990, established Tilecera Incorporated in Tennessee, USA, in a US $45 million joint venture with Italian-based Jiacobazzi Stilgres, to produce ceramic tiles
Unicord	In late 1989 purchased Bumble Bee Tuna, America's third largest tuna canner, for US $285 million
Saha-Union	Opened a factory in early 1990 in

Table 10.14 (continued)

Home country and firm	Nature of investment
	Tallapoosa, Georgia, to produce sewing yarn under its Venus brand name
Mexico	
Cementos Mexicanos	Invested more than US $800 million in 1989 for cement companies not only in Mexico but also in Texas and California, enabling the firm to become the largest in the North American cement market and the fourth largest in the world
Vitro	This, the largest glass company in Mexico, has closed a US $920 million deal to purchase the second largest US glassmaker, Anchor Glass Container Corporation
Malaysia	
Malaysia Mining Corporation	Acquired 46 per cent interest in the Australia-based Ashton Mining Limited
Magnum Corporation	Owns Phicom Limited in England which manufactures scientific instruments
Sime Darby	Acquired the UK firm Carboxyl Chemicals Limited which manufactures metallic stearates, wires and drawn lubricants and defoaments. The firm also owns 3 manufacturing companies in Australia in such industries as biodegradable detergents and engineering
Firms unknown at source	Four trading investments in North America while 2 trading companies are located both in the EC and Japan. The trading firms in the EC and North America are involved both in commodity trading and in export promotion of Malaysian products
	At least 11 investment companies owned by Malaysian MNEs in the UK which are part of large conglomerate groups such as Sime Darby and perform the important function of identifying opportunities for portfolio as well as direct investment
	There are at least 3 property investments as well as 3 diamond and coal mining activities undertaken by Malaysian MNEs in Australia
Brazil	
Copersucar	Purchased the San Francisco-based US

Table 10.14　(continued)

Home country and firm	Nature of investment
	firm Hill Brothers in 1976 primarily to serve as a marketing outlet for Brazilian products such as cocoa and sugar in the USA. This objective was not achieved since Brazilian surplus sugar was largely absorbed by the domestic programme to use sugar-based alcohol as an automobile fuel
Banco do Brasil	Established overseas banking operations as part of the export-oriented industrialisation strategy of Brazil. The competitive assets of this bank derive from its access to large assets of at least US $23 billion
Gradiente	Bought the British turntable maker, Garrard, in 1979
Securit	Entered into a joint venture to open a furniture showroom and assembly plant in the USA
Labra	Acquired a pencil factory in Portugal and Bardella bought 50 per cent of a capital goods maker in the USA in 1985
Companhía Vale do Rio Doce	Bought 25 per cent of a steel mill in the USA in 1984
Embraer	Established aircraft assembly operations in Belfast, Northern Ireland, and has linked up with Aeritalia and Aermacchi of Italy to manufacture the sophisticated AMX fighter-bomber in Brazil and Italy
Grupo Cerdau	Brazil's largest private steel producer, it agreed in October 1989 to purchase Courtice Steel of Canada for US $52 million. The acquisition seems to have materialised as Courtice Steel was recorded as being under Brazilian control in *Statistics Canada* by 1990
Various firms	Portugal has been the principal target of Brazilian direct investments in order to protect its trade interests with the EC, its most important export market
Argentina	
Firms unknown at source	A substantial proportion of Argentine direct investments are in manufacturing, in particular, oils and fats (USA), foodstuffs (Italy) and publishing (Spain)
Banco de la Nación Argentina,	Overseas expansion of these banks in

Table 10.14 (continued)

Home country and firm	Nature of investment
Argentine Banking Corporation, Banco Provincia de Buenos Aires, Banco de Cordoba	developed countries is geared to gain access to world private capital markets as well as to serve the banking needs of Latin American exporting firms

South Korea

Korean trading firms	Trading investments integrated with other firms to form giant industrial–commercial conglomerates in the developed countries to support the growth of export-oriented manufacturing investments
Bank of Korea, Bank of Seoul, Cho Hung Bank, Commercial Bank of Korea, Hanil Bank, Korea Exchange Bank, Korean First Bank	Overseas expansion in developed countries is associated with the increased significance of import-substituting manufacturing and service investments in export markets and geared to gain access to capital resources needed to finance Korea's industrial growth and to provide financial services required by Korean exporters and importers
Gold Star	Established a television plant and microwave oven plant in Alabama, USA, in order to protect and expand this export market which was limited by export quotas
Samsung Electronics	Opened a subsidiary in New Jersey, USA, in 1984, in Portugal in 1982 and in Birmingham, UK in 1987 to produce colour televisions and/or microwave ovens
Daewoo Electronics	Invested in France (microwave ovens) and USA (computers and semiconductors). In 1986, the firm acquired Cordata Technologies Incorporated in California to gain access to the distribution of personal computers in the USA
	Established a petroleum refining firm in Belgium in 1986 to procure and refine oil on the spot market and is also building a plant for the manufacture of microwave ovens in Longwy, France. With an investment projected at US $120 million, Daewoo, with the collaboration of 10 to 15 Korean manufacturers, plans to form a large-scale manufacturing complex for textile products in the USA which will

Table 10.14   (continued)

Home country and firm	Nature of investment
	include spinning and sewing factories, and dyeing plant
Pohang Iron and Steel Company	Formed a joint venture with US Steel Corporation in late 1985 for the production of sheet and tin products to escape import restrictions on POSCO's export volume to the USA. The inital investment of US $180 million in the joint venture was expected to expand to US $300 million in the late 1980s
Sung Kwang Limited	Invested US $500,000 in the USA in 1986 which represented the first case of an overseas entry to the US dye industry
Ssangyong	Resource-based investments in the developed countries such as Canada, Australia and the USA to extract uranium, bituminous coal, oil and forestry products. Ssangyong has also wholly owned foreign ventures in Japan and USA
Taiwan Firms unknown at source	Almost 85 per cent of approved Taiwan direct investments in the USA in 1985 were in the manufacturing sector, in particular, electronics and electrical appliances, chemicals, rubber and plastics. Taiwan MNEs have also invested in the food and beverage, pulp and paper products, non-metallic minerals, basic metals and metal products, and mechanical engineering sectors in the USA
Formosa Plastics Group	In 1980, the firm acquired a chemical plant of Imperial Chemical Industries (UK) in Baton Rouge, Louisiana, USA. By August 1989, the firm, together with Formosa Chemical and Fibre Corporation, had established two separate 100 per cent owned subsidiaries in the USA to produce downstream components in the manufacture of plastic products as part of a US $1.9 billion US expansion plan. Part of the investment includes the establishment of an ethylene plant in Texas estimated to cost US $400 million
Acer Incorporated	The 10th largest producer of IBM-compatible personal computers in the world, Acer has shown a

Table 10.14 (continued)

Home country and firm	Nature of investment
	determination to succeed in the USA with the acquisition of Counterpoint Computers, a manufacturer of multi-user computer systems. This firm has 11 branches in Europe and will start manufacturing PCs in Helmond, the Netherlands
Continental Engineering Corporation	Purchased American Bridge Company for more than US $200 million in early 1989
Tatung	This large Taiwan electrical appliances producer has successfully established production facilities in the UK and Luxembourg
Firms unknown at source	Service sector investments constitute almost a quarter of Taiwanese FDI in the USA and, in particular, the trading sector, to facilitate the export of Taiwanese products
50 to 60 small and midsize manufacturing firms	Operate in the 200-acre (80-hectare) Far East Industrial Park in Cork, Ireland, to establish a niche prior to the unification of the market by the end of 1992
Delta Electronic Industrial Company	Unveiled its first European expansion project in November 1989 – a $28 million, 5-year plan to establish a manufacturing facility at Inchinnan, near Glasgow, Scotland, where it will initially turn out switching power supplies
Mitac International Corporation	Also plans to build a plant in the United Kingdom
Plus and Plus Company, Mictrotek International Incorporated	Studying surveys for moving production lines to Europe
Hualon Corporation	This major producer of synthetic fibres plans to invest $4.8 million in a factory in Ireland
Bank of Communications, International Commercial Bank of China, Chang Hwa Commercial Bank	Plan to begin operations in Amsterdam
China Trust Company	Considering establishing services in London
Singapore Jack Chia-MPH and others	Established direct investments in real estate, hotels, tourism, trading and property development in Australia and North America

Table 10.14　(continued)

Home country and firm	Nature of investment
Lami-pak Industries	Established a packaging plant in the UK as well as licensing activities in other European countries
Times Publishing	Obtained equity interests in publishing companies in Australia, New Zealand and UK in order to seek new export markets and investment opportunities
Wah Chang	Invested in tungsten production in the USA
Hong Kong Semi-Tech	Paid HK $588 million in 1988 for Consumers Distributing, a chain of 90 catalogue showrooms in the USA. In 1989, the firm spent HK $2.18 billion to gain control of Singer Sewing Machines Company, owner of the Singer brand name and a vast network of sewing machine factories and consumer durable shops. In June 1990, the firm provided HK $580 million for its Toronto-based parent, International Semi-Tech Microelectronics (ISTM), to buy Singer's European business, including factories and stores in 11 countries
Bonaventure Textiles Limited, Fang Bros Knitting Limited, Shing Cheong Electronics Limited	Produce clothing and watches in the Republic of Ireland

Sources: Diaz-Alejandro, 1977; Wells, 1980; Jo, 1981; Ting and Schive, 1981; White, 1981; Villela, 1983; Lall, 1983b; Kumar and Kim, 1984; Lim, 1984; Chen, 1986; Davis, 1987; Seaward, 1987; Industrial Development Authority Ireland, 1988; Moore, 1988; Wilson, 1988; Korea Institute for Foreign Investment, 1989; Xiaoning, 1989; *Far Eastern Economic Review*, 24 August 1989; Friedland, 1990; *Business Asia*, 15 October 1990; *Asian Wall Street Journal Weekly*, 31 December 1990.

Note:　1　Excludes research-intensive export-oriented investments by Third World MNEs in developed countries which has been summarised in Table 9.11, Chapter 9.

which is expected to accommodate fifty to sixty small and medium-sized manufacturers. As shown in Table 10.14, Taiwan's computer industry is most active in European expansion, as seen by the activities of Acer (regarded as Taiwan's IBM), Delta Electronic Industrial Company, Mitac International Corporation, Plus & Plus Company and Mictrotek International Incorporated. Similarly, synthetic fibre producers such as Hualon Corporation plan to invest US $4.8 million in a factory in Ireland. Even financial institutions have joined the bandwagon of Taiwan investors in the EC. The Bank of Communications, the International Commercial Bank of China and Chang Hwa

Commercial Bank plan to begin operations in Amsterdam, while China Trust Company is considering services in London (*Asian Wall Street Journal Weekly*, 31 December 1990).

Similarly, Hong Kong textile and watch manufacturers and South Korean and Taiwanese electronics firms are establishing factories within the EC, especially in Ireland and the UK and also in Spain which has lower labour costs. Portugal has been the principal target of Brazilian direct investments, in order to protect its trade interests with the EC, its most important export market.

Apart from import-substituting manufacturing investments, another important sector of investment for MNEs from the NICs, which is now increasingly pursued by firms from the lower-income developing countries, is the trading sector. Firms from South Korea, Singapore and Taiwan have long-established trading investments in the developed countries specifically to promote exports of products in their major export markets there. Apart from trading investments, Third World firms have promoted exports through acquiring established firms to gain access to their established marketing and distribution networks and brand names. Mention has already been made in the last chapter of the acquisition by Mantrust of Indonesia and by Unicord of Thailand respectively, of the US tuna fish companies Chicken of the Sea and Bumble Bee Tuna. Similarly, Copersucar, a co-operative formed by about seventy of the largest sugar producers in Brazil, purchased Hill Brothers, a US-based firm involved in coffee processing and distribution, to provide a marketing outlet for Brazilian products such as cocoa and sugar in the USA. However, this objective was not achieved, since Brazil's surplus sugar was largely absorbed by a domestic programme to use sugar-based alcohol as an automobile fuel (Wells, 1980; Villela, 1983).

Apart from manufacturing and trading investments, other sophisticated forms of investments in the services sector, such as real estate and property development, are increasingly undertaken by firms from the NICs such as Jack Chia-MPH and other Singaporean firms in North America and Australia. A major factor in the growth of property investments by Jack Chia-MPH in Australia is the incentives offered by the Australian government. For example, in 1982 the Australian government granted the firm a waiver in the purchase of Australian property of up to A $30 million (Lim, 1984).

The banking and finance sectors are other important areas of investment for Third World MNEs. Banks from NICs such as Korea, Hong Kong, Singapore, Argentina and Brazil have established banking operations in their major export markets in the developed countries as part of the wider, outward-looking industrialisation strategies of their home countries, which have necessitated greater access to world capital markets and to banking services by their export-oriented firms. Banks from these countries, especially from Hong Kong, Singapore and Argentina, have developed competitive ownership advantages, partly as a result of the concentration of inward foreign MNEs in the banking and finance sectors of these countries which has exerted a developmental impact on local banks.

Outward direct investments by firms from the lower-income developing countries, such as the Philippines, in the developed countries are at a relatively earlier stage and are concentrated in export-promoting trade investments to support the growth of exports, especially of non-traditional products, as well as to consolidate marketing costs. Chapter 8 showed that a number of Philippine banks have also established operations in major financial centres such as Hong Kong, London and New York as well as in other developed countries but the major rationale for these investments is not to exploit any form of competitive advantages that these banks may have *vis-à-vis* banks in developed countries but to diversify their activities, acquire the latest banking technologies and practices, raise capital, capture the remittances of Filipino workers in these countries and manage the portfolio investments of their clients.

Manufacturing investments by lower-income developing countries in developed ones are generally comparatively few compared to those of the NICs, owing in part to the earlier stage of their firms' innovative activities, which does not allow for effective competition with developed country firms at the present time. Chapter 8 showed that the Philippine firm San Miguel Corporation initiated its international beer production by acquiring two breweries in the USA in the late 1930s but these ventures were divested in the late 1950s owing to the severe competition faced through the emergence of large national breweries. Similarly, although the restrictive foreign investment policies of the Indian government have resulted in the development of a strong and diverse capital goods sector, as well as a large pool of managerial and technical expertise with substantial design capabilities over a broad range of industries, Indian MNEs have been unable to make substantial direct investments in the developed countries. Perhaps the most significant are those of Kiloskar, which has successfully established diesel assembly plant and milling machinery operations in Germany.

The major exception to the generalisation that manufacturing investments in developed countries are more important to firms from the NICs than those from lower-income developing countries is the case of Malaysia. The diversity and the complexity of overseas investments by Malaysian MNEs, often in research-intensive industries such as scientific instruments, notwithstanding the non-NIC status of Malaysia, can be explained by their distinctive nature as nationalised foreign firms which are starting from a high technological base and which have substantial international investment experience.

Investments in the developed countries, of which an important component is research-intensive investment that embodies large-scale and more sophisticated technological skills, cannot be analysed effectively within the conceptual framework of the product cycle model. This model predicates that firms from developing countries have few exportable advantages relevant for competing in the more advanced countries since their main ownership advantages depend fundamentally on small-scale manufacture, substitution of local materials and other such skills. However, the more advanced capacities of some of these firms for localised technological change and the early emergence of their capacity for technological accumulation have enabled them to undertake investments in the developed countries partly to gain access to more advanced, although complementary, forms of foreign technology which could be combined with their indigenously generated technological innovation.

Hence, within the framework of growth of production and the process of technological accumulation adopted in this book, the increasing complexity and sophistication of firms' localised technological innovation as countries advance through stages of development and as firms' international production experience is broadened would lead some of the firms from more advanced and higher income developing countries, such as the NICs, which are engaged in more complex forms of outward investment, to establish successful international operations in the developed countries.

INVESTMENTS IN CONSTRUCTION SITES IN THE MIDDLE EAST AND ELSEWHERE

The preceding chapter has shown that the major Third World investors in the construction sector emerge from South Korea, Brazil, Argentina, Hong Kong, the Philippines and India. The distinctive nature of the construction investments by firms from the NICs was seen to derive from the need to have a continuous presence and to maintain their comparative advantage, which derives from their access to a skilled labour force. Contractors from developing countries strive for geographical diversification to maximise the utilisation of their country-specific advantage of skilled labour, and aim for technology transfer from contractors from the developed countries. By contrast, contractors from the developed countries strive to introduce higher forms of technological skills in their product in order to maximise the utilisation of their country-specific advantage, which lies in technological expertise. Therefore the distinctive feature of the construction investments of the developed *vis-à-vis* those the developing countries is the greater significance of technological accumulation for the international construction activities of the developed countries.

The comparatively higher forms of localised technological advantages of construction firms from the NICs and their longer experience in construction, both in domestic and foreign markets, have enabled them to broaden the geographical scope of their overseas construction activities from neighbouring developing countries (stage 1) to other developing countries that are of further psychic distance (stage 2) and at times even to developed countries (stage 3).

For example, the South Korean firm Hyundai Engineering and Construction Company began overseas construction activities in Southeast Asia in 1965 and by the 1970s this firm, as well as other Korean firms, had extended overseas construction operations to other developing countries in the Middle East, Pacific Region and Africa. By the late 1970s, the construction activities of

South Korean firms were largely concentrated in North America (43.0 per cent), the Middle East (35.2 per cent) and Southeast Asia (20.8 per cent). Hence, the geographical development of Korean overseas construction activities progressed from stage 1 to stage 3 within a span of fewer than fifteen years.

South Korean construction and consulting activities in the Middle East reflected the boom in construction in the region, particularly after 1977. Korean consulting firms in association with Korean construction firms have also expanded their international operations in the Middle East to serve the growing technological needs of Korean and local firms and to play a catalytic role in promoting joint ventures through their broadened international contacts and accumulated technological expertise. Although the Middle East continued to attract 86 per cent of the total construction orders of Korean construction firms in 1983, the construction business in these countries has since declined, due to the decrease in oil prices, the weakened comparative superiority and fierce competition from both developed and developing countries.

By comparison, the construction activities of Brazilian firms are less geographically diversified, although they have spanned not only the neighbouring Latin American countries such as Venezuela, Colombia, Peru, Chile and Uruguay but also Iraq, Mauritania, Angola, Algeria and Tanzania, where they have built hydroelectric plant, railways, expressways and university buildings. In addition, their investments in engineering activities are widely spread in Latin America, Nigeria, Tanzania, Algeria and the Middle East.

By contrast, the construction activities of firms from Argentina have continued to be dominant in neighbouring Latin American countries because of the closer psychic distance arising from the presence of ethnic ties. The most significant of these investments are the construction of railway facilities in Bolivia; large residential complexes in Venezuela; an oil pipeline in Ecuador; and recreational centres in Brazil (Katz and Kosacoff, 1983). Similarly, construction sector investments by Hong Kong firms are mostly concentrated in neighbouring countries such as Indonesia and Taiwan. In 1982, 41 per cent of the total FDI in the construction sector in Indonesia was accounted for by Hong Kong MNEs (Chen, 1981).

By contrast, the geographical development of overseas construction activities by Philippine MNEs has entered the second stage and has been important in the Middle East as well as neighbouring countries. Chapter 8 showed that as early as 1973 the Construction Development Corporation of the Philippines restored the Borobudur Temples in Indonesia, for US $9.0 million. However, since the first oil price hike in 1973–4, the Middle East has become the most significant recipient of Philippine construction investments as these countries have embarked on a massive infrastructure programme. Overseas construction there by Philippine firms continued to be favourable until the mid-1980s despite the slowdown in spending of some Middle Eastern countries.

In the early 1980s, Philippine construction investments diversified once again in neighbouring Southeast Asian countries. Ayala International, a Philippine conglomerate MNE, built *Istana*, the national palace of Brunei, at a cost of US $350 million. According to the *Guinness Book of Records* this is the largest building in the world and represents the biggest single construction project undertaken by an ASEAN firm. The firm also has construction interests in Kuala Lumpur, Malaysia, and Kota Kinabalu, Sabah, where it constructed both residential complexes and the Centrepoint hotel-bus-terminal commercial complex.

Similarly, Indian MNEs in construction and consultancy have established overseas operations primarily in the Middle East in conjunction with the promotion of various forms of technology and commodity exports which require a continuous presence in these countries (Lall, 1980b).

CONCLUSION

This chapter extended the analysis of the implications of the industrial development of Third World outward investment, with special reference to that of Asia and Latin America, on the geographical distribution of such investment. The increasing complexity of the geographical development of Third World outward investment, as a result of the increasing complexity of the sectoral pattern of their outward FDI, in turn determined by the stage of development of countries and the cumulative process of technological accumulation of firms, was described as following three sequential stages. The first stage, which is considered the most important for new international investors involved largely in resource-based investments, is directed towards investments in neighbouring and/or ethnically related territories. The second stage is production in several neighbouring and/or ethnically related territories, or in other non-ethnically related territories or developing countries, of simple forms of manufacturing and service investments. In the third and most advanced stage, the firms are better able to exploit their advantages in more complex and sophisticated manufacturing and service investments in countries further away from their home countries and neighbouring developing countries – those with further psychic distance and possibly in the developed countries.

With the exception of Hong Kong and Singapore, the general conclusion is that the geographical development of MNEs from the more advanced developing countries such as the NICs is increasingly in the third stage, compared to that of MNEs from lower-income developing countries whose international investment is at a relatively early stage. The geographical development of the latter set of countries is generally still largely concentrated within the first and second stages, with investments in neighbouring ethnically related developing countries and other non-ethnically related developing countries continuing to be important.

11 In conclusion: implications for theory and policy

THE GROWTH OF THIRD WORLD OUTWARD INVESTMENT AND THE STRUCTURAL CHANGE IN THE PATTERN OF INTERNATIONAL PRODUCTION

The emergence and growth of MNEs from developing countries constitute one of the more recent fundamental changes in the geographical pattern of international production to which the theories of the MNE are currently being adapted to provide an explanation. The emergence and growth of Third World MNEs have led to increased geographical diversity and competing sources of foreign direct investment.

Although the emergence of Third World MNEs from Latin America can be traced as far back as 1900, and those from Asia as far back as 1950, their significance and rapid growth have become noticeable only in the period since the mid-1970s and have been more pronounced in the period since the mid-1980s. As with the growth of UK firms in the nineteenth century, US MNEs in the early post-war period and, more recently, firms from Japan and Germany, the surge in the growth of international production by Third World firms based in the Asian NICs in the period since the mid-1980s may be attributed to the presence of a strong local currency as a result of a strong export position and overall balance of payments surplus. By 1989, direct investments from developing countries had increased almost twenty-fold to a conservative estimate of US $61,662 million, representing an annual average growth of 23 per cent from the 1975 level. Such an annual average growth rate of outward FDI from developing countries is almost twice as fast as that of developed countries, at 12 per cent. As a result of the high growth rates of outward FDI, the share of developing countries in the world stock of outward direct investment is estimated to have increased to at least 4.5 per cent by the end of the 1980s from a share of 1.2 per cent in 1975 and 0.8 per cent in 1960. The relatively small share accounted for by these countries relative to the already large accumulated worldwide stock of outward investment by the late 1980s is a manifestation of their recent entry and process of catching up in an era when foreign direct investment has increasingly become the new instrument of international economic integration in the world economy.[1]

Nevertheless, Hong Kong and Brazil, both newly industrialised developing countries, occupied the fourteenth and fifteenth places among the world's largest outward investors in 1983.

The general rise in the internationalisation of firms from newer source countries from both developed and developing regions (or those at lower stages of development) since the mid-1970s has brought about a structural change in the relationship between net outward investment and a country's relative stage of development as predicated by the earlier versions of the concept of the investment development cycle. These newer source countries from Germany, Japan, the smaller developed countries and the Third World have demonstrated the capacity and incentive to follow the earlier outward multinational expansion of the traditional source countries, the USA and the UK, at a much earlier stage of their national development. The general trend towards internationalisation, as demonstrated by the earlier and more rapid multinational expansion of firms from these newer source countries, has indicated that the national stage of development is no longer a good predictor of a country's overall net outward investment position.

This structural change, resulting from the general trend towards internationalisation, requires an alternative framework of analysis that considers the growth of newer sources of outward investment over time, including those at some threshold absolute lower level of development, which may acquire increasing capabilities and incentive for internationalisation and catch up with the earlier outward multinational investment expansion of the traditional source countries from the developed world. Theories predicated on relative levels of development, which is a strong feature of both the investment development cycle and product cycle models, may provide necessary but not sufficient explanations of the general trend towards internationalisation. These theories, which relate net outward investment to a country's relative level of development, will only be certain to be sustained over time if the determining factor is the country's relative stage of development, i.e. where firms at lower levels of development never really gain a major capacity for the internationalisation of their operations.

The general trend towards internationalisation, with the growth of the newer multinationals from countries at some threshold absolute lower level of development, including those at intermediate and lower stages of development, and the richer developing countries, may be better explained by absolute theories that consider the capacity of firms to internationalise based not on some relative levels of development but on a cumulative process of development in which firms gain ownership advantages over time. In this book, the generation of ownership advantages was analysed in the framework of technological competence of firms previously applied to explain the growth of MNEs from the more industrialised countries. The findings show that, with suitable qualifications to take account of the distinctive innovation of Third World firms, the framework of technological competence equally applies to explain the capacity of Third World firms to engage in international production at an earlier stage of development. The variable factor at different stages of development is not, therefore, the existence but the form of technological accumulation and competence. The theories of localised technological change and technological accumulation, which relate more to the absolute levels of development, have greater relevance in dynamic terms in explaining the internationalisation of firms than the investment development cycle and product cycle model, which relate more to relative levels of development.

However, even these theories that relate to the absolute capacity of countries and their firms to engage in outward investment are not sufficient, although necessary, to explain the generalised rise in internationalisation of firms which caused the structural change that occurred in the relationship between NOI and GNP per capita in the mid-1970s. Other reasons behind the general trend towards internationalisation depend fundamentally, although not exclusively, on the changing characteristics of the new technology paradigm as referred to by Freeman and Perez (1986), the ability to weaken the bargaining power of trade unions, and the political circumstances (Cantwell, 1987c).

THE CONCEPT OF STAGES OF DEVELOPMENT IN INTERNATIONAL PRODUCTION

More recent versions of the investment development path have argued that as well as the level of net outward investment, the character and composition of outward direct investment of a country's firms vary with the national stage of development (Dunning, 1986a, c). The early forms of foreign investment are frequently resource based or sometimes import substituting, but in each case have a quite specific location associated with a particular type of activity. However, as firms mature their outward investments evolve from a single activity or product in a particular location towards a more international perspective on the location of their different types of production activities. Such MNEs are responsible for directly organising an international division of labour within the firm in place of markets. At this stage, the character of their international production activities and their ownership advantages are also determined by firm-specific factors in addition to country-specific factors. However, home country-specific factors still dominate over firm-specific ones in determining national technological performance (Patel and Pavitt, 1991).

The speed at which various countries move through different stages of the investment development path and their direction are taken to depend on the structure of indigenous resource endowments, the extent of interaction with the rest of the world, the size of the local market, the economic system, government policy and, finally, the nature of the markets for the kind of transactions firms wish to engage in with foreign entities. These factors influence the extent and kind of ownership, location and internalisation advantages that different firms from different countries possess, and the industrial pattern of their inward and outward investments.

The general trend towards the internationalisation of business and the increasing complexity of its pattern and composition can therefore be analysed within the more general context of an investment development path. This book has examined the proposition of the changing character and composition of outward direct investment as development proceeds. The sectoral and geographical development of international production of countries, associated with innovative activities of firms, changes gradually over time in a way that to some extent can be predicted. This is because of the cumulative nature of technology and therefore the process of technological development ceases to be a random process but rather is constrained to zones closely related technologically to existing technological activities (see also Nelson and Winter, 1977, and the more recent work of Dosi and Orsenigo, 1988). The general argument was that the outward direct investment of countries follows a developmental or evolutionary course over time which is initially predominant in resource-based production or simple forms of industrial production that embody limited technological requirements in the early stages of development. At later stages of development, the sectoral path of these investments tends to evolve towards more technologically sophisticated forms of manufacturing investments. Such increasing complexity in the sectoral development of outward investment with expanding capacities for more complex technological accumulation is associated with higher stages of industrial development of home countries and broader experience of firms in international production.

The increasing complexity in the sectoral pattern of outward FDI has important implications for the geographical diversification of these investments. The stages of development approach in the analysis of the geographical distribution of Third World outward direct investment adopted in this book can be thought of in the context of Johanson's and Vahlne's (1977) *model of knowledge development and increasing foreign commitments*. In their work, the internationalisation process of the firm is regarded as following a sequential pattern in which initial foreign investments are directed to those markets with the greatest proximity in terms of psychic distance. The psychic distance at which foreign ventures are located is likely to increase directly with the complexity of the sectoral pattern of outward direct investments.

In the context of Third World MNEs in general, the increasing geographical diversification in their outward direct investment, as a result of the increasing complexity of the sectoral pattern, is described as following three sequential stages. The first stage, which is considered the most important for new international investors involved largely in resource-based and simpler manufacturing investments, is directed towards neighbouring and/or ethnically related territories. Evidence for this is shown in the relatively high share of intra-regional direct investment by developing country firms from Asia and Latin America, most of which, as predicated by the product cycle hypothesis, is located in other developing countries that are further down the pecking order than the home country. However, the evidence suggests that the importance of investments in these neighbouring and/or ethnically related territories declines as investing firms gain greater experience in international production and as their home countries progress to higher stages of industrial development.

Apart from the ethnic factor, the general trend towards internationalisation of business, with more recent outward investor firms engaging in international production at an earlier stage compared to the more traditional investors, has made investments in neighbouring developing countries that are further down the pecking order and have lower labour costs particularly important. These outward investments in export-oriented, labour-intensive manufacturing are typically undertaken by the small- and medium-sized firms in the Asian NICs that have experienced the greatest difficulty in maintaining the price competitiveness of their export products in the home country because of increasing labour costs and are forced to relocate production to countries where labour costs are lower.

The second stage entails production in several neighbouring and/or ethnically related territories or in other non-ethnically related developing countries of simple forms of manufacturing and associated service investments. Lower production and transport costs and close psychic distance, as well as the presence of favourable investment opportunities and the desire to make fuller use of regional economic integration, seem likely to explain significant direct investments within distinct regional groupings at this stage of development. As in the first stage, the technological advantages in the second stage are based on an adapted product or process innovation obtained through experience in developing countries, which can be applied to other developing countries at a lower stage of development. At the second stage, however, as home countries advance through higher stages of development and as the localised technologically innovative activities of firms become more significant, owing in part to their greater experience in international production, there are increased possibilities for economies of scale which arise out of operating subsidiaries in neighbouring and/or ethnically related territories or in non-ethnically related developing countries and enable firms to benefit more from internalisation advantages in terms of capturing economies of externalities and interdependent activities. In sum, the dynamic view adopted in this book refers to the increasing scope for international integration of foreign subsidiaries as countries advance through higher stages of development and as the localised technologically innovative activities of firms become more complex and more important. The presence of more sophisticated forms of innovation enables firms from the NICs to reduce their reliance on neighbouring and/or ethnically related territories, as they are able to exploit their advantages in countries further away from home and possibly in developed countries. In the third stage, firms with more complex technological capabilities become better able to conduct more sophisticated outward manufacturing investments in countries further away from their home base and in those further away in terms of psychic distance, including developed countries.

This book's findings suggest that the sectoral development of outward FDI from developing countries generally follows the evolutionary path of the more traditional investor countries as well as that of the newer investors from the developed world. Developing countries may be classified into groups defined with reference to the stage of industrial development, which is reflected in the industrial structure and technological complexity of their outward investment. Those belonging to the most advanced group have more complex forms of outward investment because of their greater capacity for technological accumulation, associated with higher stages of industrial development. The newly industrialised countries, defined as including South Korea, Taiwan, Singapore and Hong Kong in Asia and Mexico, Brazil and Argentina in Latin America, are therefore assigned to this most advanced group.

The cross-sectional analysis of Third World outward investment suggests that its sectoral composition is such that resource-based and simpler manufacturing investments that embody lower levels of technological accumulation are generally predominant forms for firms from developing countries at lower stages of industrial development such as those from China, India, Pakistan, the Philippines, Thailand, Colombia and Peru. The geographical development of MNEs from these countries, whose international investment is at a relatively early stage and whose innovative activities are based on production experience in developing countries, is still largely concentrated within resource-rich or neighbouring ethnically related developing countries as well as other non-ethnically related developing countries with a close psychic distance.

With the exception of the Asian city states which have followed more narrow industrial development strategies, owing to their small market size which has contributed to their firms' more limited technological competence, Third World firms from countries at more advanced stages of industrial development which have accumulated a greater amount of technological competence have an increasing capability to sustain more complex forms of manufacturing and research-linked outward investments. These more complex forms of international production often necessitate locating in countries with a further psychic distance, such as non-ethnically related developing countries as well as developed ones. The finding of increasing sectoral and geographical complexity in the pattern of international production in cross-sectional terms is further collaborated with time-series data that allow for a dynamic analysis of the sectoral and geographical evolution of the outward investments of some Third World MNEs.

Hence, with the exception of the Asian city states, the evidence confirms the initial hypothesis made about the interrelationship between the increasing complexity of the sectoral and geographical distribution of outward FDI with progressively higher stages of industrial development and broader experience in international production. Owing to the more limited technological advantages of firms from Hong Kong and Singapore and the greater importance of export-oriented, labour-intensive outward investments in neighbouring countries, their international production activities are still largely concentrated in developing countries. Their outward investments are not prompted by rapid industrialisation and the need for industrial restructuring in the home countries but by the necessity to expand productive capacity outside the limited domestic market and to overcome protectionist pressures and maintain export competitiveness. In the case of Hong Kong, apart from circumventing the limited size of the domestic market and the shortage of land and labour, there is an additional reason to diversify investment risks because of economic and political uncertainty with the impending return of the territory to the People's Republic of China in 1997.

This dynamic view of the increasing sectoral and geographical complexity of Third World outward investment as countries advance through higher stages of industrial development can be reinterpreted within the framework adopted by Dunning (1981a, 1988a) of firms possessing asset and transaction advantages. At more advanced stages, the asset advantages of Third World MNEs are supplemented by the more important transactional advantages because firms' hierarchies benefit more in terms of the common governance of separate but interrelated activities located in different countries. However, in the dynamic context adopted in this book, such asset and transactional advantages are analysed in terms of the extent to which firms engage in integrated activities across national boundaries as a feature in the growth of production within which each firm has distinct technological advantages which are difficult to transfer to, and implement by, other firms with different technological advantages.

The stages of international production of Third World MNEs in relation to the macroeconomic development theories of international production

The analysis of the rapidly evolving character and composition of outward investment from developing countries provides a Third World dimension to the macroeconomic developmental theories of international production formulated to explain outward direct investment from the developed countries. For example, Ozawa (1990) has described Japanese outward direct investment as following distinct stages of development. The distinctive characteristics of Japanese FDI from 1950 to the early 1970s in resource-based and labour-intensive or technologically standardised industries in developing countries belonged to the first two phases of Japanese overseas activity. Since the late 1960s, as a result of industrial restructuring in Japan, overseas investment has evolved to the third, import-substituting or recycling, stage of international production. This is centred on the mass production of assembly-based, consumer durables in high-income countries such as the USA and Europe, supported by a network of subcontractors. By the mid-1970s the transformation of Japanese outward FDI towards more complex manufacturing activities in industrialised countries such as the USA and parts of Europe had become well established (Franko, 1983). Since the early 1980s, a fourth phase of Japanese FDI can be discerned. This is characterised as the flexible manufacturing of highly differentiated goods, involving the application of computer-aided designing (CAD), computer-aided engineering (CAE) and computer-aided manufacturing (CAM) (Ozawa, 1990).

Although most Third World MNEs are unlikely to develop as fast as their Japanese counterparts, who can today draw on frontier technologies, the trend is towards a persistent upgrading in the types of activity for which these firms are responsible. The phases of development suggested by Ozawa for Japanese MNEs are thus relevant to the study of Third World multinationals. The first two phases of Japanese outward investment from the 1950s until the early 1970s in resource-based and labour-intensive manufacturing activities in developing countries are relevant in explaining the early stages of the growth of outward investment from the newer investors based in the developing countries. In the early stages, as domestic firms in developing countries upgrade their domestic production activities, resource-based and less technologically sophisticated manufacturing activities are transferred to countries at an earlier stage of development with lower levels of technological capacities and production costs. In the later stages, more sophisticated manufacturing activities are transferred abroad, even to the developed countries. The current overseas activities of the newly industrialised countries in the developed countries, although based on a less research-intensive form of technological innovation than Japanese MNEs, can nevertheless be described as being in the early phases of the import-substituting or surplus recycling form of international production.

EXPLANATIONS OF GROWTH OF THIRD WORLD MULTINATIONALS: THE ROLE OF OWNERSHIP ADVANTAGES

Post-war developments in the nature of international production necessitated a shift in emphasis to an analysis of ownership advantages based on the theories of the firm and the growth of the firm. The Hymer theory of the growth of the firm, in which the major cause of international operations is seen to derive from the possession of oligopolistic forms of ownership advantages arising from the presence of firms of unequal abilities, has very limited explanatory power when applied to the growth of Third World MNEs that are hardly involved in competition under oligopolistic market structures. Hymer describes ownership advantages in the context of Bain-type advantages of established firms such as those that arise from scale economies, knowledge advantages, distribution networks, product diversification and credit advantages. Other theorists in the industrial organisation tradition describe ownership advantages as the possession of superior product and production technology, superior management skills and preferential access to production inputs (Mason, Miller and Weigel, 1975); or firm-specific know-how and other intangible income-producing proprietary assets resulting from past investments in product or process R&D and/or investments in advertising and other promotional techniques which enable the firm to engage in some form of product differentiation (Hirsch, 1976); or technology, product differentiation, scale economies and skills (Lall, 1980a). These ownership advantages, which create barriers to entry to rival firms and increase the market power of the firm, and/or create a temporary monopoly power, are not significant explanatory factors for the multinational investment expansion of Third World firms that are at an earlier stage of industrial development but have nevertheless gained the capacity and incentive to initiate international production activities. As a result of their less complex forms of ownership advantages, the catching up process of Third World firms in international production largely occurs within the later stages of the *product cycle* (Vernon) or *industry technology cycle* (Magee) as the industry moves over time from less competitive to more competitive market structures and from less standardised to more standardised products as a result of diffusion and changes in product or process technology.

This book has focused, in particular, on the role of newly emerging technological factors in explaining firms' international production activities and the distinctive nature in which these factors have found expression in the growth as well as the changing sectoral and geographical composition of international production. Although the developmental course of the most recent outward investors from the Third World has been faster than the more mature multinationals from Europe, the USA and Japan, the forms of innovation of firms from the Third World differ from those of the industrialised countries because of the different stages of their industrial development. The analysis of innovation of Third World firms requires an empirical investigation of technology creation beyond R&D and international patenting activities. Hence, although the existence of innovation within a firm determines the growth of international production, especially in manufacturing, the changing form of innovation as opposed to its existence is the variable factor in stages of development. The concept of rapidly emerging technological capabilities or technological competence is therefore a useful means of analysing the international growth of manufacturing firms from quite different environments, and at different stages of technological development and capacity.

In the early stages of technological development, ownership advantages tend to be more country specific and, in the case of resource-based firms, associated in part with the abundance or scarcity of natural resources in the home country. In either case, firms gain accumulated expertise in the exploitation and processing of natural resources present in the home or host country. The early stages of technological development of these Third World MNEs are broadly parallel with the early forms of technological ownership advantages of US and British firms in the wood processing and metal and coal processing industries respectively, which were acquired because of the abundant availability of timber and coal in the USA and the UK (Rosenberg, 1976).

However, as home countries advance through progressively higher stages of industrial development, and as firms accumulate greater experience in international investment, the technological embodiment of their outward direct investment activity becomes more significant and more complex. This is seen in part in research-intensive investments in centres of innovation in the developed countries through which firms gain access to more advanced although complementary forms of foreign technology which can be adapted and integrated with their indigenously generated technology.

This analytical framework has historical antecedents in the technology gap theory of trade of Posner (1961), and also in the work of Leontief (1954) and Johnson (1958) who emphasised the important role of technological factors in explaining the pattern of US trade. Leontief, in particular, pointed to the embodiment of higher skills in the pattern of US export products while Johnson referred to the presence of a slower rate of innovation in Europe than in the USA in explaining the existence of a persistent dollar shortage in Europe. The presence of a technological gap as an important factor in explaining patterns of trade became even more apparent in the early 1960s with the work of Posner, who pioneered the technology gap theory of trade. His work, which emphasised the different rates of innovation and learning among different firms and countries, provides the basis for this book's analysis of the unique nature of technological characteristics as well as the obstacles to further technological development of Third World firms.

This book has attempted to address the central issue of the relationship between the pattern of domestic development, the emergence of competitive or ownership advantages on the part of indigenous firms and the growth of outward investments in particular sectors. The main argument is that the significance and complexity of technological innovation of firms from developing countries are largely determined by, and in turn affect, the stage of industrial development of their home countries and the cumulative process by which they build upon their unique technological experience with increasing experience in international investment. The existence and accumulation of technological capabilities are an important determinant as well as effect of the pattern and growth of their international production activities.

In the earlier stages of development, the nature of international production may be described in the context of the product cycle model as well as the Kojima theory grounded on the neoclassical Hecksher-Ohlin principle of comparative advantage (or costs). These models are particularly helpful in explaining the relocation abroad of production of comparatively disadvantaged (or marginal) industries as locational advantages accrue in host countries at an earlier stage of development. Such relocation of production is undertaken because the domestic firms still have the technological and organisational advantages associated with the lower technology and more labour-intensive production activities which can be more profitably exploited in a foreign environment with lower levels of technological capacities and production costs. The relocation of sunset industries, described by declining international competitiveness in production in the home country, to more competitive foreign locations may be an alternative to their managed decline in the home country. These theories therefore present a useful explanatory framework within which to view the growth of international production of countries undergoing rapid growth such as Japan, Germany and the newly industrialising countries.

The analysis of the technological advantages of Third World MNEs across a wide range of industries suggests a rapid pace of industrial development accompanied by the existence and accumulation of indigenous technological capability, especially in those sectors in which indigenous firms have accumulated production experience. The analysis of the emerging technological capabilities of Third World firms extended beyond the static neoclassical application of Kojima's comparative based model, which considered foreign production in the narrow context of sustaining a gradual process of managed decline of mature or technologically standardised industries in the home country through their relocation abroad. In this book, the process of international production itself is considered an important means by which to meet the need to sustain the dynamic process of domestic industrial restructuring in the home country, which will be crucial in the development of future comparative advantage. This can be undertaken both directly, through specialisation and concentration in the home country of more highly value-added, technologically intensive industries that offer greater opportunities for growth with the relocation of sunset industries abroad; and indirectly, through the infusion of new and more advanced technology brought about through access to complementary foreign technologies abroad. Such a strategy enables dualistic development to be made on an international scale.

The existence and accumulation of indigenous technological capability therefore become an important cause and effect of the growth of international production networks and the process of domestic industrial restructuring. The existence of indigenous technological capability influences the ability to engage in international production, which in turn influences the accumulation of further technological capabilities which enable firms to upgrade domestic industry towards more technologically intensive activities and to engage in more sophisticated forms of outward investment.[2]

In this context of international production as an important instrument of domestic industrial restructuring, exceptions are found once again in the cases of Hong Kong and Singapore. In the case of Hong Kong, outward investments are not part of a government-directed industrial policy, perhaps because of the uncertainties of 1997. Whatever the reasons for the lack of active government support for outward investments in Hong Kong, which is characteristic of the other Asian NICs, the result has been that Hong Kong firms have invested outside their national boundaries motivated by the generation of quick returns rather than long-run growth based on planning and research and development (ESCAP, 1986) or fulfilling any macroeconomic development objectives for their home country. This explains why the development of their technological capabilities is confined to fairly narrow and defined areas of activity which limits the growth of their future comparative advantage in higher technologically intensive industries (Fransman, 1984; Chen, 1988; Lall, 1990). Although Singapore is in some respects different from Hong Kong, because of the more interventionist and paternalistic role of the state and its lower manufacturing value-added, it does share many common features with Hong Kong in that investments in education and R&D, and the breadth of its industrial and technological base, are inferior to those of the other Asian NICs such as South Korea and Taiwan (see Pang and Lim, 1989).

The narrow industrial base of Hong Kong is seen by Lall (1991) as an added pressure to internationalise, as opposed to countries that have a broader scope for industrial diversification. The argument is based on the premise that countries that specialise in a wider range of industries can retain a strong production and export base through a continual process of domestic industrial restructuring in the home country. However, although the pressure to internationalise at a much earlier stage may be greater for Hong Kong firms, owing to weak local technological capabilities and the narrow scope for diversification into other industrial sectors in the home country, there may be difficulty in sustaining competitive advantages in labour-intensive manufacturing in the longer term as predicated by the product cycle model owing *inter alia* to the growth of firms from lower-income developing countries which will gain relatively greater comparative advantage in the simpler forms of manufacturing activities over time. In an era of technological competition in international industries, comparative advantage must be directed towards and sustained in activities associated with more sophisticated forms of innovation, including R&D, which offer greater and wider technological opportunities for growth.

The cases of the Asian city states and other lower-income developing countries show that apart from technological advantages, managerial experience, training and international orientation are also significant determinants of international production competitiveness of Third World MNEs (see Lecraw, 1991, for case study on Indonesia). The rather limited competitive advantages of Hong Kong firms in the area of technology are somehow sustained through their large reservoir of entrepreneurial and marketing know-how and access to global distribution channels. The strengths of Hong Kong MNEs in marketing and distribution as well as management and entrepreneurship, as opposed to technological accumulation considerations, render the product cycle model relevant to explain the growth of these firms.

THE ROLE OF TECHNOLOGICAL ADVANTAGES AND THE THEORIES OF THIRD WORLD MULTINATIONALS

The process of independent technological accumulation on the part of Third World MNEs therefore necessitates the formulation of theories of the growth of Third World MNEs that recognise that the role of innovation within firms has been distinctive and increasing, and has become more sophisticated as international production has developed. Although the evolving sectoral pattern of Third World outward investment has been faster and bears a certain resemblance to the changing sectoral composition of the international production of firms from the developed countries during an earlier period, the innovative capacities of Third World firms are distinguished from those of the more mature multinationals from the developed countries because of their different stages of industrial development and a process of technological change that is firm specific, cumulative, differentiated and path dependent. Hence, although an underlying link exists between the increasing importance and complexity of technological accumulation as greater experience in international investment is acquired and increasing sophistication in the industrial and geographical pattern of outward direct investment, the significance and form of the technological accumulation vary between the technological leaders of the USA, the UK, Japan and Germany and the simpler and less research-intensive technology creation of Third World MNEs. The MNEs of the USA, the UK, Japan, Germany and other industrialised countries are frequently technological leaders which have accumulated greater production experience and have more significant investments in R&D in research- and technologically intensive industries. By comparison, Third World MNEs rely to a greater extent upon simpler and less research-intensive forms of technology creation. Hence, although a quicker pace in the evolutionary trend of the sectoral pattern of Third World outward investment is observed, these firms do not exhibit the same rapid pace of technological advancement as firms from the more developed countries.

The distinctive innovative activities of firms based in Third World countries provide evidence that they have followed a technological course that is to some extent independent and a function of their own unique learning experience. Owing to their earlier stage of technological development, the distinct innovative capacities of Third World MNEs are not necessarily limited to descaling and increasing the labour intensity of sophisticated technologies or specialisation in the production of lower-skill and lower technologically-intensive sectors. They also extend to innovations that are based on lower levels of research, size, technological experience and skills and aim to achieve improvements by modernising older techniques, including outdated foreign technology. These firms undertake technological improvements in products, processes or equipment which may represent a revival or modernisation of older technologies, including technology considered outdated in the developed countries. Without a strong reliance on R&D activity, a greater part of their innovatory capacity has been given to production engineering, learning by doing and using, organisational capabilities and, as shown in Chapter 6, the assimilation of imported technology. However, there are a steadily rising number of examples of Third World multinational enterprises that have become genuinely innovative through their progression to the early stages of R&D activities, even though such activities have tended to be less scientifically refined and have not generally involved frontier technologies.

With higher rates of innovation through investments in production engineering and in some cases R&D, the relative technological latecomers from the Third World are increasingly able to catch up with the technological leaders from the more developed countries. Some of these Third World firms have gradually acquired innovative capacities in fields in which technological opportunities are arising faster and stand to benefit from a system of technological interrelatedness by drawing on the more advanced but complementary technologies of established firms from the developed countries. The further expansion of these firms through more complex development of R&D capabilities may be expected as these countries advance through higher stages of industrial development and as their firms acquire greater experience in international production. The future technological course in the direction of R&D is also greatly influenced by indigenous efforts in developing countries to strengthen the national system of innovation through the establishment of R&D laboratories, science parks and technological research institutions, overseas investments in research and efforts to seek more advanced but complementary foreign technologies through various modalities. The emergence of the NICs' firms' capacity for more sophisticated innovative activities, compared with the lower-income developing countries, explains their greater ability to catch up with the older and more established MNEs.

The rapid technological development of Third World multinationals provides the *raison d'être* for the re-examination of the theories currently proposed to explain their growth. The scope of innovation of Third World MNEs differs in the two main schools of thought, i.e. between the framework of the product cycle model articulated by Wells and the model of localised technological change advanced by Lall. In the former, the growth of external market demand and differences in relative factor prices primarily determine the extent and form of innovation and product development as well as the location of production by local firms. But in the model of localised technological change, and in the more general theory of technological accumulation and competence used in this book, the existence and continuation of technological accumulation become the primary determinants of international production by local firms.

Within the overall framework of technological competence, the findings of this book suggest that the product cycle model is most applicable for Third World firms at the earliest stages of development in international production that have accumulated production experience in developing countries. For countries at a rather more advanced stage, whose firms have developed a broader range of technological advantages, including the ability to develop an independent and distinctive technological capacity, the theory of localised technological change assumes increasing importance. In addition, for the more advanced newly industrialising countries that are showing early signs of the capacity for R&D the theory of technological accumulation is acquiring greater significance in explaining their firms' outward investment. The truly innovative capacities of firms from the NICs are expected to become increasingly important and complex as their countries advance through progressively higher stages of industrial development and as they acquire greater experience in international production.

The product cycle model

The dynamic theory of Third World outward investment formulated in this book recognises the relevance of the product cycle model in analysing the Schumpeterian process by which the innovative activities of technological leaders in the more industrialised countries are subsequently imitated by a competitive fringe of Third World firms which catch up with the technological leaders by increasing productivity growth and output. The process of catching up on the part of Third World firms is manifested through the imitation of innovative activities during the maturation and standardisation phases of product and process development.

In particular, the process of catching up is envisaged during the fourth stage in the product life cycle, as the location of production is shifted to Third World countries because relative factor prices dictate that capital-intensive means of production be replaced by labour-intensive means, and especially that higher-skilled labour be replaced by less-skilled or unskilled labour. A cost-determined equilibrium regulates the shift of these lower-skilled and unskilled labour stages of production of standardised products to developing countries where labour costs are lowest and where incomes begin to catch up.[3]

The growth of Third World firms may be conveniently viewed from a Schumpeterian perspective which essentially sees the internationalisation of firms from developing countries as a stage in the product life cycle or technological gap model. These theories propose that the international expansion of countries through trade and international production falls into a pecking order depending on their ability to produce a particular product, which reflects their different capacities for technological innovation and differences in production costs. In the product cycle model in particular the unique generation and development of innovation of firms is determined by differences in the national environment of their home countries. While American firms innovate by way of developing high-income and labour-saving products, and European firms innovate raw-material saving and capital-saving technologies, the expansion of Third World MNEs is based on unique innovation derived from the adaptation of foreign technology to the circumstances of smaller plant, smaller firm size, greater labour intensity and the development of special skills in the maintenance, repair and supply of spare parts for second-hand machinery, as well as a technological advantage in their ability efficiently to utilise locally available natural resources rather than imports, and other advantages in terms of low salary payments to managers who are none the less adept at organising in developing country conditions. To the extent that these firms have unique technological advantages, these advantages are limited to the imitation and adaptation of foreign technology in the production of standardised goods at the tail end of the product and industry life cycle, in accordance with the requirements of Third World markets and production conditions. As a result, these firms are fundamentally engaged in industries characterised by low expenditures on R&D and low product differentiation.

Since the ownership advantages of Third World MNEs described by the product cycle model have low or no barriers to entry, the predication is that developing country affiliates of US or European MNEs may imitate and copy the technological improvements achieved by Third World firms in the course of adapting an essentially foreign technology and product development. Even new MNEs in other developing countries may gain the capacity to copy or adopt the competitive advantages of the current Third World outward investors and replace those firms whose advantages have eroded.

Empirical investigation confirms that some domestic firms in developing countries, especially those at early stages of outward investment, have innovative capabilities that respond to the unique conditions of their home markets and result in the creation of entirely new products or processes or a modification of existing products or processes. The technological development of Third World MNEs engaged in outward direct investment in mature, standardised and often resource-based industries is strongly linked to the technological course of firms from developed countries, which is assimilated and adapted to suit Third World markets and production conditions. The adaptations pertain mainly to the scale of operations, factor intensity, lower production costs including factor costs, and the use of locally available raw materials, equipment and other product and process design, which represent accumulated production experience and learning by doing and using. These outward investments by Third World MNEs whose home countries are at earlier stages of industrial development and whose firms are at an early stage in their international production activities are frequently undertaken in neighbouring and/or ethnically related territories. This lends support to the product life cycle hypothesis which predicates that Third World MNEs can best exploit their competitive advantages in other developing countries, at least as far as their initial international investments are concerned.

The major strength of the product cycle model lies in explaining the role of non-technological factors as important ownership advantages of firms in the later stages of the product life cycle where the nature of the product demands that accessibility to market information is greater and competition is largely, if not solely, on the basis of price. The search for the lowest cost source of supply therefore becomes the priority of investor firms. At this stage, a firm's main ownership advantages are based on marketing and distribution, unlike the earlier stages where they were based on the firm's ability to engage in technological innovation. These other forms of ownership advantage become useful in explaining not only the initial growth of MNEs from lower-income developing countries at early stages of industrial development, whose firms have limited technological advantages, but also the sustained growth of MNEs from Hong Kong and Singapore which, despite their more advanced stage of national development as measured by GNP per capita, also have limited technological advantages. The model therefore provides a useful explanatory framework to analyse the international production of mature and standardised products by firms both from developing countries at early stages of development and from Asian city states (especially Hong Kong) that have rather more international investment experience and have retained competitive advantages in the foreign production of mature and standardised products.

In addition, the role of financial factors in explaining the growth of international production may be explained by the product cycle model. The model deals with the process by which technologically innovative firms become significant exporters in the first stage of foreign market servicing and then become outward direct investors at a later stage when locational advantages, helped by a strong local currency, favour international production. The presence of a strong local currency as a result of a strong export position and overall balance of payments surplus may provide the impetus for the growth of outward direct investment by locally based innovative firms (Cantwell, 1991b). This process effectively characterised the growth of UK firms in the nineteenth century, US MNEs in the early post-war period and firms from Japan and Germany more recently (Cantwell, 1989b). This growth process also holds true for some Third World firms based in the Asian NICs in the period since the mid-1980s.

However, where there are competing innovations among firms from the follower country in which the process of catching up is determined by firms' inherent potential for innovation rather than limited to their scope for imitating the innovation of the technological leader, then the PCM is on weaker ground. Such a framework of analysis, which specifies the market as the prime determinant of innovation, assigns a narrower scope for innovation to Third World firms whose technological course is restricted to the imitation and adaptation of foreign technology in accordance with the requirements of Third World markets and production conditions. More recent theories show that innovative activities are not determined entirely by current or anticipated market signals as predicated by the PCM but also by evolving technologies (Atkinson and Stiglitz, 1969; Mowery and Rosenberg, 1979). These theories, which allow wider scope for the analysis of innovative activities, offer better explanations for the way in which firms from developing countries develop the capacity to follow an *independent* or *localised* technological trajectory or evolutionary pattern through progressively higher stages of industrial development and greater experience in international production. The relegated importance of the firm and industry to that of the product and the market constitutes a major limitation on the effectiveness of the product cycle model in explaining patterns of innovative activities in firms and countries (Pavitt, 1988).

The inability of Third World MNEs to sustain ownership advantages over time in the product cycle model emerges precisely from the specification of the market as the prime determinant of innovation. The predication of the model that new MNEs from other developing countries may gain the capacity to copy or adopt the competitive advantages of the current Third World outward investors over time and replace those firms whose advantages have eroded has limited explanatory power in the light of the recent developments in international production and, in particular, the increasing sources of outward investment from other developing countries. Indeed, as a result of their greater experience in international production and the cumulative nature of technology development, the earlier sources of outward investment have developed more complex and more significant forms of competitive ownership advantage, helped by the increasing importance of their innovative activities. By contrast, the competitive ownership advantages of the newer sources of outward investment are much simpler because their innovative activities are at an early stage. However, these advantages may be expected to become more complex and more sophisticated as experience in international production is broadened. These findings disagree with the contention of the product cycle model that Third World MNEs have no capacity to sustain their competitive ownership advantages over time.

Perhaps another important criticism of the PCM in explaining the growth of Third World MNEs is that recent forms of international production, especially those undertaken by the NICs in the developed world in more technologically complex sectors, embodying large-scale and more sophisticated technological skills, cannot be analysed effectively within the conceptual framework of the model. This is because the model predicates that firms from developing countries have few exportable advantages relevant for competing in the more advanced countries since their main ownership advantages are taken to depend fundamentally on small-scale manufacture, substitution of local materials and other such skills.

More importantly, research-intensive investments by Third World MNEs in the developed countries to gain access to technology created in a particular location cannot be analysed within the Schumpeterian framework of the PCM, which lays down a chain of events starting with innovation in the leading country and leading to a technology transfer and diffusion to a competitive fringe of firms in the host country. This book has presented evidence that firms from developing countries, like those from the developed world, may engage in foreign direct investment or licensing activities as a form of backward technology transfer in order to gain access to complementary foreign technology which is generated in a particular location but which these firms can combine with their indigenous technological innovation.

Within the framework of growth of production and the process of technological accumulation adopted in this book, the increasing complexity and sophistication of localised technological innovation of firms from more advanced and higher-income developing countries which are engaged in more complex forms of outward investment have enabled them to establish successful international operations in the developed countries. Indeed, the more advanced capacities of some of these firms for independent technological change, and the early emergence of their capacity for R&D, have enabled firms from the NICs to undertake investments in the developed countries, partly to gain access to more advanced although complementary forms of foreign technology which could be combined with their indigenously generated technological innovation.

The model of localised technological change

The steadily rising number of examples of Third World multinational enterprises that have become genuinely innovative necessitates a broader theoretical explanation of their indigenous technological capabilities. The rapid evolution in the pattern of their activities has meant that theories may have to take account of the possibility that these firms can also embark on independent technological accumulation. The theories have to recognise the increasing role of innovation within firms, which becomes more sophisticated as international production develops. As with the growth of international production established by industrialised country MNEs in the manufacturing sector, the growth of Third World MNEs has become increasingly dependent on the existence of innovation within the firm.

Within the current theories of the growth of Third World multinationals, Lall's theory of localised technological change, an idea which is implicit in the more general theory of technological accumulation, suggests that technology consists of very specific knowledge of particular products, processes and markets and that firms follow a technological course that is to some extent independent and a function of their own unique learning experience (see Lindbeck, 1981). In particular, the technological course of Third World firms is considered independent from that of the more developed countries as a result of indigenous technological creation, adaptation of foreign technology or specialisation in techniques which were used at an earlier stage in the developed countries.

The model therefore provides a valuable theoretical tool for examining the distinctive nature of proprietary or firm-specific advantages of developing country firms *vis-à-vis* developed country firms. Unlike the product cycle model, the theory stresses that the unique ownership advantages of firms from developing countries may not necessarily derive from the imitation and adaptation of foreign technology in accordance with the requirements of Third World markets and production conditions. The capability of a firm in a developing country to innovate for a unique proprietary advantage conforms to the evolutionary theory of technical change which regards such change as localised at the firm level, path dependent and partly irreversible.

Owing to the earlier stage of technological development, the distinctive advantages of Third World firms derive from their ability to innovate on essentially different lines from those of the more advanced countries, i.e. innovations that are based on lower levels of research, size, technological experience and skills, and to achieve improvements by modernising an older technique, including outdated foreign technology. Hence, while the competitive assets of firms from developed countries are derived from frontier technologies and sophisticated marketing, those from developing countries are derived from widely diffused technologies, special knowledge of marketing relatively undifferentiated products or special managerial or other skills.

Through modification of the static, conventional Hecksher-Ohlin trade theory into a dynamic theory of trade and investment for developing countries, Lall's theory of localised technological change allows for changing dynamic comparative advantages of developing countries that are increasingly capable of producing goods which embody high levels of technology, skills and scale. These technical changes are in part brought about by technological and organisational innovations introduced by inward foreign direct investment in developing countries to take advantage of low factor costs and in part by the indigenous technological and marketing innovation undertaken by local firms.

The main conclusion that can be drawn from the analysis of the technological development of Third World MNEs in Chapter 6 is that these firms have the capability to follow an independent or localised technological trajectory in which imported technology is integrated with the development of indigenous technology in a manner consistent with their dynamic growth path of changing comparative advantage. The nature of technology as firm specific, differentiated and often tacit means that integration is more successful when foreign-generated technology is complementary to the path taken by indigenous innovation, and can therefore be beneficially integrated and adapted within a unique programme of technology generation. The more complementary the technology is to current firm technology, the lower are the costs relative to the benefits of adoption and integration. Such a view of the technological advantages of Third World firms, which encompasses not only the simple imitation and adaptation of innovation carried out previously in a more advanced country by a competitive fringe of Third World firms but also a further set of related although differentiated innovations as a result of the unique technological experience of Third World firms, is consistent with the views of Atkinson and Stiglitz (1969), Rosenberg (1976), Nelson and Winter (1977, 1982) and Stiglitz (1987). Indeed, the innovative activities undertaken by firms are triggered and guided not only by evolving market signals, as predicated by the product cycle model, but also by evolving technologies (Atkinson and Stiglitz, 1969; Rosenberg, 1982; Pavitt, 1988).

Due to the localised and irreversible nature of technological change, the distinctive innovative capacities of Third World MNEs cannot be readily replicated by the affiliates of industrialised country MNEs in developing countries that have progressed to new technologies, since there are high costs in efficiently reproducing or transferring older technologies. For example, if MNEs from industrialised countries attempt to replicate the technological advantages of Third World MNEs in small-scale manufacture, they lose the advantages of accumulated experience in large-scale manufacture (Wells, 1978).

The theory of technological accumulation

Both the theory of localised technological change and the more general theory of technological accumulation propose that firms follow distinctive technological trajectories in the course of innovation, which are at least to some extent independent of one another. The theory of technological accumulation as firm specific, cumulative and differentiated advanced by Pavitt has previously been applied to explain the historical evolution of the sectoral pattern of comparative advantage in innovative activity of national groups of firms in industrialised countries (Cantwell, 1990) and in work on competition between rival technologies (Arthur, 1988, 1989). These innovative activities, especially in R&D, have been found to be statistically significant determinants of differences in export and productivity performance (Soete, 1981b; Fagerberg, 1988) and also of internationalisation of business among the major OECD countries (Vernon, 1974, 1979; Cantwell, 1989a).

Evidence of the dynamic changes in the area of Third World outward investment is provided by some of the newly industrialising countries whose outward investment is increasingly associated with early stages of R&D activity. Some of these countries have advanced beyond the lower-income developing countries, whose outward investment still embodies innovative activities limited to the imitation and adaptation of foreign technology to suit Third World production conditions as well as the generation of independent or localised technological change.

Some Third World MNEs now operate in sectors whose products are far from standardised, and their activities are helping to generate a fresh pulse of innovation in some cases. Indeed, there are new multinationals operating in the Third World, particularly in the newly industrialised countries, that are genuinely and uniquely innovative, and in some cases discussed in this book these firms have progressed to early stages of R&D activities, in part through investments in research facilities in the developed countries. Although these firms are still at a relatively early stage of development, their abilities in technological accumulation continue to expand. Their further outward expansion in more technologically sophisticated activities may be expected to increase as their home countries advance through higher stages of development and as they acquire greater experience in international production. The emergence of these firms' capacity for more complex technical changes explains their greater ability to catch up more rapidly with the older and more established MNEs from the more developed countries.

For these firms from the NICs, with the exception of the Asian city states, the theory of technological accumulation is increasingly becoming a more useful framework within which to explain the expansion of their production networks across national boundaries. However, the argument is extended to consider the fact that the degree and form of firms' innovative activities are in turn determined by the stage of industrial development of the home country. Countries at a lower stage of development such as the developing countries have the capability for simpler forms of innovation, compared to industrialised countries at a advanced stage of development whose innovative activities are becoming increasingly important and complex. Hence, the difference between MNEs from developed and developing countries, and indeed between MNEs from the NICs and lower-income developing countries, lies in the form and complexity of their innovation.

The theoretical conclusion

Perhaps the most important theoretical conclusion that could be drawn from the analysis of the evolutionary pattern of the nature of Third World outward investment is that the concept of technological competence of firms has general relevance as an explanation of the emergence and growth of Third World outward investment. The findings of this book suggest that the process of technological accumulation characteristic of the growth of international production of firms from the industrialised countries also finds application as an important determinant and effect of the pattern and growth of the international production activities of Third World MNEs. The growth of outward direct investment is a cumulative process, where the form of technological accumulation rather than its existence varies with the stage of development.

The two prevailing schools of thought on Third World MNEs can therefore be applied to the outward direct investment of countries at different stages of technological development or alternatively to the outward direct investment of firms within countries over time. In the earlier stage, the product cycle model provides a useful framework within which to analyse the internationalisation of firms whose technological capacities are limited to the assimilation and adaptation of innovative activities of technologically leading firms in accordance with the requirements of Third World markets and production conditions (Wells, 1986). The model is also useful in explaining the role of organisationally based technology creation such as managerial and marketing techniques, as well as financial factors, in the growth of Asian NICs. In the later stages, as home countries advance through progressively higher stages of development and as their firms acquire greater experience in international investment, the concepts of localised or independently generated technological change and eventually technological accumulation through R&D assume greater significance. This allows for an independent or differentiated technological path on the part of Third World firms, which contrasts with the more limited scope of technological innovation embodied in the product cycle model. The latter model cannot explain the different patterns of innovative activities in firms and countries apart from those determined by market demand factors. In fact, firms' innovative activities are triggered and guided both by evolving market signals and evolving technologies.

The essence of the dynamic evolutionary approach towards explaining the growth of new sources of outward investment from the Third World therefore lies in the shift from the product cycle model towards the theories of localised technological change and technological accumulation as developing countries advance through progressively higher stages of development and as firms acquire greater experience in international production.

THE GROWTH OF THIRD WORLD MNEs AND THE ROLE OF FOREIGN TECHNOLOGY

This book has argued that even in the Asian NICs that have developed some national innovative capabilities in R&D, by and large the main sources of growth of outward direct investment from developing countries in the catch-up phase of technological development are found in their ability to access technology from abroad through diffusion and in their indigenous capacity to exploit the benefits of technology created abroad. Research expenditure by developing countries geared to the achievement and improvement of international competitiveness are therefore related more to technology importation than to autonomous research spending on domestic R&D.

This book has demonstrated how, in certain cases, foreign technology has been an agent of growth through supporting dynamic changes in comparative advantages of firms and industries in developing countries towards higher value-added and higher technology intensive industries, in accordance with policies implemented by governments in those countries. In particular, the growth of more complex sectors of industry associated with more advanced stages of development is reflected in the growth of international production of local firms. The transfer from abroad of more advanced but complementary technologies has helped some local firms in developing countries to strengthen their technological capabilities and to generate more competitive forms of ownership advantages, necessary in carrying out their international production activities.

Perhaps the most shining example of the success with which MNEs from a developing country have effectively absorbed foreign technology in the course of technological development is provided by South Korea. The assimilation of foreign technology, especially in higher technologically intensive industries, was seen as part of the overall technological efforts of Korean firms to achieve productive efficiency and to compete effectively through increased market shares in overseas markets (Nishimizu and Robinson, 1984). South Korean MNEs' aggressive pursuit of foreign technology through licences, outright purchase, joint ventures and other agreements, reverse engineering and feedback from their foreign buyers makes South Korea unique among the developing countries in its commitment to the development of its indigenous product models and state-of-the-art process technology (see also Euh and Min, 1986; Porter, 1990).

The analysis of the growth of MNEs from developing countries in the late twentieth century on the basis of international diffusion of foreign technology has historical antecedents in the growth of MNEs from Europe and the USA in the nineteenth century and Japan and other developed countries earlier in the twentieth century. Inward investment, along with other modalities of foreign technology transfer, fulfilled a similar developmental upgrading role in the internationalisation of firms from the UK, Germany, Italy and, to a lesser extent, Japan in the immediate post-war period.

Historically, a developmental linkage was established between US MNEs and some technologically innovative indigenous firms in the UK and the advanced Continental European nations as local firms in the host countries enhanced their technological ownership advantages either through the development of their own indigenous technology or the importation of American technology (Dunning and Cantwell, 1982). The large increase in direct investments originating from West Germany and Japan is similarly a reflection of the developmental industrial upgrading of these countries, associated with stronger rates of technological innovation of indigenous firms complemented by access to foreign technology. Inward import-substituting FDI acted as a stimulus to indigenous technological innovation in West Germany, while licensed foreign technology acted as a corresponding stimulus in Japan. Japanese firms initially responded to competition in international markets through export growth, which in turn acted as a spur to further international competitive performance in the form of FDI. The rapidly evolving comparative advantages of Japanese MNEs have been assisted by a continuous process of industrial restructuring and economic growth in Japan based on the production of more technology-intensive products, in part through the introduction of the most advanced foreign technologies in the form of licensing (Ozawa, 1985). Such continuous assimilation and adaptation of foreign technology, channelled primarily through licensing agreements, have resulted in Japan's export of technology, particularly in connection with, and in support of, their overseas direct investments (Ozawa, 1974).

Chapter 6 analysed the role of foreign technology in encouraging the accumulation of indigenous technological capacity and growth of production of Third World MNEs. In common with the technological development of the more established outward investors, inward foreign manufacturing investments in both import-substituting and export-oriented industries have exerted a large developmental influence on the internationalisation of some of these countries, especially the Asian NICs. Indeed, a common feature of the internationalisation of firms from the Asian NICs is *inter alia* the concentration of import-substituting and export-oriented FDI at an earlier period, which has perpetuated a virtuous cycle by generating development spill-over effects to technologically innovative indigenous firms that have developed the capacity to respond independently to the competition offered by foreign MNEs through linkages and competition (see Findlay, 1978). The most important source of productive efficiency of foreign investment in developing countries lies in the competitive pressure generated by foreign firms on domestic firms (Blomstrom, 1986).

Through the presence of adaptive foreign affiliates and the integration of their unique technological characteristics with local systems in their host countries, some of these local firms succeeded in catching up in the process of international economic expansion through trade and foreign direct investment. As in the earlier stages of international production of firms from the more industrialised countries, the imposition of tariff barriers in export markets provided opportunities for outward FDI by developing countries, especially in the case of Asian NICs, which enabled firms to expand their network of productive activities across national boundaries.

The increasing technological capacities of some local firms in developing countries, enhanced by the developmental role of FDI and other modalities of foreign technology transfer, are reflected not only in the increasing degree of internationalisation – from exports to direct investment – but also in changing patterns in the composition of such international economic activities. The pattern of foreign economic involvement, from exports of certain manufactured goods in the initial phase of international expansion followed, or accompanied by, outward direct investment and changing composition, that characterised the growth of foreign investment by firms from Germany and Japan in the immediate post-war period, can therefore be applied to the more recent growth of FDI by firms from the NICs.

The modalities of foreign technology transfer for Third World MNEs

Foreign direct investment is the major modality of technology transfer in most developing countries (Findlay, 1978; Lall, 1983d), especially in sectors characterised by product differentiation resulting from research and development, the use of patents, brand names, trade marks, and intensive promotion and marketing, or where the absorptive capacity of the recipient country is low (UNCTC, 1987). However, for developing countries that have assumed rather more restrictive policies towards MNEs and/or higher forms of local technological capabilities, contractual resource flows or non-equity forms of technology transfer which represent arm's length international exchange of the intermediate product have become much more important modalities for the transfer of technology. Apart from inward foreign direct investment, this book has offered some examples of strategic alliances or cooperative arrangements with foreign firms in the form of joint ventures, licensing agreements, turnkey plant, technical assistance, subcontracting arrangements and other forms of non-equity investments, and forms of foreign technology transfer which have also become important complementary modalities for the transfer of more advanced and frequently non-proprietary technology in developing countries (Lenae, 1985; Perlmutter and Heenan, 1986). Yet another important modality for the transfer of foreign technology to developing countries is effected by the research-intensive outward investments of developing countries to take advantage of local technological competence in a particular location. These research-intensive outward investments enable firms to widen the scope of their technological specialisation through the development of related skills and technologies.

The main determinants of the form of technology transfer are the technological content of industry operations, the extent of barriers to entry, the degree of competition, and the bargaining power and policies of host countries (UNCTC, 1987). In any case, whether through the modality of foreign direct investment or contractual resource flows, or non-equity forms of technology transfer, the MNE is the agent of dynamic comparative advantage, in new technology generation as well as in mature technology diffusion (Lall, 1979).

In the country case studies analysed in this book, MNEs from Asia have been shown to rely on foreign direct investment, and subcontracting and supplier contracts, in fostering competitiveness in the lower technology-intensive sectors of the electronics industry. The analysis of the pattern of technology development, transfer and diffusion of firms in the microelectronics industry of Asian NICs shows the increasingly important role of foreign direct investment in the development of indigenous technological capability, in part through on-the-job training (Hakam and Chang, 1988; Hock, 1988b). For example, the growth in offshore assembly plant of the US, Japanese and European semiconductor manufacturers in Singapore, Hong Kong, Taiwan and South Korea has spurred the emergence and probable future development of the domestic semiconductor industry in these countries (Davis and Hatano, 1985).[4] The importance of foreign direct investment as a modality for foreign technology transfer and diffusion has been more important for the Asian NICs, except for South Korea where the evidence indicates that foreign direct investment has been an important source of technology in only a few sectors, primarily in chemicals, electronics and petroleum refining (Westphal, Rhee and Pursell, 1984).

In addition, the linkages that some companies in the Asian NICs have with foreign MNEs, especially with respect to sourcing of components, parts and semi-manufactured products, have helped to build their technological competitiveness. For example, a number of locally owned subcontracting companies in Hong Kong was established on the basis of contracts concluded with firms based in the smaller developed countries which provided equipment as well as technical and marketing assistance (Davis and Hatano, 1985; Hock, 1988a; Oman, 1988).

These foreign direct investments, and subcontracting and supplier contracts, have enabled local firms in Asia to strengthen their technological innovation, particularly in electronics production, *inter alia* through their access to technical information in open computer architecture that has enabled reverse engineering and imitation in hardware and software design to be carried out. The production of copies, clones and IBM-compatible computers may be attributed to efforts in this direction by firms from Asia. Taiwan in particular accounted for 10 per cent of the world's personal computer market in 1988 (Liu, 1988). Strengthened technological innovation in Asian countries has given some of their firms the capacity to undertake technology exports and outward direct investments.

Foreign direct investment has also been important as a modality of foreign technology transfer in the Latin American NICs and in particular Brazil, which has the greatest reliance on foreign direct investment and foreign licensing agreements to promote the development of localised technological advantages and production experience, especially in the capital goods sector (Dahlman and Sercovich, 1984). In the lower-income developing countries such as Malaysia, the modality of foreign direct investment has transferred technology mostly through foreign and on-the-job training and the presence of resident expatriate and visiting foreign consultants. Local firms have demonstrated fairly high capabilities to absorb and adapt new technology including new management systems because of their relatively abundant supplies of skills and technical knowledge (Ariff and Lim, 1984; Lee, 1989). During the 1980s the comparative advantage of Malaysia evolved in the electronics industry, enabling the country to become the third largest exporter of semiconductor devices in the world after Japan and the United States (Goldstein, 1989).[5] This was sustained by the ability of Malaysia to attract investment from US and Japanese semiconductor firms in line with the New Economic Policy of 1970 to industrialise the economy, based at least initially on the attraction of foreign capital (Ehrlich, 1988).

However, more complex and specialised technological import requirements of developing countries have been met through contractual resource flows or non-equity forms of technology transfer which represent arm's length international exchange of technology in the form of joint ventures, licensing, purchase of equipment, technical assistance, and other forms of strategic alliances and cooperative relationships with foreign firms. Arm's length international transactions in technology through joint ventures and liberal licensing agreements with foreign based multinational corporations have been particularly important in the case of South Korea, in comparison to other Asian NICs, partly because of earlier government policy which regulated the inflow of foreign direct investment (UNCTC, 1986; Enos and Park, 1988). Joint ventures played a major innovative role in the establishment of facilities to produce basic petrochemicals and derivative synthetic fibres and resins, an important example being polyester fibre and yarn (Westphal, Rhee and Pursell, 1984). Apart from licensing and joint ventures, the central diversification strategy adopted by South Korean firms is similar to that in Taiwan which is based on the utilisation of foreign technology, especially in managing supplier contracts, technology agreements and outright purchase of equipment. These technological alliances with foreign firms have helped to change their comparative advantages dynamically through the acquisition of additional capabilities and new technologies especially in research-intensive industries and have also assisted in the acquisition of marketing and management know-how (Jo, 1981; Westphal, Kim and Dahlman, 1984; Euh and Min, 1986; Crawford, 1987).

These contractual technology resource flows have also been important in Latin American NICs, particularly Argentina where licensing contracts covering both joint and domestic ventures have proved to be the general means of rapid and sure access to standard information technology and to supplies for the assembly and possible manufacture of electronics products. The strategy of foreign participation in the acquisition of new information technology has proved worth while in the sense of ensuring rapid entry which allows the generation of localised technology to be quickly accomplished (Inter-American Development Bank, 1988 Report). Similarly, more complex technological import requirements in the later stages of Brazilian industrialisation were also largely met through licensing and technical agreements with foreign producers (Guimaraes, 1986).[6] The joint ventures that Brazilian MNEs established with foreign firms in the developed countries have enabled their firms to expand their activity in high-technology sectors. Efforts to promote the development of indigenous production capacities in computers, especially in mini- and microcomputers and peripherals, have been sustained through licensing with smaller computer firms that have been prepared to supply technology through arm's length transactions (Oman, 1988). The emergence of outward-looking Brazilian firms in the area of information technology and the important role of foreign licensing agreements may be attributed to these firms' acquisition of technological capacities largely through a coherent national policy embodied in the National Informatics Law of 1984, which imposed government limitations on inward foreign direct investment and effectively imposed a policy of market reserve of the computer industry for locally owned firms.[7]

The importance of these arm's length technology transfer modalities of foreign technology transfer is also seen in the lower-income developing countries. For example, in China joint ventures with foreign firms have been important sources of new foreign technology and management skills since 1979 when China promulgated the Law on Joint Ventures (Tai, 1987; Huasheng and Kerkhofs, 1988). In addition, foreign technology in Malaysia has mainly been accessed through technical assistance and joint venture agreements as well as management contracts, mainly from Japan, the USA and the UK (Jegathesan, 1990). Technology transfer through joint ventures and purely technical collaboration agreements (or licensing agreements) with foreign firms in the Philippines is often associated with new business ventures involving fairly complex technology, requiring close supervision and control and/or marketing tie-ups.

The costs of foreign technology transfer

Apart from the costs incurred in the importation of foreign technology, MNEs from developing countries, like firms based in the more industrialised countries, have made considerable efforts in facilitating indigenous technological change, and have incurred significant costs in assimilation. The process of technology transfer is even more costly between firms in countries at different stages of development (Rosenberg, 1976). In developing countries, in particular, a number of social, organisational and economic characteristics either preclude or make highly uneconomic any attempt at straightforwardly replicating 'from-the-shelf' technology previously used in developed countries.

Among these distinctive characteristics of Third World technology compared to that of the more developed, industrialised countries are smaller domestic markets, higher rates of tariff protection, weaker competitive atmosphere, stronger distortions in technical information, market imperfections, shortages of skills and more dramatic levels of uncertainty, etc. (Katz, 1984b). The costly process of technology transfer between developed and developing countries is evident *inter alia* in terms of difficulties in communication and training owing to lower levels of entrepreneurship, education or absorptive capacities, and the need for more specialised and updated technology (Slaybaugh, 1981; Lall, 1984a).

The path of technological development and diffusion in developing countries is thereby significantly influenced by the sustained presence of a technological gap in the world economy brought about by differences in the quality and scale of commitment to R&D, and the clustering of technical innovations and dynamic economies of scale across countries (Posner, 1961; Dosi and Soete, 1988). The extent of convergence or divergence in inter-firm and inter-national technological capabilities is explained by the evolutionary nature of innovation which is determined by the interplay of science-related opportunities, country-specific and technology-specific institutions and the nature and intensity of economic stimuli. This may help to explain why, despite the rapid development of advanced technology and its diffusion between industrialised countries, a significant technological lag continues to exist in developing countries. The work of Hufbauer (1965) on the longer imitation lags in firms from large developing countries as opposed to large, medium or small firms from the industrialised countries has shown that the level of development is a much more crucial determinant of the technological progress of firms and countries.

Apart from the presence of the gap in international technological capabilities, the path of technological development and diffusion is also dictated by the stability of the sectoral pattern of comparative advantage held by each national group of firms in the medium term (Dosi and Orsenigo, 1988). This means that firms and countries are locked in to a specific course of technological development which is conditioned by the prevailing techno-economic paradigm and only changes when shifts occur in the paradigm as a result of a radical restructuring of the fields of technological opportunities that offer the fastest growth (Freeman and Perez, 1988; Cantwell, 1990).[8]

The ability of firms and countries to engage in higher technologically intensive industries whose growth is dependent on research and development crucially determines their position in technological competition, not only in international trade as shown by Soete (1981b) but also in a whole range of international economic activities. Since technology determines cross-country variations in economic performance in international industries, firms and countries that are more likely to succeed in world markets are those able to engage in the development and design of new products, as well as the improvement of old products and the manufacturing technology by which such products are made. By contrast, firms and countries that remain in industries described by abundant natural resource endowment or technological maturity and low research intensity face the danger of being 'locked in' to a low growth development pattern and at the same time 'locked out' of technologies that offer greater potential for higher growth. This fact raises a number of policy issues for developing countries, including the need to keep pace with the continuous process of attaining indigenous technological learning capacities to prevent a further widening of the technological gap.

The windows of opportunity for Third World MNEs

Periods of changes in the techno-economic paradigm represent excellent windows of opportunity for developing countries to break out of the vicious circle of technological and economic dependency and specialisation in technologically mature products and industries, and to catch up effectively in the process of technological advancement. Apart from the innovative efforts in R&D sourced through the network of the firm, which may be limited for Third World companies, this may require access to foreign technology in the form of licensing and/or other collaboration arrangements with organisations responsible for pioneering the new paradigm (Chesnais, 1988; Teece, 1988).

Developing countries may be better able to catch up in the production of new products emerging from the new technological paradigm if their firms have not been 'locked in' to old technology systems and the concomitant infrastructure, which are difficult to change. The relative importance of the various components of the national system of innovation increase or decrease over the life cycle of the technology, with the phases of introduction and maturity having the lowest thresholds of entry (Perez and Soete, 1988). Immediate adoption of new technology not only implies a significant contribution to the monopoly profits of the supplying innovator but also a commitment to a technology which might well undergo significant improvements in the not too distant future (Soete, 1982). In any event, early investment in new technologies is important *inter alia* to gain first mover advantages and to prevent a 'lock out' from technological paths that might be useful in the development of areas of comparative advantage and specialisation (Freeman, 1988b; Cohen and Levinthal, 1989).

The rate of leapfrogging and catching up is faster where there are initial capabilities in closely related types of activity (Cantwell, 1991a), which in turn are determined by the national system of innovation covering such things as, *inter alia*, the supply of entrepreneurial skills and scientific and technical knowledge involved in the transfer of technology related to know-why as opposed to know-how capabilities; a reasonable level of knowledge necessary for the assimilation and eventual commercialisation of the imported technology; favourable locational advantages related to the general infrastructure; and other economic and institutional conditions exogenous to the firm. Among these conditions are the development of the local capital goods sector which functions as source and locus of technology capacity; social and political changes in the international community, such as increased protectionism in the developed countries, which determine the future development of comparative advantage; and the specific technological and industrial strategies of particular countries (Freeman, 1988b; Perez, 1988; Perez and Soete, 1988; Unger, 1988).

The development of the national system of innovation to sustain the accumulation, expansion and self-realisation of local capital and technology ensures that dependent capitalism, described by Cardoso (1973) *et al.*, is unlikely to occur. For example, the extent to which the technological development of indigenous Philippine firms has been favourably affected by the presence of foreign MNEs during the period of ISI has to some extent been constrained by the absence of government policies to promote further local development of the national system of innovation. The establishment of the national system of innovation, including that of local capital goods industries, is instrumental in the promotion of local technological development, in the enhancement of the bargaining power of local firms in relation to the importation of foreign technology and in the improvement of localised technological change (Stewart, 1977, 1979; Mitra, 1979; Pack, 1981).

In general, the findings of this book confirm that those developing countries with more developed national systems of innovation, such as the NICs, have stood to gain from the effective assimilation of advanced but complementary foreign technologies, which are useful in the development of areas of comparative advantage and specialisation, and have greater opportunities for growth. In these cases, foreign technology has strengthened innovation and growth of production by indigenous firms. For example, with the effective assimilation of foreign technologies combined with relatively high levels of domestic technological capacities by local firms in the field of consumer electronics, Korea became, by the late 1980s, the second largest producer of capital- and skill-intensive products such as video recorders and microwave ovens after Japan (Mody, 1990).

Conversely, in countries with undeveloped national systems of innovation, such as those in lower-income developing countries in Asia, Latin America and Africa, foreign technology has had a less favourable impact on the development of production experience and indigenous technological capacity of local firms. Perhaps the only exceptions to this generalisation are indigenous firms from Kenya, and to a lesser extent Zimbabwe and the Ivory Coast, which have achieved a high level of industrialisation through their access, respectively, to Indian, British and French skills and technology. A direct consequence of such technology transfer was the growth of overseas operations by a number of large local business houses in Kenya (Lall, 1987).

The stage of technological development of Third World MNEs has fundamentally determined the precise nature of foreign technology transfer that is necessary in the promotion of indigenous technological development and economic growth. In general, countries at a more elementary stage of technological development import disembodied technology in the form of know-how or production engineering which consists of minor adaptations to assimilated technology to conform to local sales, product mix and raw materials.

On the other hand, countries at a more advanced stage of technology development, such as the NICs, increasingly seek to acquire acquisition of embodied technology such as technical information, drawings, tools, machinery, process information, specifications and patents drawn from inventions that emerge from basic research. This represents a more advanced stage of technology development that leads to the acquisition of know-why capabilities, reflecting a deeper understanding of the nature of the underlying process and product technologies and leading to substantial adaptation, improvement and even replacement by new products or processes (Ozawa, 1981; Lall 1985b). Know-why capabilities embody skills in basic design and R&D and are referred to as the foundations of technological paradigms which enable the production and development of new industries and products along the trajectory anticipated for the new technology.

These imports of embodied foreign technology have promoted know-why capabilities, especially in the NICs that have developed advanced technological skills and industrial experience. However, such know-why technological capabilities require continuous injections of fresh foreign technologies to keep pace with rapid developments abroad (Lall, 1983d). This is especially important in sectors characterised by faster growth of technological change which allow leaders to increase their technological superiority relative to weaker firms trying to catch up (Cantwell, 1991a). These technological capabilities in R&D are increasingly becoming a more important determinant of technology exports and international production than know-how capabilities. However, their efforts in R&D are generally distinguished from those of the more advanced countries by their concentration on minor innovation, design imitation and assimilation, raw material adaptation, process down-scaling, equipment modification, slight product change, upgrading of components, product diversification and the like. These technical efforts reflect more advanced capabilities than production engineering (Lall, 1984a) but are nevertheless a reflection of the nature of innovation as an incremental process in their gradual progression to related and more advanced technological activities (Pavitt, 1988).

The high capabilities of some firms in developing countries to assimilate more complex foreign technologies to suit their particular technological needs has in part led Warren (1973) to argue that technological dependence does not arise from imperialist monopoly or other forms of domination but from the capabilities of indigenous firms to assimilate foreign technology. These capabilities increase with the advance of commercialisation and industrialisation and with the acquisition of education and experience, including bargaining experience (Warren, 1980).

POLICY: THE NEED FOR THE DEVELOPMENT OF NATIONAL SYSTEMS OF INNOVATION

Given that foreign technology is desirable to the extent that a dynamic growth path of changing comparative advantage and rapid industrialisation is achieved, the correct development strategy must concentrate on having the conditions required to undertake the difficult and complex process of effectively blending and assimilating foreign technology with local technological efforts in order to achieve the production efficiency and competitive technologies necessary for world competition. Foreign technological efforts are required to enable developing countries to acquire deeper and more sophisticated technological capabilities in know-why, such as those required in basic design and new product/process development (Lall, 1984a), and therefore there is a need for a selective policy towards the import of foreign technology (Stewart, 1984). Although active technological strategies, through government intervention to reduce access to readily available foreign technologies, are favoured where domestic know-how and know-why can be competitively developed, the further development of know-why requires keeping pace with rapid technological developments in the advanced industrialised countries. The lessons learned from the interventionist policies of the Indian government to achieve technological self-reliance through the development of a diverse and fairly sophisticated base in industrial technologies are that these policies also led large sectors of domestic industry to technological obsolescence and inefficiency (Lall, 1982a, 1983b, 1983d, 1984d, 1985a). Studies have shown that in cases where Indian firms have imported technology, their indigenous R&D activities have been encouraged or enhanced (see, for example, Katrak, 1989). Similar arguments apply to the protectionist policies of the Brazilian government in the development of indigenous information technology, which have led to the emergence of an inefficient and technologically obsolescent computer industry.[9]

Since governments and the suppliers of capital and technology, such as the MNEs, are the predominant sources of external influence on technological change in the early stages of technological development, governments in developing countries have an important role in fostering the international competitiveness of local firms (see Kim, 1980; Rohlwink, 1987). The role of public-sector participation in the development of national systems of innovation is crucial since long-term and consistent policies are needed to stimulate the necessary continuous interaction between producers, users and external sources of technology (Gregersen, 1988).[10]

However, developing countries have to realise that technology transfer from abroad does not constitute the only solution to problems of technological underdevelopment. The increasing emergence of technological protectionism in the advanced industrialised countries has prompted the Asian NICs in particular to hasten the domestic development of technical, engineering and scientific manpower, partly through the enactment and enforcement of legislation on the protection of intellectual property rights, the promotion of inward investment in high technology industries and access to foreign technology through licensing and other contractual means of foreign technological transfer. These technological efforts are in addition to efforts in research-linked investments in centres of innovation in the industrialised countries (Chia, 1989).

Hence, although foreign technology may play a vital role, policies have to be nationally defined and oriented to set in motion a dynamic process of technological accumulation consistent with the effective and cumulative development of an indigenous technological capacity (Bienefeld, 1984). Developing countries themselves have to initiate the process of technological transformation by creating effective demand for foreign technologies through their inherent technological knowledge and capacities to select, adopt, transplant, diffuse and develop the technology according to their unique technological requirements (UNCTAD, 1984). The establishment of indigenous technological capacity is also important *inter alia* in order to reduce technological dependence on advanced countries, which tends to involve loss of local control over many aspects of production, lessens a country's ability to bargain effectively on the terms of technology transfer and exerts adverse effects in terms of self-respect and self-reliance (Stewart, 1984). For example, South Korea's indigenous efforts to achieve technological development were made partly to decrease import dependence in precision components such as mini-bearings, video tape recorder heads, drums and other electronic parts that had to be imported from Japan. More importantly, the current efforts of South Korea to develop indigenous technology are made to increase their bargaining position in response to the emerging reluctance of their main source of technology, the United States, to transfer proprietary technology. Japan has always been reluctant to transfer technology because of fears of developing potential competitors (Berney, 1985; Clifford, 1988; Evans and Tigre, 1989; Matsuura, 1989; Johnstone, 1990b; *Korea Business World*, April 1990).

As part of their strategies to catch up with the West, Japan and the NICs have not only emphasised the development of labour-intensive technologies but also more modern capital- and research-intensive technologies, or what Blumenthal and Teubal (1975) refer to as 'future-oriented technologies', whose use may be unjustified by existing factor proportions but which may prove to be more efficient in view of future requirements. Generally, recent policy objectives of Third World countries concerning industrial restructuring have included incentives for foreign firms to assist in technological upgrading of local firms through investments in higher value-added technologically intensive industries. These countries seek to change their comparative advantage dynamically as regards the local design, production and marketing of products that embody more advanced technology. Evidence of some of their efforts at local development in order to strengthen national systems of innovation in higher technologically intensive industries is given in Table 11.1.

A significant although neglected aspect of the formulation and implementation of technology development policies in developing countries, which has led to limited transfer and diffusion, pertains to the stipulations regarding the use, transfer and regulation of imported technologies (Patarsuk, 1989). In the area of foreign direct investment in particular, an important distinction has to be made between policies geared to existing investments and policies geared to attract new investment. Policies geared to existing investments should ensure greater interaction with local firms and have a more positive impact on industrialisation through *inter alia* greater local downstream processing activities. On the other hand, policies to attract new investments which would enable greater integration and linkages with local firms should be undertaken and incentives for these kinds of investment extended in the form of tax *et al.* benefits.

The extent of technological development of some lower-income developing countries has been constrained by the absence of government policies to promote further technological development of the different elements of the national system of innovation. This lack of emphasis on the development of the national system of innovation has enabled foreign technologies to perpetuate a vicious circle of cumulative decline by stifling domestic competition through their increased market power, as predicated by Hymer, which leads to the reinforcement of these countries' technological underdevelopment, as predicated by Newfarmer, Bornschier and Chase-Dunn, and other neo-Marxist economists. For example, the low levels of indigenous technological capabilities of Thai firms in higher technologically intensive industrial sectors such as petroleum refining, tin smelting and electronic equipment have resulted in the total control of these sectors by foreign MNEs (see Phiphatseritham, 1982). In addition, the absence of local Thai firms equipped with the technology needed in the production of local materials and parts for the electrical products and equipment industries has resulted in the importation of these materials and parts by foreign MNEs (see also Tambunlertchai and McGovern, 1984). The high import dependence of this industry has meant that few or no linkage effects have been established by foreign MNEs with local firms, which has led to a dependent state of industrial development in Thailand (Prasartset, 1988). This book has also shown that the low level of technological capabilities of Philippine firms has been a particularly important factor in the limited success of government local-content programmes in the capital- and technology-intensive sectors, such as motor vehicles and electrical appliances, which were intended to foster the development of small industries through the establishment of subcontracting networks with large foreign firms (Hill, 1985). The crucial factor in the difficult process of transfer of skills to local personnel by foreign firms in such industries as electrical appliances and iron foundries has been the limited level of Philippine capabilities, confined, for example, to areas of manual dexterity such as the operation of machine tools, sheet-metal work or welding as opposed to more sophisticated operations such as foundries or annealing. In addition, the

Filipino work ethic has placed quality control and pride in workmanship low
down in the order of importance (Odaka, 1984).

By contrast, in some of the larger LDCs such as Brazil, Mexico, Argentina,
Malaysia, India, Pakistan and South Korea the presence of determined government intervention has led to the development of national systems of
innovation which have enabled local firms to reduce their technological dependence and increase their technological interdependence with foreign
MNEs. As a result of government policies to reduce technological dependence
through the promotion of local technological development, particularly in the
intermediate and capital goods industries, and sector-specific policies towards
FDI, indigenous firms in these countries have been able to adapt and modify
foreign technologies to suit their own technological innovation and to produce
technologically sophisticated products.

The unique features of international production of developing countries
analysed in this book have shown that Third World multinationals perform
both active and passive roles in the rapidly changing international division of
labour in the current world economy. The author's prediction is that as
developing countries advance through progressively higher stages of industrial
development, as their firms achieve greater international production experience and as windows of opportunity are opened during changes in the
prevailing techno-economic paradigm, the technological embodiment of their
future outward investments will become more significant and more complex.
Subject to structural impediments to technological development such as the
presence of gaps in international technological capabilities and the stability in
the sectoral pattern of comparative advantage held by each national group of
firms in the medium term, the newer investors from the lower-income developing countries may increasingly adopt the technological sophistication of the
NICs in the future, aided by government policies that help to develop the
national system of innovation. On the other hand, the MNEs from the NICs
may increasingly be expected to adopt the even greater technological sophistication of MNEs from the developed countries. Meanwhile, newer sources of
FDI are expected to emerge from the present lower-income developing countries, as these advance in their stage of development and as indigenous firms
increasingly acquire the technological capacity to catch up with the dynamic
and cumulative process of international production.

Table 11.1 Evidence of local technological development efforts in developing countries: the strengthening of national systems of innovation

Modality	Efforts in technological development
Training and education	
South Korea	Training and education have been crucial in the technological development of South Korea. For example, an important factor behind the growth of competitive domestic producers of semiconductors is growth in local university training in the scientific and engineering fields and repatriation of technical talent
Protection	
✓ South Korea	The Korean industrial policy implemented in 1973 promoted the growth of the industry in part through infant industry protection and the assurance of a guaranteed market. The resultant trade pressures from the USA and the attainment of competitive abilities by Korean firms resulted in the reversal of the previous infant industry protection policy for consumer electronics in 1987
✓ Latin America	In more recent years, owing to concerns that foreign technology might be inappropriate for these countries, special policies designed to promote indigenous technological development have become more characteristic of development policies, in addition to the role of foreign private investment and trade in technology. The strengthening of indigenous technology increases the capacity of countries and firms to assimilate new knowledge and is considered an effective instrument in achieving technological change and rapid industrialisation in the region
Science towns and research institutions	
Singapore	Government and the private sector have established science towns to promote technical research
Taiwan	The Industrial Technology Research Institute (ITRI) was established by the Taiwan government in 1973 to develop new ideas, or license advanced technology from abroad. The Electronics Research and Science Organisation (ERSO) emerged as an offshoot of ITRI and has

Table 11.1 (continued)

Modality	Efforts in technological development
	been tremendously successful in the establishment of domestic semiconductors and computer memory chipmaking ability to support Taiwan's personal computer industry. The overall objective of ERSO is to develop generic technologies and through direct connections with local academic research departments, both ITRI and ERSO lend technical support to the private sector through the transfer of technologies for modest fees to local companies, primarily small- and medium-sized businesses, since Taiwan has few conglomerate firms with the financial capacity to spend heavily on R&D
South Korea	The research institutions that are funded in part or wholly by the Korean government and industry are the Korean Advanced Institute of Science and Technology, the Korean Electronics and Telecommunications Research Institute (ETRI), the Korean Automotive Systems Research Institute and the Korean Biogenetics Research Institute, which supplements university research but whose efforts are almost exclusively directed to industry[1]
Formation of conglomerates	
South Korea	Like the Japanese conglomerate groups (*zaibatsu* or *keiretsu*) that enjoy economies of scale comparable to the major MNEs based in the developed countries, the emergence of a few conglomerate groups (*chaebol*) has been enhanced by the interaction of state policy and private entrepreneurship. These conglomerate groups allow major investments in high technology industries *inter alia* through their ability to raise large amounts of capital more efficiently, their greater ability to bear risks through cross-subsidisation of production lines, their long-term incentives to upgrade product quality and their economies of scale and scope in manufacturing and R&D. These groups represent some of the institutional factors that have led to the faster technological development of

Table 11.1 (continued)

Modality	Efforts in technological development
	the electronics industry in South Korea compared to Taiwan. The major *chaebols* that have accumulated experience in the local manufacture of consumer electronics products have been able to engage in exports even in the USA where they have been able to sustain competition through the sale of personal computer clones that replicate standard computer architecture but sell on the basis of lower costs
Financial and tax incentives	
Singapore	The government offers tax incentives, venture capital funds and low interest loans to local firms to upgrade their technological capacities
Taiwan	The government is providing start-up capital for small firms as well as exemption from tariff payments on certain equipment
South Korea	The South Korean government, like those of Taiwan and Singapore, provides substantial amounts of low interest capital for companies
R & D	
South Korea	Although their scientific and technical resources are still modest, an important factor in the growth of the local electronics industry of South Korea in general is expenditure on R&D. The major research efforts on South Korea in electronics, telecommunications and automobiles are a positive indication of efforts to boost the competitiveness of some of the most important industries. Particularly heavy investments in electronics R&D are evident. R&D allotment for 1990 was 133 billion won or 3 per cent of total expected revenue for 1990. This is expected to increase gradually to 5 per cent by 1996.

Sources: Bell, Ross-Larson and Westphal, 1984; Davis and Hatano, 1985; Berney, 1985; Brody, 1986; Kim, 1986; Kotkin, 1987; Inter-American Development Bank, 1988 Report; Karlin, 1988; Yu, 1988; Hock, 1988b; Henderson, 1989; Saghafi and Davidson, 1989; Evans and Tigre, 1989; Henderson, 1989; *The Economist*, March 1989; Porter, 1990; Mody, 1990.
Note: Besides R&D, the Korean Institute of Electronics Technology (KIET) provides

Table 11.1 (continued)

contractual services to private industry in the areas of design, wafer fabrication and testing. These services are effected through contract manufacturing or contract testing arrangements which the institute concludes with foreign-based MNEs through which it gains access to their product and process technologies and computerised testing know-how in return for a guarantee of a reliable and cost competitive supply of integrated circuits. Among the foreign firms that have provided assistance to local firms in their high technology industrial park in Gumi are Hewlett-Packard, AMI and Fairchild which have helped train Korean personnel in process technology, circuit design and layout, and hardware and software design, respectively. In addition, KIET has arrangements with RCA to perform circuit testing on contract in its semiconductor assembly plants in Taiwan and Malaysia in exchange for testing know-how. Other foreign-based MNEs with which KIET has conducted negotiations to supply various types of semiconductor design, manufacturing and/or testing expertise include Siemens, Philips, Western Digital Corporation, Terradyne, Data General and Digital Equipment (UNCTC, 1986).

Annex 1
Outward investments by Philippine MNEs in the trading and banking sectors

OUTWARD INVESTMENTS IN THE TRADING SECTOR

The earlier stages of Philippine economic development, characterised by heavy reliance on the primary sector, necessitated the development of some form of service sector investments to support resource-based industrialisation. The Philippine government embarked on an export promotion drive, especially for non-traditional products, during the 1970s through the establishment of large trading companies patterned on the Japanese and Korean models. Some of the largest Philippine conglomerate firms and banks which were called upon by the government to form trading companies and establish overseas branches are as follows: Allied Transnational Export Import Corporation, Ayala Corporation, Delta International, First Philippine Trading Corporation, Herdis International Trading Corporation, Lepanto-Filsyn-Tondeña Trading Corporation, Marsteel Consolidated Incorporated, Rustan Investment and Development Corporation, Ultra International Trading Corporation, United Coconut Planters Bank and Universal Robina Corporation. These trading companies had two major objectives. The first was to promote exports of existing traditional and non-traditional products and to consolidate marketing efforts in identifying new export opportunities in order to reduce marketing costs. A secondary objective was to acquire raw materials and services for the Philippine economy (Lim, 1984).

OUTWARD INVESTMENTS IN THE BANKING AND FINANCE SECTOR

Substantial direct investments have also been undertaken by Philippine firms in the banking sector in such major export markets as the USA and Hong Kong. The importance to Philippine banks of overseas branches is manifested in the net worth of their overseas subsidiaries, which is often equal to, if not larger than, that of the parent firm.

Tables 8.6 and 8.7 showed that half of all Philippine direct investment in the USA in 1980 was directed towards the banking and finance sector with a total value of at least US $36.0 million. By 1990, the value of such investments had increased further to US $44.0 million. The relative importance of perceived competitive assets of Philippine banks *vis-à-vis* both indigenous banks and foreign banks from developed countries in host countries is tabulated in Table A1. The Philippine banks were asked to rank several potential competitive assets according to their declining relative order of importance, with one being the most important. The different ranks achieved by the various competitive assets were then averaged to give an overall impression of the relative importance of each competitive asset for all Philippine banks included in the survey.

The most important perceived competitive assets of Philippine banks included in the survey with respect to other banks in host countries are better knowledge about banking in developing countries, access to lower-cost skilled banking personnel and better management skills, banking and financial expertise. The respectable levels of banking and financial experience acquired by Philippine banks are important firm-specific advantages brought about by the reasonably well-developed Philippine capital market and the availability of education and training facilities both in the Philippines and abroad, to which Filipino nationals have access. Indirect government intervention may also be a major country-specific variable that has favourably affected the generation of ownership-specific advantages of Philippine banks which have led to their domestic and overseas expansion. Philippine foreign investment policy stipulates that no more than 30 per cent foreign ownership is allowed in the banking sector, with the exception of offshore banking units. Another important point that should be considered in analysing the overseas investments of Philippine banks is the massive financial outflows since the 1980s, owing largely to the intensifying political and economic uncertainties in the country. Some of the Philippine banks had to manage the large-scale outflow of portfolio investments of their valued clients through the establishment of banking networks abroad. The main ownership advantage of these banks is therefore their close relationship with, and the trust of, clients who wish to transfer their funds abroad.

Notwithstanding their perceived competitive assets, as summarised in Table A1, the main explanation for the FDI of Philippine banks in the main financial centres is not that these banks have ownership advantages *vis-à-vis* other banks which can be exploited abroad but that they have to establish foreign operations to manage their clients' funds effectively, to develop the latest banking technologies and practices and to diversify their activities, all of which necessitate operation in the international financial centres. A review of some Philippine firms in the banking sector that have established international operations is provided below.

The Philippine National Bank (PNB) is a state-owned bank which possesses unique ownership-specific advantages *vis-à-vis* privately owned banks. In particular, PNB is able to negotiate for funds at lower costs because of its access to relatively larger financial resources. The bank established overseas branches in New York and London in 1917 and 1969 respectively and in the early 1980s opened two more overseas branches, in Houston, Texas, and in

Beijing, China, as well as representative offices in Toronto and Vancouver. The opening of the overseas branch in Beijing is especially significant as this makes PNB the first ASEAN bank to establish an office there.

The desire to exploit retail banking skills is a major reason for the establishment of Philippine overseas bank branches and subsidiaries. Bancom, at one time a leading Philippine financial institution with US links, ventured to Hong Kong and neighbouring Southeast Asian countries to provide money market and investment banking services. Special mention has to be made of the establishment of branches by Philippine banks in the Middle East to service the banking needs of Filipino workers resident in those countries. The Philippine National Bank handles the biggest share of remittances from Filipino workers in the Middle East.

The main thrust of the international operations of the Philippine Commercial International Bank (PCIBank), formerly the Philippine Commercial and Industrial Bank, is aggressively to pursue the remittances of Filipino workers overseas. In Hong Kong, for example, PCIBank uses mass-media campaigns to inform Filipino residents of the requirement to remit at least 70 per cent of their earnings. PCIBank has similarly penetrated the Middle East to persuade Filipino workers to remit their earnings through the banking system. The bank has two agencies, in New York and Los Angeles, which operate as branches with limited banking operations; four representative offices, in Houston, Madrid, Frankfurt and London, and a deposit-taking subsidiary, PCI Capital Asia Limited, based in Hong Kong.

The Manila Banking Corporation, which used to be another of the country's largest commercial banks, acquired universal banking status in 1982.[1] It established a wholly owned subsidiary commercial bank in Los Angeles, California, in 1983 with an initial capitalisation of US $5 million, the minimum paid-up capital required under California's banking laws. Besides providing traditional banking services, this branch also functioned as an information centre, providing US firms and financial institutions with data on business opportunities in the Philippines.

Similar reasons have compelled Metropolitan Bank and Trust Company to establish international subsidiaries, namely, the International Bank of California in Los Angeles and the First Metro International Investment Company Limited in Hong Kong, as well as offices in Taipei and Guam. In addition, Equitable Banking Corporation is the country's first commercial bank to establish a branch office outside the Philippines, with one based in Hong Kong.

In 1982 the Bank of the Philippine Islands (BPI) became the seventh commercial bank in the Philippines to achieve universal banking status. The branch networks of BPI have been strengthened as a result of their universal bank status and the acquisition of a domestic bank, the Commercial Bank and Trust Company, through a merger in 1981. Since then, BPI has concentrated efforts on consolidating holdings and expanding resources and expertise to support its universal banking thrust. Since the early 1980s, BPI has been gearing up for the acquisition of seven foreign financial institutions in which the bank has a minority interest. Some of these financial institutions are Inter-Finance Bhd in Kuching, Sarawak, and AF Capital Sdn Bhd in Kuala Lumpur, Malaysia; the newly established PT Indon-Ayala Leasing Company in Jakarta, Indonesia, and Ayala International Finance Limited (AIFL) in Hong Kong. Most, if not all, of these foreign financial institutions were owned by Ayala Investment and Development Corporation (AIDC), a Philippine parent firm. On 1 March 1986, BPI acquired AIDC and in the process also acquired its foreign subsidiary, Ayala International Finance Limited (AIFC) in Hong Kong, which was renamed BPI International Finance Limited.

In addition, in the mid-1980s, BPI acquired full control of the New York-based Asian International Bank (AIB) which was originally established in 1981 by BPI and five other major Philippine banks, namely Far East Bank & Trust Company, China Banking Corporation, Pacific Banking Corporation, Consolidated Bank & Trust Company and Equitable Banking Corporation. These five Philippine commercial banks have sold their stakes in AIB to BPI.

Unlike other Philippine universal banks that have merged with investment houses to acquire the necessary expertise to handle non-banking lines, Allied Banking Corporation hopes to develop its indigenous banking expertise. It has one of the most extensive overseas operations among local commercial banks. Allied Bank has a deposit-taking subsidiary, Allied Capital Resources Limited, in Hong Kong, an offshore banking office in Bahrain and representative offices in London, Singapore, Sydney and Tokyo. Other Philippine banks that wish to acquire universal banking status have established overseas subsidiaries to meet the 500 million Philippine peso (US $54.5 million) capitalisation requirement. The establishment of Manila CBC Finance Incorporated and Manila CBC Finance (Hong Kong) Limited in 1980 by China Banking Corporation represents efforts to meet universal banking regulations.

Table A1 Perceived competitive assets of Philippine banks with respect to other banks in the host country

Perceived asset	Mean
Better knowledge about banking in developing countries	1.50
Access to lower-cost banking personnel	1.50
Better management skills, banking and financial expertise	2.50

Source: Questionnaires.

Annex 2
Service sector investments of Third World MNEs

Outward direct investments in the service sector also follow development patterns. The general expectation is that within particular service sectors, developing countries at a more advanced stage of development, such as the NICs, have accumulated higher forms of ownership advantages as a result of their longer experience in international investment.

The pattern of outward investment by developing countries in the service sector is closely associated with the earlier pattern of inward investment in the service sector in these countries. The increased significance of trade investments by developing countries engaged in outward investments in the resource-based sector is closely related to the earlier pattern of firms in the trading, transport and communications sectors that were established by developed countries in the developing countries prior to the Second World War. These service-oriented investments were established to support the resource-based investments and, particularly, to facilitate the transfer of unprocessed natural resources from their source in the developing countries to their final markets in the developed countries.

The pattern of growth of these service investments in recent years is also clearly linked to the relocation of simpler forms of manufacturing activities by export-oriented MNEs from the more developed countries in the NICs since 1960. This has prompted the development of direct investments in trade and construction activities to support these export-oriented manufacturing investments as well as the rapid growth of the industrial sector in these countries. The establishment of Japanese trading firms or *sogo-shosha* in commercial and distribution channels in the developed countries, which have been instrumental in the achievement of the export-led growth of Japan, is an excellent case in point.

As these NICs embarked on their own resource-based and export-oriented manufacturing investments in the 1970s and 1980s, there was an accompanying growth of outward investment in trading and construction activities. However, although the role of such trading activities in the earlier resource-based investment was to promote the export of unprocessed natural resources from their source to their final markets in the home country, such a role is increasingly geared to promoting the export of manufactured products from the downstream and horizontal manufacturing outlets abroad, owing to the greater importance of outward direct investments in the manufacturing sector.

The distinctive nature of service investments by the NICs compared to the lower-income developing countries is also reflected in their higher forms of ownership advantages in more sophisticated and rationalised forms of services such as banking and finance. These investments are partly a result of the concentration of inward FDI in the banking sector in some of these countries, particularly Hong Kong and Singapore, which has fulfilled a tutorial role to local firms in the Kojima sense and which enabled local firms to accumulate ownership advantages which could be exploited abroad.

This section analyses the ownership advantages of different Third World MNEs in the construction, banking and trading sectors as related to their stage of development.

EXPORT-ORIENTED SERVICE INVESTMENTS IN THE TRADING SECTOR

Export-oriented service investments in the trading sector are investments geared largely to supporting the growth of export-oriented manufacturing investments and/or resource-based investments. Japanese trading firms or *sogo-shosha* have supported the export-led growth of the Japanese economy and, in particular, their export-oriented manufacturing and resource-based investments through their intermediary role in the commercial and distributive channels in developed countries such as those in Europe and North America.

These trading investments are especially important in countries whose industrial exports are growing and face intense competition in export markets such as the NICs. For example, investment in on-site trading, warehousing and distribution channels to serve overseas markets accounted for 13.3 per cent of total Korean FDI in 1984, and even by the late 1980s such sectors still accounted for at least 10 per cent. The establishment of eleven licensed general trading firms is considered as a first step towards the internationalisation of export-oriented Korean manufacturing firms in as much as a number of these firms have developed a sufficient and flexible capacity for home-based production and have access to adequate funds to purchase on-site service facilities although they are still relatively weak in marketing skills.

These Korean trading firms are integrated with other firms to form giant industrial–commercial conglomerates modelled along the lines of Japanese *sogo-shoshas* and receive various incentives from the government. These incentives include priority on international bidding for government projects involving more than US $500,000, favourable export and import financing terms and the right to import raw materials (Jo, 1981; Cho, 1982).

Similarly, a majority of Singaporean FDI outside Asia is confined to trading. Investment in the trading sector is the second most important form of investment for Singaporean FDI in Malaysia and the Philippines after manufacturing. However, investments by Singaporean MNEs have been most significant in Hong Kong where most of these firms operate trading, finance or investment-holding companies. The importance of trading investment in Hong Kong is enhanced by the chartering and marine service companies of Pan Electrical Industries.

That lower-income developing countries like the Philippines are following the pattern of the developed countries in the earlier period as well as that of the NICs in the more recent period and are themselves engaging in investments in the trading sector was shown in the last chapter. The twelve trading companies, patterned on the Japanese and Korean models and established by the largest Philippine conglomerate firms in response to Philippine government policy, operated overseas branches in the traditional export markets such as the USA and Japan in order to fulfil the objectives of promoting exports of existing traditional and non-traditional products, identifying new export opportunities and acquiring raw materials needed for industrialisation in the Philippines.

RATIONALISED SERVICE INVESTMENTS IN THE BANKING SECTOR

Rationalised forms of service investments are the most sophisticated forms of service investment. These forms of service investments, such as those in the banking and finance, property, and other service sectors apart from trading and construction, have been especially important areas of service investments for firms from the NICs which have developed sufficient ownership advantages to be exploited abroad in the form of direct investments.

Investment in the banking and finance sector has been especially important for the NICs, especially Hong Kong and Singapore, partly as a result of the concentration of inward foreign investment in the banking and finance sectors in these countries, which has enabled them to become centres of financial activity not only within Asia but also within the developing world. The dynamic interaction between foreign and local banks may have enabled local banks to accumulate ownership advantages relevant to banking which could be exploited through the establishment of international banking networks.

Banking sector investments by Hong Kong banks have been concentrated mainly in Singapore, Malaysia, Thailand, the UK and the USA (Wu and Wu, 1980). The investment by Hong Kong firms in the service sectors, and particularly those in banking, property development, hotels, shipping and tourism, is largely attributed to increasing uncertainties over the long-term future of Hong Kong (see also Liu,1990). Nevertheless, outward investment by Hong Kong banks has been comparatively limited owing in part to the expansion of British banks such as Hong Kong Shanghai Bank, and to a lesser extent US banks such as Citibank, in Asia, which had met many of the international banking requirements of the overseas Chinese communities throughout Southeast Asia long before Hong Kong banks had generated sufficient ownership advantages to establish their own international banking networks.

Similarly, the overseas expansion of Korean banks is associated with the increased significance of export-oriented manufacturing and service investment in export markets. Apart from gaining access to capital resources needed to finance Korea's industrial growth, these overseas banks have been instrumental in providing the financial services required by Korean exporters and investors.

NICs in Latin America, such as Argentina and Brazil, have also established substantial overseas investments in the banking sector. A major factor in the international expansion of Latin American banks, both public and private, lies in the relatively recent outward-looking policies of the major semi-industrialised Latin American countries which have led to an increased use of the world private capital markets, as well as the growing banking requirements of exporting firms. In the case of Argentina, the concentration of inward FDI in the banking and finance sector may have played a developmental role in upgrading the ownership advantages of local banks which enabled these banks to establish their international banking networks.

The state-owned Banco de la Nacion Argentina has established eighteen overseas branches throughout Latin America and six representative offices, in San Francisco, Chicago, New York, Paris, London and Tokyo. Other private banks such as Galicia, Credito Argentina, Ganadero and Español together comprise the Argentine Banking Corporation in New York, which was incorporated with a net value of US $2.5 million. Banco Provincia de Buenos Aires, Banco de Cordoba, Banco Frances and Banco Rio (also private banks) have also established overseas branches (Katz and Kosacoff, 1983).

Similarly, the state-owned Banco do Brasil has established overseas banking activities and has been listed among the top fifty international banks. The major competitive advantage of this bank derives from access to huge assets of at least US $23 billion, outstripping Chemical Bank, Lloyds and Bank of Tokyo (Diaz-Alejandro, 1977). The expansion of Brazilian banks abroad is a reflection of their objectives to borrow on the international capital markets and to support the growing exports by Brazilian firms to other Latin American countries (Guimaraes, 1986).

Investment in the banking sector has also been quite significant for the lower-income developing countries. In Latin America, banks from Venezuela and Colombia have established overseas operations in other Latin American countries and in some cases in active competition with foreign banks. Philippine banks also feature in overseas banking investments, but the main explanation for FDI by these banks in the financial centres such as the USA and Hong Kong does not derive from the possession of particular ownership advantages *vis-à-vis* US or Hong Kong banks but from the need to manage the portfolio investments of their clients effectively, to develop the latest banking technologies and practices, and to diversify activities, which necessitates banking operations in these financial centres.

Bibliography

Aggarwal, R. (1987) 'Foreign Operations of Singapore Industrial Firms: A Study of Emerging Multinationals from a Newly Industrializing Country', in M. Dutta (ed.) *Asia-Pacific Economies: Promises and Challanges, Research in International Business and Finance*, vol. 6, part B, Greenwich, Connecticut: JAI Press.

Aggarwal, R. and Agmon, T. (1990) 'The International Success of Developing Country Firms: Role of Government-Directed Comparative Advantage', *Management International Review* 30(2): 163–80.

Agmon, T. and Lessard, D. (1977) 'Investor Recognition of Corporate International Diversification', *Journal of Finance*, September.

Akashah, S. E. (1987) 'Innovation versus Transfer of Technology: A Case Study of R&D in a Petroleum-Producing Country', *International Journal of Technology Management* 2 (2): 249–62.

Aliber, R. Z. (1970) 'A Theory of Direct Foreign Investment', in C. P. Kindleberger (ed.) *The International Corporation: A Symposium*, Cambridge, Massachusetts: MIT Press.

Aliber, R. Z. (1971) 'The Multinational Enterprise in a Multiple Currency World', in J. H. Dunning (ed.) *The Multinational Enterprise*, London: George Allen & Unwin.

Althuser, A., Anderson, M., Jones, D., Roos, D. and Womack, J. (1984) *The Future of the Automobile: Report of MIT's International Automobile Program*, Cambridge, Massachusetts: MIT Press/George Allen & Unwin.

Amin, S. (1974) *Accumulation on a World Scale: A Critique of the Theory of Underdevelopment*, two volumes, New York: Monthly Review Press.

Andersen, E. S. and Lundvall, B.-Å. (1988) 'Small National Systems of Innovation Facing Technological Revolutions: An Analytical Framework', in Christopher Freeman and Bengt-Åke Lundvall (eds) *Small Countries Facing the Technological Revolution*, London: Pinter Publishers.

Andrade, M. C. de (1967) *Espaço, Polarizaço e Desenvolvimento: A Teoria dos Polos de Desenvolvimento e a Realidade Nordestina*, Recife: Centro Regional de Administração Municipal.

Andrianov, V. (1990) 'NICs of Asia: High-Tech Products Priority', *Far Eastern Affairs* 6.

Ariff, M. and Lim, C. P. (1984) 'Foreign Investment in Malaysia', paper submitted to the Commonwealth Secretariat, London, October.

Armour, H. O. and Teece, D. J. (1980) 'Vertical Integration and Technological Innovation', *Review of Economics and Statistics* 62.

Arrow, K. J. (1962) 'Economic Welfare and the Allocation of Resources for Invention', in National Bureau of Economic Research, *The Rate and Direction of Inventive Activity: Economic and Social Factors*, Princeton, New Jersey: Princeton University Press.

Arthur, W. B. (1988) 'Competing Technologies: An Overview', in G. Dosi, C. Freeman, R. R. Nelson, G. Silverberg and L. L. G. Soete (eds) *Technical Change and Economic Theory*, London: Frances Pinter.

Arthur, W. B. (1989) 'Competing Technologies, Increasing Returns, and Lock-In by Historical Events', *The Economic Journal* 99, March.

Atkinson, A. B. and Stiglitz, J. E. (1969) 'A New View of Technological Change', *The Economic Journal* 79, September.

Aydin, N. and Terpstra V. (1981) 'Marketing Know-how Transfers by Multinationals: A Case Study in Turkey', *Journal of International Business Studies*, Winter.

Baba, Y. (1987) 'Internationalisation and Technical Change in Japanese Electronics Firms or Why the Product Cycle Doesn't Work', Mimeo, Science Policy Research Unit, University of Sussex, January.

Badulescu, P. (1991) 'International Technological Knowledge Differences and Economic Growth Comparisons: USA versus West Germany and Sweden versus Norway, 1963–1988', *Applied Economics* 23(1B): 263–82.

Bain, J. S. (1956) *Barriers to New Competition*, Cambridge, Massachusetts: Harvard University Press.

Baker, S. (1989) 'Mexico's Giants March North', International Business Section, *Business Week*, 13 November: 63–7.

Baran, P. A. (1960) *The Political Economy of Growth*, New York: Prometheus.

Baran, P. A. (1969) *The Longer View: Essays Toward a Critique of Political Economy*, New York: Monthly Review Press.

Baran, P. and Sweezy, P. M. (1966) *Monopoly Capital*, New York and London: Monthly Review Press.

Barkin, D. (1981) 'Internationalization of Capital: An Allernative Approach' *Latin American Perspectives* 8, Summer and Autumn: 156–61.

Bautista, L. R. (1990) 'Joint Venture Agreements and Technology Transfer: The Philippine Experience', in UNCTAD, *Joint Ventures As A Channel for the Transfer of Technology*, New York: United Nations.

Bell, M. (1984) 'Learning and The Accumulation of Industrial Technological capability in Developing countries', in M. Fransman and K. King (eds) *Technological Capability in the Third World*, New York: St Martin's Press.

Bell, M., Ross-Larson, B. and Westphal, L. (1984) 'Assessing the Performance of Infant Industries', *Journal of Development Economics*, September–October, 101–28.

Berney, K. (1985) 'The Four Dragons Rush to Play Catch-up Game', *Electronics Week*, 6 May.

Beveridge, A. J. (1900) 'Our Philippine Policy', *Congressional Record*, 9 January, 704–11. Reprinted in D. B. Schirmer and S. R. Shalom (eds) (1987) *The Philippines Reader: A History of Colonialism, Dictatorship, and Resistance*, Boston: South End Press.

Bhatt, M. C. and Dalal, K. L. (n. d.) 'Indian Enterprises Abroad', Mimeo, the Indian Council for Research on International Economic Relations, New Delhi.

Bienefeld, M. (1984) 'International Constraints and Opportunities', in M. Fransman and K. King (eds) *Technological Capability in the Third World*, New York: St Martin's Press.

Blomstrom, M. (1986) 'Foreign Investment and Productive Efficiency: The Case of Mexico', *Journal of Industrial Economics* 35 (1), September.

Blumenthal, T. and Teubal, M. (1975) 'Factor Proportions and Future Oriented Technology: Theory and an Application to Japan', Mimeo, Tel Aviv University, David Horowitz Institute.

Blumenthal, T. (1979) 'A Note on the Relationship between Domestic Research and Development and Imports of Technology', *Economic Development and Cultural Change* 27(2), January.

Bornschier, V. and Chase-Dunn, C. (1985) *Transnational Corporations and Underdevelopment*, New York: Praeger Publishers.

Brody, H. (1986) 'Taiwan: From Imitation to Innovation', *High Technology* 6 (11), November.

Buckley, P. J. and Casson, M. C. (1976) *The Future of the Multinational Enterprise*, London and Basingstoke: Macmillan.

Buckley, P. J. and Enderwick, P. (1985) 'Manpower Management in the Domestic and International Construction Industry', Mimeo, University of Bradford Management Centre and University of Belfast Department of Economics.

Buckley, P. J. (1983) 'New Theories of International Business: Some Unresolved Issues', in M. C. Casson (ed.) *The Growth of International Business*, London: George Allen & Unwin.

Buckley, P. J. (1985) 'A Critical View of the Theories of the Multinational Enterprise', in P. J. Buckley and M. C. Casson (eds) *The Economic Theories of the Multinational Enterprise*, London: Macmillan.

Buckley, P. J. (1988) 'The Limits of Explanation: Testing the Internalisation Theory of the Multinational Enterprise', *Journal of International Business Studies* 19(2), Summer: 18–93.

Cahill, M. (1989) 'Product Development Leadership Will Be A Major IE Role in the '90s', *Industrial Engineering* 21(8), August.

Caillods, F. (1984) 'Education, Organisation of Work and Indigenous Technological Capacity', in M. Fransman and K. King (eds) *Technological Capability in the Third World*, New York: St Martin's Press.

Callis, H. G. (1942) 'Foreign Capital in Southeast Asia', Mimeo, International Secretariat, Institute of Pacific Relations, New York.

Cantwell, J. A., Corley, T. A. B. and Dunning, J. H. (1986) 'An Exploration of Some Historical Antecedents to the Modern Theory of International Production', in G. Jones and P. Hertner (eds) *Multinationals: Theory and History*, Farnborough: Gower.

Cantwell, J. A. and Tolentino, P. E. E. (1990) 'Technological Accumulation and Third World Multinationals', University of Reading Discussion Paper in International Investment and Business Studies, no. 139.

Cantwell, J. A. (1985) Review of R. S. Newfarmer (ed.) 'Profits, Progress and Poverty: Case Studies of International Industries in Latin America', *The Economic Journal*, December.

Cantwell, J. A. (1987a) 'The Contribution of Recent Foreign Direct Investment in Services to a Changing International Division of Labour', paper prepared for the UNCTC meeting on TNCs in Services, New York, July.

Cantwell, J. A. (1987b) 'The Role of Foreign Direct Investment in Development in Africa', paper submitted to the World Bank, April.

Cantwell, J. A. (1987c) 'The Determinants of the Internationalisation of Firms', Mimeo, University of Reading.

Cantwell, J. A. (1987d) 'The Reorganisation of European Industries After Integration: Selected Evidence on the Role of MNE Activities', *Journal of Common Market Studies* 26(2), December: 127–51.

Cantwell, J. A. (1987e) 'A Dynamic Model of the Post-War Growth of International Economic Activity in Europe and the US', University of Reading Discussion Paper in International Investment and Business Studies, no. 104, Summer.

Cantwell, J. A. (1987f) 'Technological Competition and Intra-Industry Production in Europe', University of Reading Discussion Paper in International Investment and Business Studies, no. 106.

Cantwell, J. A. (1989a) *Technological Innovation and Multinational Corporations*, Oxford: Basil Blackwell.

Cantwell, J. A. (1989b) 'The Changing Form of Multinational Enterprise Expansion in the Twentieth Century', in A. Teichova, M. Levy-Leboyer and H. Hussbaum (eds) *Historical Studies in International Corporate Business*, Cambridge: Cambridge University Press.

Cantwell, J. A. (1990) 'Historical Trends in International Patterns of Technological Innovation', in J. Foreman Peck (ed.) *New Perspectives on the Late Victorian Economy*, Cambridge: Cambridge University Press.

Cantwell, J. A. (1991a) 'The Technological Competence Theory of International Production and its Implications', in D. McFetridge (ed.) *Foreign Investment, Technology and Economic Growth*, Toronto: University of Toronto Press.

Cantwell, J. A. (1991b) 'A Survey of Theories of International Production', in C. Pitelis and R. Sugden (eds) *The Nature of the Transnational Firm*, London and New York: Routledge.

Cardoso, F. H. (1973) 'Associated Dependent Development: Theoretical and Practical Implications', in A. Stepan (ed.) *Authoritarian Brazil: Origins, Policies and Future*, New Haven, Connecticut: Yale University Press.

Cardoso, F. H. and Enzo, F. (1979) *Dependency and Development* (translated by M. M. Urquidi), Berkeley, California: University of California Press.

Carlsson, B. (1987) 'Reflections on Industrial Dynamics: The Challenges Ahead', *International Journal of Industrial Organisation* 5(2), June.

Casson, M. C., Barry, D., Foreman-Peck, J., Hennart, J. F., Horner, D., Read, R. A. and Wolf, B. M. (1986) *Multinationals and World Trade: Vertical Integration and the Division of Labour in World Industries*, London: George Allen & Unwin.

Casson, M. C. (1979) *Alternatives to the Multinational Enterprise*, London: Macmillan.

Casson, M. C. (1986) 'Foreign Investment and Economic Welfare: Internalising the Implementation of Threats', Mimeo, University of Reading.

Casson, M. C. (1987) *The Firm and the Market: Studies in Multinational Enterprise and the Scope of the Firm*, Oxford: Basil Blackwell.

Casson, M. C. (ed.) (1983) *The Growth of International Business*, London: George Allen & Unwin.

Castro, A. A. (1969) 'Import Substitution and Trade Promotion: Trade and Development', Discussion Paper, Institute of Economic Development and Research, School of Economics, University of the Philippines, no. 69–10, 27 June.

Caves, R. E. (1971) 'International Corporations: The Industrial Economics of Foreign Investment', *Economica* 38, February.

Chandler, A. D., Jr (1980) 'The Growth of the Transnational Industrial Firm in the United States and the United Kingdom', *Economic History Review*, series 2, vol. 33.

Chandler, A. D., Jr (1986) 'The Evolution of Modern Global Competition', in M. E. Porter (ed.) *Competition in Global Industries*, Boston, Massachusetts: Harvard Business School Press.

Chen, C.-H. (1986) 'Taiwan's foreign direct investment', *Journal of World Trade Law* 20(6), November/December: 639–64.

Chen, E. K. Y. (1981) 'Hong Kong Multinationals in Asia: Characteristics and Objectives', in K. Kumar and M. G. McLeod (eds) *Multinationals from Developing Countries*, Lexington, Massachusetts: D. C. Heath and Company.

Chen, E. K. Y. (1983a) *Multinational Corporations, Technology and Employment*, New York: St Martin's Press.

Chen, E. K. Y. (1983b) 'Multinationals from Hong Kong', in S. Lall (ed.) *The New Multinationals*, Chichester: John Wiley and Sons.

Chen, E. K. Y. (1988) 'The Role of TNCs in Hong Kong's Economic Development', paper prepared for the International Conference on the Changing Role of TNCs in Asia–Pacific Development, Sydney, Australia, 14–15 July.

Chesnais, F. (1988) 'Multinational Enterprises and the International Diffusion of Technology', in G. Dosi, C. Freeman, R. R. Nelson, G. Silverberg and L. L. G. Soete (eds) *Technical Change and Economic Theory*, London: Frances Pinter.

Chia, S. Y. (1989) 'Asian NIEs as Traders and Investors', paper presented at the Conference on the Future of the Asia–Pacific Economies (FAPE III), Bangkok, 8–10 November.

Chilcote, R. H. (1984) *Theories of Development and Underdevelopment*, Boulder, Colorado, and London: Westview Press.

Cho, D. S. (1982) 'Anatomy of Korean General Trading Company', Mimeo, Seoul National University.

Clemente, W. A. and Bautista, E. (1978) 'The Cost of Technology Transfer: The Case of the Philippine Drug Industry', in W. Clemente, F. Bacuñgan and F. Laxa (eds) 'Multinational Corporations in the Philippines', Mimeo, Technology Resource Center, Republic of the Philippines.

Clifford, M. (1988) 'Breakneck Expansion: South Korea's Booming Electronics Industry Still Relies on Imports, *Far Eastern Economic Review* 139(1), 7 January: 42–3.

Clifford, M. (1989) 'Thanks for the Memories: Hitachi Helps South Korea's Goldstar become a Microchip Major', *Far Eastern Economic Review* 145(34), 24 August.

Coase, R. H. (1937) 'The Nature of the Firm', *Economica* 4(4).

Cohen, B. S. (1975) *Multinational Firms and Asian Exports*, New Haven, Connecticut: Yale University Press.

Cohen, W. M. and Levinthal, D. A. (1989) 'Innovation and Learning: The Two Faces of R&D', *The Economic Journal* 99(397), September.

Colman, D. and Nixson, F. (1986) *Economics of Change in Less Developed Countries*, Oxford: Philip Allan Publishers.

Colson, F. (1985) 'New Perspectives on the Brazilian Computer Industry: 1985 and Beyond', *Multinational Business* 4.

Commons, J. R. (1934) *Institutional Economics: Its Place in Political Economy*, Madison: University of Wisconsin Press. Reprinted 1951, New York: Macmillan Press.

Contractor, F. J. and Lorange, P. (eds) (1988) *Cooperative Strategies in International Business*, Lexington, Massachusetts: Lexington Books.

Contractor, F. J. (1979) 'The Cost of Technology Transfers in Overseas Licensing', Ph.D. Dissertation, University of Pennsylvania.

Cooper, C. and Sercovich, F. C. (1970) 'The Channels and Mechanisms for the Transfer of Technology from Developed to Developing Countries', UNCTAD Paper TD/D/AC 11/5.

Corporate Information Center, National Council of Churches of Christ in the USA (1973) 'The Philippines: American Corporations, Martial Law and Underdevelopment', *IDOC* 57, International/North American Edition, November.

Cowling, K. and Sugden, R. (1987) *Transnational Monopoly Capitalism*, Brighton: Wheatsheaf Books.

Crane, G. (1990) 'DRAM Wars', *Korea Business World*, January: 10–16.

Crawford, M. H. (1987) 'Technology Transfer and the Computerisation of South Korea and Taiwan – Part I: Developments in the Private Sector', *Information Age* 9(1), January: 10–16.

Cummings, B. (1987) 'The Origins and Development of the Northeast Asian Political Economy: Industrial Sectors, Product Cycles and Political Consequences', in F. Deyo (ed.) *The Political Economy of the New Asian Industrialism*, Ithaca: Cornell University Press.

Dahlman, C. J. and Sercovich, F. C. (1984) 'Exports of Technology from Semi-Industrial Economies and Local Technological Development', *Journal of Development Economics* 16.

Dahlman, C. J. (1984) 'Foreign Technology and Indigenous Technological Capability in Brazil', in M. Fransman and K. King (eds) *Technological Capability in the Third World*, New York: St Martin's Press.

Dahlman, C. J. (1989) 'Technological Change in Industry in Developing Countries', *Finance and Development* 26(2), June: 13–15.

Das, S. (1987) 'Externalities and Technology Transfer Through Multinational Corporations: A Theoretical Analysis', *Journal of International Economics* 22(1–2), February.

Davis, D. (1987) 'Direct Investment Abroad by Developing Countries', paper prepared for the UNCTC Fourth Survey on the Transnational Corporations in World Development.

Davis, W. E. and Hatano, D. G. (1985) 'The American Semiconductor Industry and the Ascendancy of East Asia', *California Management Review* 27(4), Summer.

Debes, C., Holstein, W. J. and Javetzki, B. (1987) 'India Locks in on a High-Tech Economy', *Business Week* 3013 (Industrial/Technology Edition), 24 August:80D–E.

Demsetz, H. (1969) 'Information and Efficiency: Another Viewpoint', *Journal of Law and Economics* 12, April.

Desai, A. V. (1984) 'Achievements and Limitations of India's Technological Capability', in M. Fransman and K. King (eds) *Technological Capability in the Third World*, New York: St Martin's Press.

Desai, M. (1987) 'Comments on Sukhamoy Chakravarty: Marxist Economics and Contemporary Developing Economies', *Cambridge Journal of Economics* 11(2), June.

Diaz-Alejandro, C. F. (1977) 'Foreign Direct Investment by Latin American Firms', in T. Agmon and C. P. Kindleberger (eds) *Multinationals from Small Countries*, Cambridge, Massachusetts: MIT Press.

Diokno, R. (1946) 'Roxas Violates the Constitution', *Amerasia*, 10(6), December: 75–8. Reprinted in D. B. Schirmer and S. R. Shalom (eds) (1987) *The Philippines Reader: A History of Colonialism, Dictatorship, and Resistance*, Boston: South End Press.

Dore, R. (1984) 'Technological Self-Reliance: Sturdy Ideal or Self-Serving Rhetoric', in M. Fransman and K. King (eds) *Technological Capability in the Third World*, New York: St Martin's Press.

Dos Santos, T. (1970) 'Dependencia Económica y Alternativas de Cambio en América Latina', *Revista Mexicana de Sociología* 32, March–April: 417–63.

Dosi, G. and Orsenigo, L. (1988) 'Coordination and Transformation: An Overview of Structures, Behaviours and Change in Evolutionary Environments', in G. Dosi, C. Freeman, R. R. Nelson, G. Silverberg and L. L. G. Soete (eds) *Technical Change and Economic Theory*, London: Frances Pinter.

Dosi, G. and Soete, L. (1988) 'Technological Change and International Trade', in G. Dosi, C. Freeman, R. Nelson, G. Silverberg and L. Soete (eds) *Technical Change and Economic Theory*, London: Pinter Publishers.

Dosi, G. (1982) 'Technological Paradigms and Technological Trajectories', *Research Policy* 11(3), June.

Dosi, G. (1984) *Technical Change and Industrial Transformation*, London: Macmillan.

Dosi, G. (1988) 'The Nature of the Innovative Process', in G. Dosi, C. Freeman, R. R. Nelson, G. Silverberg and L. L. G. Soete (eds) *Technical Change and Economic Theory*, London: Frances Pinter.

Douglass, G. K. (1966) 'Innovation and International Trade', Mimeo, Claremont, California.

Doz, Y. (1986) *Strategic Management in Multinational Companies*, Oxford: Pergamon Press.

Drucker, P. F. (1974) 'Multinationals and Developing Countries: Myths and Realities', *Foreign Affairs* 53, October.

Dunning, J. H. and Cantwell, J. A. (1982) 'Inward Direct Investment from the US and

Europe's Technological Competitiveness', University of Reading Discussion Paper in International Investment and Business Studies, no. 65, September.

Dunning, J. H. and Cantwell, J. A. (1987) *Directory of Statistics of International Investment and Production*, London: Macmillan and New York: New York University Press.

Dunning, J. H. and Cantwell, J. A. (1990) 'The Changing Role of Multinational Enterprises in the International Creation, Transfer and Diffusion of Technology', in F. Arcangeli, P. A. David and G. Dosi (eds) *Technology Diffusion and Economic Growth: International and National Policy Perspectives*, Oxford: Oxford University Press.

Dunning, J. H. and Pearce, R. D. (1985) *The World's Largest Industrial Enterprises 1962–1983*, Aldershot, Hants: Gower.

Dunning, J. H. and Rugman, A. M. (1985) 'The Influence of Hymer's Dissertation on the Theory of FDI', *American Economic Review*, May.

Dunning, J. H. (1981a) *International Production and the Multinational Enterprise*, London: George Allen & Unwin.

Dunning, J. H. (1981b) 'Explaining Outward Direct Investment of Developing Countries: In Support of the Eclectic Theory of International Production', in K. Kumar and M. G. McLeod (eds) *Multinationals from Developing Countries*, Lexington, Massachusetts: D. C. Heath and Company.

Dunning, J. H. (1982) 'Explaining the International Direct Investment Position of Countries: Towards a Dynamic or Developmental Approach', in J. Black and J. H. Dunning (eds) *International Capital Movements*, London: Macmillan.

Dunning, J. H. (1983a) 'Changes in the Level and Structure of International Production: The Last 100 Years', in M. C. Casson (ed.) *The Growth of International Business*, London: George Allen & Unwin.

Dunning, J. H. (1983b) 'Market Power of the Firm and International Transfer of Technology', *International Journal of Industrial Organisation* 1(1).

Dunning, J. H. (1986a) 'The Investment Development Cycle and Third World Multinationals', in K. M. Khan (ed.) *Multinationals of the South*, London: Frances Pinter.

Dunning, J. H. (1986b) *Japanese Participation in British Industry*, London: Croom Helm.

Dunning, J. H. (1986c) 'The Investment Development Cycle Revisited', *Weltwirtschaftliches Archiv* 4 (122).

Dunning, J. H. (1988a) 'The Eclectic Paradigm of International Production: A Restatement and Some Possible Extensions', *Journal of International Business Studies* 19(1), Spring: 1–31.

Dunning, J. H. (1988b) 'The Theory of International Production', *The International Trade Journal* 3.

Dunning, J. H. (1988c) 'International Business, the Recession and Economic Restructuring', in N. Hood and J.-E. Vahlne (eds) *Strategies in Global Competition*, London: Croom Helm.

Economist, The (1989) 'Investment by Taiwan: The Embarassment of Riches', 25 March.

Ehrlich, P. (1988) 'Malaysia: Fabrication Host Spot?', *Electronic News* 18, 6 June.

Eisemon, T. O. (1984) 'Insular and Open Strategies for Enhancing Scientific and Technological Capacities: Indian Educational Expansion and its Implications for African Countries', in M. Fransman and K. King (eds) *Technological Capability in the Third World*, New York: St Martin's Press.

Elliott, R. F. and Wood, P. W (1981) 'The International Transfer of Technology and Western European Integration', in R. G. Hawkins and A. J. Prasad (eds) *Technology Transfer and Economic Development*, Research in International Business and Finance, vol. 2, Greenwich, Connecticut: JAI Press.

Emmanuel, A. (1972) *Unequal Exchange: A Study of the Imperialism of Trade*, New York: Monthly Review Press.

Engineering News Record (1981) 'Joint Ventures Win Big Contracts', 30 April.

Enos, J. L. and Park, W. H. (1988) *The Adoption and Diffusion of Imported Technology: The Case of Korea*, London: Croom Helm.

ESCAP/UNCTC (1986) 'Hong Kong Transnational Corporation Investments in Developing Asian Economies', *Asia–Pacific TNC Review*, January.

ESCAP/UNCTC (1987) 'Transnational Corporations and Electronics Industries of ASEAN Economies', UNCTC Current Studies Series A, no. 5, ESCAP/UNCTC Joint Unit on Transnational Corporations, New York.

ESCAP/UNCTC (1988) *Transnational Corporations from Developing Asian and Pacific Economies*, Bangkok: ESCAP/UNCTC Joint Unit on Transnational Corporations.

Espenshade, A. V. (1955) 'Investment in the Philippines: Conditions and Outlook for the United States', study prepared for the US Department of Commerce.

Euh, Y-D and Min, S. H. (1986) 'Foreign direct investment from developing countries: the case of Korean firms', *The Developing Economies* 14, June.

Evans, P. B. and Tigre, P. B. (1989) 'Going Beyond Clones in Brazil and Korea: A Comparative Analysis of NIC Strategies in the Computer Industry', *World Development* 17(11).

Fagerberg, J. (1988) 'Why Growth Rates Differ', in G. Dosi, C. Freeman, R. Nelson, G. Silverberg and L. Soete (eds) *Technical Change and Economic Theory*, London: Pinter Publishers.

Felix, D. (1964) 'Monetarists, Structuralists and Import Substituting Industrialisation: A Critical Appraisal', in W. Baer and I. Kerstenetzky (eds) *Inflation and Growth in Latin America*, Homewood, Illinois: Irwin.

Findlay, R. (1978) 'Relative Backwardness, Direct Foreign Investment, and the Transfer of Technology: A Simple Dynamic Model', *Quarterly Journal of Economics* XCII(1), February.

Flowers, E. B. (1976) 'Oligopolistic Reactions in European and Canadian Direct Investment in the United States', *Journal of International Business Studies* 7.

Ford, A. and Clifford, M. (1987) 'Hyundai reined in: South Korean carmaker penalised for dumping in Canada', *Far Eastern Economic Review* 138(51) December 17: 128–9.

Forestier, K. (1989) 'Going multinational', *Asian Business* 25(4): April 14.

Frank, A. G. (1966) 'The Development of Underdevelopment', *Monthly Review* 18, September: 17–31.

Frank, A. G. (1975) *On Capitalist Underdevelopment*, Bombay: Oxford University Press.

Frank, A. G. (1978a) *Dependent Accumulation and Underdevelopment*, London: Macmillan.

Frank, A. G. (1978b) *World Accumulation, 1492-1789*, New York: Monthly Review Press.

Frank, A. G. (1979) *Mexican Agriculture, 1521-1630: Transformation of the Mode of Production*, Cambridge: Cambridge University Press.

Franko, L. G. (1976) *The European Multinationals*, London: Harper and Row.

Franko, L. G. (1983) *The Threat of Japanese Multinationals*, Chichester: John Wiley & Sons.

Fransman, M. (1984) 'Some Hypotheses Regarding Indigenous Technological Capability and the Case of Machine Production in Hong Kong', in M. Fransman and K. King (eds) *Technological Capability in the Third World*, New York: St Martin's Press.

Freeman, C., Clark, J. and Soete, L. L. G. (1982) *Unemployment and Technical Innovation*, London: Pinter Publishers.

Freeman, C. and Perez, C. (1986) 'The Diffusion of Technical Innovations and Changes of Techno-Economic Paradigm', paper prepared for the Conference on Innovation Diffusion, Venice, Italy, 17–22 March.

Freeman, C. and Perez, C. (1988) 'Structural Crises of Adjustment: Business Cycles and Investment Behaviour', in G. Dosi, C. Freeman, R. Nelson, G. Silverberg and L. Soete (eds) *Technical Change and Economic Theory*, London: Pinter Publishers.

Freeman, C. (1988a) 'Preface to Part II: Evolution, Technology and Institutions', in G. Dosi, C. Freeman, R. Nelson, G. Silverberg and L. Soete (eds) *Technical Change and Economic Theory*, London: Pinter Publishers.

Freeman, C. (1988b) 'Technology Gaps, International Trade and the Problems of Smaller and Less-Developed Economies', in C. Freeman and B.-Å. Lundvall (eds) *Small Countries Facing the Technological Revolution*, London: Pinter Publishers.

Friedland, J. (1990), 'Singer in Harmony: Hong Kong Firms Aim to capitalise on Brand Name', *Far Eastern Economic Review* 149(3), 27 September: 56.

Friedman, W. and Kalmanoff, G. (eds) (1961) *Joint International Business Ventures*, New York: Columbia University Press.

Fröbel, F., Heinrichs, J. and Kreye, O. (1980) *The New International Division of Labour*, Cambridge: Cambridge University Press.

Frondizi, S. (1957) *La Realidad Argentina: Ensayo de Interpretación Sociolÿgica*, 2nd edn, 2 volumes, Buenos Aires: Praxis.

Galang, J. (1988) 'Hereditary Defects: Manila is Burdened with Past Industrial Planning', *Far Eastern Economic Review* 140 (16), 21 April: 76.

Galbraith, J. K. (1967) *The New Industrial State*, London.

GATT (1990) *International Trade, 1989–90*, vol. I, Geneva: GATT.

Germidis, D. (1977) 'Technology Transfer, Regional Co-operation and Multinational Firms', in D. Germidis (ed.) *Transfer of Technology by Multinational Corporations*, vol. 1, Paris: OECD Development Centre.

Giddy, I. H. (1978) 'The Demise of the Product Cycle Model in International Business Theory', *The Columbia Journal of World Business* 13(1), Spring.

Global Trade (1989) 'The Lee Iacocca of Korea', 109(6), June.

Goldstein, C. (1989) 'Chips of Change: Electronics Transforms the Face of Malaysia's Industry', *Far Eastern Economic Review* 145(36), 7 September: 98–9.

Goldstein, C. (1990) 'Switch Watch: Hong Kong Firms Hail Role Change', *Far Eastern Economic Review* 149(38), 20 September: 93–4.

Graham, E. M. and Krugman, P. (1989) *Foreign Direct Investment in the United States*, Washington, DC: Institute for International Economics.

Graham, E. M. (1975) 'Oligopolistic Imitation and European Direct Investment', Ph. D. Dissertation, Harvard Graduate School of Business Administration.

Graham, E. M. (1978) 'Transatlantic Investment by Multinational Firms: A Rivalistic Phenomenon?', *Journal of Post-Keynesian Economics* 1(1), Autumn.

Graham, E. M. (1985) 'Intra-Industry Direct Investment, Market Structure, Firm Rivalry and Technological Performance', in A. Erdilek (ed.) *Multinationals as Mutual Invaders: Intra-Industry Direct Foreign Investment*, London: Croom Helm.

Gregersen, B. (1988) 'Public Sector Participation in Innovation Systems', in C. Freeman and B.-Å. Lundvall (eds) *Small Countries Facing the Technological Revolution*, London and New York: Pinter Publishers.

Grosse, R. (1985) 'An Imperfect Competition Theory of the MNE', *Journal of International Business Studies* 16, Spring.

Guimaraes, E. A. (1986) 'The Activities of Brazilian Firms Abroad', in C. Oman (ed.) *New Forms of Overseas Investment by Developing Countries: The Case of India, Korea and Brazil*, Paris: OECD.

Hakam, A. N. and Chang, Z.-Y. (1988) 'Patterns of Technology Transfer in Singapore: The Case of the Electronics and Computer Industry', *International Journal of Technology Management* 3(1–2).

Hamel, G. and Prahalad, C. K. (1988) 'Creating Global Strategic Capability', in N. Hood and J. Vahlne (eds), *Strategies in Global Competition*, London: Croom Helm.

Harrigan, K. R. (1987) 'Strategic Alliances: Their New Role in Global Competition', *Columbia Journal of World Business* 22(2), Summer: 67–9.

Helleiner, G. K. (1975) 'The Role of Multinational Corporations in the Less Developed Countries' Trade in Technology', *World Development* 3(4), April.

Henderson, J. (1989) *The Globalisation of High Technology Production*, London and New York: Routledge.

Hiemenz, U. (1987) 'Foreign Direct Investment and Industrialization in ASEAN Countries', *Welwirtschaftliches Archiv* 123(1).

Hill, H. (1985) 'Subcontracting, Technological Diffusion, and the Development of Small Enterprise in Philippine Manufacturing', *Journal of Developing Areas* 19(2), January.

Hill, H. and Pang, E. F. (1988) 'The State and Industrial Restructuring: A Comparison of the Aerospace Industry in Indonesia and Singapore', *ASEAN Economic Bulletin* 5(2): 152–68.

Hirsch, S. (1967) *Location of Industry and International Competitiveness*, Oxford: Oxford University Press.

Hirsch, S. (1976) 'An International Trade and Investment Theory of the Firm', *Oxford Economic Papers* 28, July.

Hirschman, A. O. (1968) 'The Political Economy of Import-Substituting Industrialisation in Latin America', *Quarterly Journal of Economics* LXXXII(1), February.

Hock, T. L. (1988a) 'Hong Kong: Upmarket, Hi-Tech is New Beep in Electronics', *Asian Finance* 14(3), 15 March: 36–43.

Hock, T. L. (1988b) 'Singapore: Hi-tech Vision Turns Industry Wheels', *Asian Finance* 14(9), 15 September: 92–5.

Hong Kong Industry Department (1987) *Overseas Investment in Hong Kong Manufacturing Industries*, Government of Territory of Hong Kong.

Hood, N. and Young, S. (1979) *The Economics of the Multinational Enterprise*, London: Longman.

Hornell, E. and Vahlne, J. E. (1972) 'The Deciding Factors in the Choice of Subsidiary Sales Company as a Channel for Exports', *Acta Universitatis Usaliensis* 6.

Horst, T. (1972) 'Firm and Industry Determinants of the Decision to Invest Abroad: An Empirical Study', *Review of Economics and Statistics* 54, August.

Howe, G. N. (1981) 'Dependency Theory, Imperialism, and the Production of Surplus Value on a World Scale', *Latin American Perspectives* 8, Summer and Autumn: 82–102.

Huasheng, Z. and Kerkhofs, M. (1988) 'System 12 Technology Transfer to the People's Republic of China', *International Journal of Technology Management* (Switzerland) 3(1–2): 204–11.

Hufbauer, G. C. (1965) *Synthetic Materials and the Theory of International Trade*, London: Duckworth.

Hufbauer, G. C. (1970) 'The Impact of National Characteristics and Technology on the Commodity Composition of Trade in Manufactured Goods', in R. Vernon (ed.) *The Technology Factor in International Trade*, New York: Columbia University Press.

Hymer, S. and Rowthorn, R. (1970) 'Multinational Corporations and Industrial Organisation: The Non-American Challenge', in C. P. Kindleberger (ed.) *The International Corporation: A Symposium*, Cambridge, Massachusetts: MIT Press.

Hymer, S. H. (1960) 'The International Operations of National Firms: A Study of Direct Foreign Investment', Ph. D. Dissertation, Massachusetts Institute of Technology. Published 1976 Cambridge, Massachussetts: MIT Press.

Hymer, S. H. (1972) 'The Multinational Corporation and the Law of Uneven

Development', in J. N. Bhagwati (ed.) *Economics and the World Order*, New York: Macmillan.

Industrial Development Authority Ireland (1988) *Overseas Companies in Ireland*, March.

Inter-American Development Bank (1988) Economic and Social Progress in Latin America: 1988 Report, Special Section: Science and Technology, Washington, DC.

Japan Economic Planning Agency (1987) *Economic Survey of Japan, 1986–87* (in Japanese), Tokyo: Economic Planning Agency.

Jegathesan, J. (1990), 'Factors Affecting Access to Technology Through Joint Ventures', in UNCTAD *Joint Ventures As A Channel for the Transfer of Technology*, New York: United Nations.

Jenkins, R. O. (1979) 'The Export Performance of Multinational Corporations in Mexican Industry', *Journal of Development Studies* 15(3), April.

Jenkins, R. O. (1984) *Transnational Corporations and Industrial Transformation in Latin America*, London: Macmillan.

Jenkins, R. O. (1987) *Transnational Corporations and Uneven Development*, London: Methuen.

Jenkins, S. (1954) *American Economic Policy Toward the Philippines*, Stanford: Stanford University Press, 34–7.

Jiang, F. (1990) 'Fresh Pastures: Hong Kong Firms Shift Investment Away From China', *Far Eastern Economic Review*, 20 September: 92–3.

Jo, S.-H. (1981) 'Overseas Direct Investment by South Korean Firms: Direction and Pattern', in K. Kumar and M. G. McLeod (eds) *Multinationals from Developing Countries*, Lexington, Massachusetts: D. C. Heath and Company.

Johanson, J. and Matson, L.-G. (1988) 'Internationalisation in Industrial Systems – A Network Approach', in N. Hood and J.-E. Vahlne (eds) *Strategies in Global Competition*, London: Croom Helm.

Johanson, J. and Vahlne, J. E. (1977) 'The Internationalisation Process of the Firm – A Model of Knowledge Development and Increasing Foreign Market Commitments', *Journal of International Business Studies* 8(1), Spring/Summer.

Johnson, B. (1988) 'An Institutional Approach to the Small-Country Problem', in C. Freeman and B-Å. Lundvall (eds) *Small Countries Facing the Technological Revolution*, London: Pinter Publishers.

Johnson, H. G. (1958) *International Trade and Economic Growth*, London: Allen & Unwin.

Johnson, H. G. (1970) 'The Efficiency and Welfare Implications of the International Corporation', in C. P. Kindleberger (ed.) *The International Corporation: A Symposium*, Cambridge, Massachusetts: MIT Press.

Johnson, H. G. (1977) 'Economic Benefits', in H. R. Hahlo, J. G. Smith and R. W. Wright (eds) *Nationalism and the Multinational Enterprise: Legal, Economic, and Managerial Aspects*, Netherlands: A. W. Sijhoff.

Johnstone, B. (1990a) 'Research and Innovation: The Need to Think Big', *Far Eastern Economic Review* 149(31), 2 August: 54.

Johnstone, B. (1990b) 'Bargaining Chips', *Far Eastern Economic Review* 149(29), 19 July: 26.

Joseph, A. (1989) 'Japan's Foreign Direct Investment in India's Industrial Development', paper prepared for the United Nations Centre on Transnational Corporations, February.

Juilland, M.-J. (1986) 'Asian money – and More', *Venture* 8(11), November.

Julius, D. (1990) *Global Companies and Public Policy*, London: Pinter Publishers.

Jun, J. W. (1989) 'Korean Overseas Investment: Patterns, Characteristics and Strategic Behaviours', paper prepared for the ESCAP/UNCTC Joint Unit on Transnational Corporations.

Kaldor, N. (1934) 'The Equilibrium of the Firm', *The Economic Journal* 44, March.

Kang, J. (1990) 'Companies for the 21st Century', *Korea Business World,* September.
Kaplinsky, R. (1984) 'Trade in Technology – Who, What, Where and When?', in M. Fransman and K. King (eds) *Technological Capability in the Third World,* New York: St Martin's Press.
Karlin, B. (1988) 'Singapore's Plan to Offer More Than Cheap Assembly', *Electronic Business* 14(10), 15 May: 124, 126.
Karlsson, C. (1988) 'Corporate Families to Handle Galloping Technology', in N. Hood and J.-E. Vahlne (eds) *Strategies in Global Competition,* London: Croom Helm.
Katrak, H. (1989) 'Imported Technologies and R&D in a Newly Industrialising Country: The Experience of Indian Enterprises', *Journal of Development Economics* 31(1), July: 123–39.
Katz, J. and Ablin, E. (1978) 'De La Industria Incipiente y La Exportación de Tecnología: La Experiencia Argentina en la Venta Internacional de Plantas Industriales y Obras de Ingeniería', Programa BID/CEPAL de Investigaciónes en Temas de Ciencia y Tecnología, Buenos Aires, April.
Katz, J. and Kosacoff, B. (1983) 'Multinationals from Argentina', in S. Lall (ed.) *The New Multinationals,* Chichester: John Wiley & Sons.
Katz, J. M. (1984a) 'Domestic Technological Innovations and Dynamic Comparative Advantage', *Journal of Development Economics* 16.
Katz, J. M. (1984b) 'Comparative Advantages of Latin American Metalworking Industries', in M. Fransman and K. King (eds) *Technological Capability in the Third World,* New York: St Martin's Press.
Katz, J. (1978) 'Technological Change, Economic Development and Intra and Extra Regional Relations in Latin America', Buenos Aires: IDB/ECLA, Working Paper no. 30.
Kay, N. (1983) 'Multinational Enterprise: A Review Article', *Scottish Journal of Political Economy* 30.
Keesing, D. (1979) 'World Trade and Output of Manufactures: Structural Trends and Developing Countries' Exports', World Bank Staff Working Paper, no. 316.
Ken, O. (1977) 'Bataan Export Processing Zone: Its Development and Social Implications', in *Japan–Asia Quarterly Review*, Free Trade Zones and Industrialisation of Asia, Special Issue, Tokyo.
Khanna, S. (1984) 'TNCs and Technology Transfer: Contours of Dependence in the Indian Petrochemical Industry', Indian Institute of Management, Working Paper, no. 83, Calcutta, India, July.
Kim, L. and Lee, H. (1987) 'Patterns of Technological Change in a Rapidly Developing Country: A Synthesis', *Technovation* 6(4), September.
Kim, L. S. (1986) 'New Technologies and their Economic Effects: A Feasibility Study in Korea', paper commissioned by the United Nations University for a project under the direction of Prof. C. Cooper, Maastricht, Netherlands: University of Limburg, 10 October.
Kim, L. (1980) 'Stages of Development of Industrial Technology in a Developing Country', *Research Policy* 9(3): 254–77.
Kindleberger, C. P. (1969) *American Business Abroad: Six Lectures on Direct Investment,* Cambridge, Massachusetts: MIT Press.
King, K. (1984) 'Science, Technology and Education in the Development of Indigenous Technological Capability', in M. Fransman and K. King (eds) *Technological Capability in the Third World,* New York: St Martin's Press.
Kirchbach, F. (1981) 'Economic Policies towards TNCs: the Experience of ASEAN countries', Ph. D. Dissertation, University of Regensburg, July.
Kirkland, R. I. Jr (1988) 'Entering a New Age of Boundless Competition', *Fortune* 117(6), 14 March: 4–8.
Knickerbocker, F. T. (1973) *Oligopolistic Reaction and the Multinational Enterprise,* Boston: Harvard University Press.

Knickerbocker, F. T. (1976) 'Market Structure and Market Power Consequences of Foreign Direct Investment by Multinational Companies', Occasional Paper, no. 8, Washington: Center for Multinational Studies.

Kobrin, S. J. (1979) 'Multinational Corporations, Sociocultural Dependence, and Industrialization: Need Satisfaction or Want Creation', *Journal of Developing Areas* 13(2), January:109–25.

Kogut, B. (1983) 'Foreign Direct Investment as a Sequential Process', in C. P Kindleberger and D. B. Audretsch (eds) *The Multinational Corporation in the 1980s*, Cambridge, Massachusetts: MIT Press.

Kogut, B. (1984) 'The Benefit of Being Global: Profiting from Environmental Volatility', Mimeo.

Kojima, K. and Ozawa, T. (1985) 'Toward a Theory of Industrial Restructuring and Dynamic Comparative Advantage', *Hitotsubashi Journal of Economics* 26(2), December.

Kojima, K. (1973) 'A Macroeconomic Approach to Foreign Direct Investment', *Hitotsubashi Journal of Economics* 14.

Kojima, K. (1975) 'International Trade and Foreign Investment: Substitutes or Complements?', *Hitotsubashi Journal of Economics* 16.

Kojima, K. (1978) *Direct Foreign Investment: A Japanese Model of Multinational Business Operations*, London: Croom Helm.

Kojima, K. (1982) 'Macroeconomic versus International Business Approach to Direct Foreign Investment', *Hitotsubashi Journal of Economics* 23(1), June.

Kojima, K. (1989) 'Theory of Internalisation by Multinational Corporations', *Hitotsubashi Journal of Economics* 30(2), December.

Kojima, K. (1990) *Japanese Direct Investment Abroad*, Tokyo: International Christian University Social Science Research Institute.

Koo, B.-Y. (1984) 'Outward Investment by Korean Firms: Their Forms and Characteristics', paper presented at the OECD meeting on collaborating researchers in New Forms of Investment in Developing Countries, Paris, June.

Koo, B.-Y. (1985) 'Korea', in J. H. Dunning (ed.) *Multinational Enterprises, Economic Structure and International Competitiveness*, Chichester: John Wiley & Sons. *Korea Business World*, April 1990.

Korea Institute for Foreign Investment (1989) 'A Study on Korean Overseas Investment', Mimeo, May.

Korean Development Bank Research Department (1985) *Industry in Korea 1984*, Korea: Korean Development Bank.

Kotkin, J. (1987) 'The Real Threat from Asia', *Inc.* 9(1), January: 23–6.

Kraar, L. (1987) 'Korea's Big Push Has Just Begun', *Fortune* 115(6), 16 March: 72–6.

Kucway, C. (1988) 'Automotive Marketing: Hyundai Learns from its Canadian Errors', *Advertising Age* 59(3), 25 July: S20.

Kumar, K. and Kim, K. Y. (1984) 'The Korean Manufacturing Multinationals', *Journal of International Business Studies*, Spring/Summer.

Kumar, K. (1981) 'Multinationalisation of Third World Public-Sector Enterprises', in K. Kumar and M. G. McLeod (eds) *Multinationals from Developing Countries*, Lexington, Massachusetts: D. C. Heath and Company.

Kuznets, S. (1968) *Toward a Theory of Economic Growth*, New York: W. W. Norton.

Kwag, D.-H. (1987) 'Korea's Overseas Investment', *Monthly Review*, Korean Exchange Bank, September.

Kwon, Y. C. and Ryans, J. K., Jr (1987) 'Dynamic Competitive Position between Foreign Joint Venture and NICs' Indigenous Firms: A Case Study in South Korea', *Foreign Trade Review* 22, July/September.

Lall, S. (1975) 'Is Dependence a Useful Concept in Analysing Underdevelopment?', *World Development* 3(11 and 12), November/December.

Lall, S. (1977) 'Transfer Pricing in Assembly Industries: A Preliminary Analysis of the Issues in Malaysia and Singapore', Mimeo, Commonwealth Secretariat, London.

Lall, S. (1979) 'The International Allocation of Research Activity by US Multinationals', *Oxford Bulletin of Economics and Statistics* 41, November.

Lall, S. (1980a) 'Monopolistic Advantages and Foreign Involvement by US Manufacturing Industry', *Oxford Economic Papers* 32(1), March.

Lall, S. (1980b) 'The Export of Capital from Developing Countries: The Indian Case', paper presented to the International Economics Study Group Conference, September.

Lall, S. (1981) *Developing Countries in the International Economy: Selected Papers*, London and Basingstoke: Macmillan.

Lall, S. (1982a) *Developing Countries as Exporters of Technology*, London and Basingstoke: Macmillan.

Lall, S. (1982b) 'The Export of Capital from Developing Countries: India', in J. Black and J. H. Dunning (eds) *International Capital Movements*, London: Macmillan.

Lall, S. (1982c) 'Technological Learning in the Third World: Some Implications of Technology Exports', in F. Stewart and J. James (eds) *The Economics of New Technology in Developing Countries*, London: Pinter Publishers.

Lall, S. (1983a) 'The Rise of Third World Multinationals from the Third World', *Third World Quarterly* 5(3), July.

Lall, S. (1983b) 'Multinationals from India', in S. Lall (ed.) *The New Multinationals*, Chichester: John Wiley & Sons.

Lall, S. (1983c) 'The Theoretical Background', in S. Lall (ed.) *The New Multinationals*, Chichester: John Wiley & Sons.

Lall, S. (1983d) 'Trade in Technology by a Slowly Industrialising Country: India', paper prepared for the Conference on International Technology Transfer: Concepts, Measures and Comparisons organised by the US Social Science Research Council in New York, 2–3 June.

Lall, S. (1984a) 'Transnationals and the Third World: Changing Perceptions', *National Westminster Bank Quarterly Review*, May.

Lall, S. (1984b) 'India', in S. Lall (ed.) 'Exports of Technology by Newly Industrialising Countries', *World Development* 12, May–June.

Lall, S. (1984c) 'South–South Economic Co-operation and Global Negotiations', in J. Bhagwati and J. G. Ruggie (eds) *Power, Passions and Purpose: Prospects for North–South Negotiations*, Cambridge, Massachusetts: MIT Press: 287–322.

Lall, S. (1984d) 'India's Technological Capability: Effects of Trade, Industrial, Science and Technology Policies', in M. Fransman and K. King (eds) *Technological Capability in the Third World*, New York: St Martin's Press.

Lall, S. (1985a) 'India', in J. H. Dunning (ed.) *Multinational Enterprises, Economic Structure and International Competitiveness*, Chichester: John Wiley & Sons.

Lall, S. (1985b) 'Multinationals and Technology Development in Host Countries', in S. Lall (ed.) *Multinationals, Technology and Exports*, London: Macmillan.

Lall, S. (1987) 'Multinationals and the Industrialization of the Least Developed Countries', paper prepared for the Federation Internationale des Universites Catholiques Conference on Development Objectives, Economic National Policies and Multinationals, Porto Alegre, Brazil, 22–6 March.

Lall, S. (1991) 'Developing Country TNCs and their Impact on Home Countries', technical paper for the United Nations Centre on Transnational Corporations, preliminary draft, February.

Lamb, C. (1991a) 'A Protectionist Virus in Brazil's Computer Plans', *Financial Times*, 23 July 1991.

Lamb, C. (1991b) 'Collor Wins Victory on Information Technology', *Financial Times*, 27 June 1991.

Laxa, F., Cardenas, E., Federizon, R. and Gesmundo, M. (1978) 'The Philippine

Automotive Industry', in W. Clemente, F. Bacuñgan and F. Laxa (eds) 'Multinational Corporations in the Philippines', Mimeo, Technology Resource Center, Republic of the Philippines.

Lecraw, D. T. (1977) 'Direct Investment by Firms from Less Developed Countries', *Oxford Economic Papers* 29(3), November: 442–57.

Lecraw, D. T. (1981) 'Internationalisation of Firms from LDCs: Evidence from the ASEAN Region', in K. Kumar and M. G. McLeod (eds) *Multinationals from Developing Countries*, Lexington, Massachusetts: Lexington Books.

Lecraw, D. (1985) 'Singapore', in J. H. Dunning (ed.) *Multinational Enterprises, Economic Structure and International Competitiveness*, Chichester: John Wiley & Sons.

Lecraw, D. (1991) 'Third World MNEs Once Again: the Case of Indonesia', Mimeo.

Lee, C.-S. (1989) 'The Overseas Experience', *Korea Business World*, October.

Lee, G. B. (1989) 'Transnational Corporations in Malaysia: Visible but Unrevealing', paper prepared for the International Seminar on Industrialization and Development: Focus on ASEAN, Bangkok, Thailand, 21–6 August.

Lee, J., Bae, Z.-T. and Choi, D.-K (1988) 'Technology Development Processes: A Model for a Developing Country with a Global Perspective', *R&D Management (UK)* 18(3), July: 235–50.

Lenae, H. (1985) 'Joint Ventures as Vehicles of Technology Transfer to the Third World', Ph. D. Dissertation, New York University.

Lenin, V. I. (1967) *Selected Works*, three volumes, Moscow: Progress Books.

Leontief, W. W. (1954) 'Domestic Production and Foreign Trade: The American Capital Position Reexamined', *Economía Internationale* 7(1), February.

Lessard, D. R. (1976) 'The Structure of Returns and Gains from International Diversification', in N. Elton and M. Gruber (eds) *International Capital Markets*, Amsterdam: North-Holland.

Lessard, D. R. (1977) 'International Diversification and Direct Foreign Investment', in D. K. Eiteman and A. I. Stonehill, *Multinational Business Finance*, Massachusetts: Addison-Wesley.

Lessard, D. R. (1979) 'Transfer Prices, Taxes and Financial Markets', in R. Hawkins (ed.) *The Economic Effects of Multinational Corporations*, Greenwich, Connecticut: JAI Press.

Liang, N. (1990) 'Competitive Strategies of Nations: A New Typology with Empirical Verification', paper presented at the Annual Conference of the Academy of International Business, Toronto, Ontario, Canada, 11–14 October.

Lim, D. (1983) 'Fiscal Incentives and Direct Foreign Investment in Less Developed Countries', *Journal of Development Studies* 19.

Lim, L. Y. C. and Pang, E. F. (1982) 'Vertical Linkages and Multinational Enterprises in Developing Countries', *World Development* 10(7), 585–95.

Lim, M. H. (1984) 'Survey of Activities of Transnational Corporations from Asian Developing Countries', paper submitted to ESCAP/UNCTC Joint Unit on Transnational Corporations, Bangkok, March.

Lindbeck, A. (1981) 'Industrial Policy as an Issue in the Economic Environment', *The World Economy* 4, 391–406.

Linder, S. B. (1961) *An Essay on Trade and Transformation*, New York: Wiley.

Lindsey, C. W. (1983) 'In Search of Dynamism: Foreign Investment in the Philippines Under Martial Law', *Pacific Affairs* 56, Autumn.

Lindsey, C. W. (1985) 'The Philippine State and Transnational Investment', in R. B. Stauffer (ed.) *Transnational Corporations and the State*, Sydney, NSW: University of Sydney Transnational Corporations Research Project.

Liu, P. (1988) 'Emperor of Taiwan's Computer Industry', *Electronic Business* 14(20), 15 October: 96–8.

Liu, Y.-H (1990) 'Relocation of Corporation Domicile: Case of Hong Kong', paper presented at the Annual Conference of the Academy of International Business, 11–14 October.

Lodge, H. C. (1898) 'Interview with President McKinley', *Boston Evening Transcript*, 3 June. Reprinted in D. B. Schirmer and S. R. Shalom (eds) (1987) *The Philippines Reader: A History of Colonialism, Dictatorship, and Resistance*, Boston: South End Press.

Luostarinen, R. (1978) 'The Impact of Physical, Cultural and Economic Distance on the Geographical Structure of the Internationalisation Pattern of the Firm', FIBO Working Paper, no. 1978/2, Helsinki School of Economics.

Maex, R. (1983) 'Employment and Multinationals in Asian Export Processing Zones', International Labour Office Working Paper, no. 26, Geneva.

Magee, S. P. (1977a) 'Multinational Corporations, the Industry Technology Cycle and Development', *Journal of World Trade Law* 11(4), July–August pp. 297–321.

Magee, S. P. (1977b) 'Technology and Appropriability Theory of the Multinational Corporation', in J. Bhagwati (ed.) *The New International Economic Order The North–South Debate*, Cambridge, Massachusetts: MIT Press.

Magee, S. P. (1981) 'The Appropriability Theory of the Multinational Corporation', *The Annals of the American Academy of Political and Social Science*, November.

Maitland, A. (1989) 'Aided by a "Reverse Brain Drain"', *Financial Times*, 10 October.

Mansfield, E. and Romeo, A. (1979) 'Technology Transfer to Overseas Subsidiaries by US-Based Firms', Working Paper, Department of Economics, University of Pennsylvania.

Mansfield, E. M. (1968) *Industrial Research and Technological Innovation*, New York: Norton.

Mansour, M. B. (1981) 'Definitional Issues in Technology Transfer: Channels, Mechanisms and Sources', in R. G. Hawkins and A. J. Prasad (eds) *Technology Transfer and Economic Development, Research in International Business and Finance*, vol. 2, Greenwich, Connecticut: JAI Press.

Marcussen, H. S. and Jens, E. T. (1982) *Internationalization of Capital–Prospects for the Third World: A Re-examination of Dependency Theory*, London: Zed Press.

Marx, K. (1943) *Articles on India*, Bombay: People's Publishing House.

Marx, K. (1976) *Capital*, Harmondsworth, Penguin.

Mason, R. H., Miller, R. R. and Weigel, D. R. (1975) *The Economics of International Business*, New York: Wiley.

Mason, R. H. (1980) 'A Comment on Professor Kojima's Japanese Type versus American Type of Technology Transfer', *Hitotsubashi Journal of Economics* 20(2), February.

Matsuura, N. F. (1989) 'Japanese Direct Investment and Control: Case of the NIEs', paper presented at the Annual Meeting of the Academy of International Business, Singapore, November.

McCormick, J. (1990) 'Taiwan's New Chip Makers Struggle for Survival', *Electronic Business*, 16(18), 17 September: 87–90.

McManus, J. C. (1972) 'The Theory of the Multinational Firm', in G. Pacquet (ed.) *The Multinational Firm and the Nation State*, Toronto: Collier-Macmillan.

Ministry of Labour and Employment (MOLE), Republic of the Philippines (1982) Briefing Paper for the President's visit to Saudi Arabia, 11 February.

Miranda, C., Jr (1989) 'Transnational Corporations in the ASEAN Region: The Philippines', in Report and Summary of Proceedings of the International Seminar on Industrialization and Development: Focus on ASEAN, Bangkok, Thailand, 21–6 August.

Mitra, J. D. (1979) 'The Capital Goods Sector in LDCs: A Case for State Intervention?', World Bank Staff Working Paper, no. 343, Washington, DC.

Mi-yong, A. (1990) 'Shaking the Foundation', *Korea Business World* 6(9), September.

Mody, A. (1990) 'Institutions and Dynamic Comparative Advantage: The Electronics Industry in South Korea and Taiwan', *Cambridge Journal of Economics* 14.

Mohri, K. (1979) 'Marx and "Underdevelopment"', *Monthly Review* 30, April: 32–42.

Monkiewicz, J. and Boguslaw, S. (1986) 'Transnational Corporations from Developing Countries and the Exports of Technology', Working Paper, no. 14, World Economy Research Institute Central School of Planning and Statistics.

Moore, J. (1988) 'Apple of Taiwan's Eye', *Far Eastern Economic Review* 141(28), 14 July: 62.

Moore, J. (1990) 'The Upstart Taipans Cash = Rich Taiwanese Scour Asia for Opportunities', *Far Eastern Economic Review* 148(6), 19 April: 84–6.

Morris, S. (1987) 'Trends in Foreign Direct Investment from India', *Economic and Political Weekly*, 7 and 14 November.

Mowery, D. and Rosenberg, N. (1979) 'The Influence Of Market Demand Upon Innovation: A Critical Review Of Some Recent Empirical Studies', *Research Policy* 8.

Moxon, R. W. (1974) 'Offshore Production in Less Developed Countries', *The Bulletin*, New York: Institute of Finance, New York University, July.

Nagaoka, S. (1989) 'Overview of Japanese Industrial Technology Development', Industry and Energy Department Working Paper, no. 6, World Bank, March.

Nakarmi, L. and Neff, R. (1989) 'Korea's Powerhouses Are Under Siege', International Business Section, *Business Week*, 20 November.

Nayyar, D. (1978) 'Transnational Corporations and Manufactured Exports from Poor Countries', *The Economic Journal* 88 (349), March.

Negandhi, A. R. and Palia, A. P. (1989) 'Alternative Approaches to Development of Computer Technologies: A Comparison of India, Japan and Singapore', in E. Kaynak and K.-H. Lee (eds) *Global Business: Asia–Pacific Dimensions*, London: Routledge.

Nelson, R. R. and Winter, S. G. (1977) 'In Search of Useful Theory of Innovation', *Research Policy* 6(1), January: 36–76.

Nelson, R. R. and Winter, S. G. (1982) *An Evolutionary Theory of Economic Change*, Cambridge, Massachusetts: Harvard University Press.

Newfarmer, R. (1980) *Transnational Conglomerates and the Economics of Dependent Development*, Greenwich, Connecticut: JAI Press.

Newfarmer, R. (ed.) (1985) *Profits, Progress and Poverty: Case Studies of International Industries in Latin America*, Notre Dame: University of Notre Dame Press.

Nishimizu, M. and Robinson, S. (1984) 'Trade Policies and Productivity Change in Semi-Industrialised Countries, *Journal of Development Economics* 16.

Odaka, K. (1984) 'Skill Formation in Development: Case Studies of Philippine Manufacturing', *Hitotsubashi Journal of Economics* 125, 105–23.

Ohkawa, K. and Rosovsky, H. (1973) *Japanese Economic Growth: Trend Acceleration in the Twentieth Century*, Stanford: Stanford University Press.

Ohmae, K. (1985) *Triad Power: The Coming Shape of Global Competition*, New York: The Free Press.

Ohmae, K. (1987) 'The Triad World View', *Journal of Business Strategy* 7, 8–19.

Ohmae, K. (1989) *The Borderless World*, New York: The Free Press.

Oman, C. (1988) 'Cooperative Strategies in Developing Countries: The New Forms of Investment', in F. J. Contractor and P. Lorange (eds) *Cooperative Strategies in International Business*, Lexington, Massachusetts: Lexington Books.

Ozawa, T. (1971) 'Transfer of Technology from Japan to Developing Countries', United Nations Institute for Training and Research Report no. 7, New York.

Ozawa, T. (1974) *Japan's Technological Challenge to the West, 1950–74: Motivation and Accomplishment*, Cambridge, Massachusetts: MIT Press.

Ozawa, T. (1977) 'Japan's Resource Dependency and Overseas Investment', *Journal of World Trade Law* 2(1), January–February.

Ozawa, T. (1979a) *Multinationalism, Japanese Style: The Political Economy of Outward Dependency*, Princeton, New Jersey: Princeton University Press.

Ozawa, T. (1979b) 'International Investment and Industrial Structure: New Theoretical Implications from the Japanese Experience', *Oxford Economic Papers* 31, March.

Ozawa, T. (1981) 'Technology Transfer and Japanese Economic Growth in the Postwar Period', in R. G. Hawkins and A. J. Prasad (eds) *Technology Transfer and Economic Development, Research in International Business and Finance*, vol. 2, Greenwich, Connecticut: JAI Press.

Ozawa, T. (1982) 'A Newer Type of Foreign Investment in Third World Resource Development', *Rivista Internazionale di Scienze Economiche e Commerciali* 29(12), December.

Ozawa, T. (1985) 'Japan', in J. H. Dunning (ed.) *Multinational Enterprises, Economic Structure and International Competitiveness*, Chichester: John Wiley & Sons.

Ozawa, T. (1990) 'Europe 1992 and Japanese Multinationals: Transplanting a Subcontracting System in the Expanded Market', in B. Bürgenmeier and J. L. Mucchielli (eds) *Multinationals and Europe 1992*, London: Routledge.

O'Brien, P. *et al.* (1979) 'Direct Foreign Investment and Technology Exports Among Developing Countries: An Empirical Analysis of the Prospects for Third World Cooperation', paper for the UNIDO Joint Study on Industrial Co-operation, Vienna, January.

Pack, H. (1981) 'Fostering the Capital-Goods Sector in LDCs', *World Development* 9(3), March.

Palloix, C. (1975) *L'Internationalisation du Capital*, Paris: François Maspero.

Palloix, C. (1977) 'The Self-Expansion of capital on a World Scale', *Review of Radical Political Economy* 9, Summer: 1–28.

Pang, E. F. and Hill, H. (1991) 'Technology Exports from a Small, Very Open NIC: The Case of Singapore', *World Development* 19(5): 553–68.

Pang, E. F. and Komaran, R. V. (1984) 'Hong Kong and Singapore Multinationals: A Comparison', preliminary draft on Research Project on New Forms of Investment in Developing Countries, Phase II, OECD, May.

Pang, E. F. and Komaran, R. V. (1985) 'Singapore Multinationals', *Columbia Journal of World Business* 20(2), Summer: 35–43.

Pang, E. F. and Lim, L. Y. C. (1989) 'High-Tech and Labour in the Asian NICs', *Labour and Society* 14: 1–15.

Pang, E. F. (1984) 'Foreign Investment and the State in Singapore', paper prepared for the Commonwealth Secretariat, September.

Park, Y. C. (1989) 'The Little Dragons and Structural Change in Pacific Asia', *The World Economy* 12(2), June: 125–61.

Parry, T. G. (1980) 'The Role of Transnational Corporations in the Developing ESCAP Region', Center for Applied Economic Research Paper, no. 10, September.

Parry, T. G. (1981) 'The Multinational Enterprise and Two-Stage Technology Transfer to Developing Nations', in R. G. Hawkins and A. J. Prasad (eds) *Technology Transfer and Economic Development, Research in International Business and Finance*, vol. 2, Greenwich, Connecticut: JAI Press.

Parry, T. G. (1985) 'Internalization as a General Theory of Foreign Direct Investment: A Critique', *Weltwirtschaftliches Archiv* 121, September.

Pasinetti, L. L. (1981) 'International Economic Relations', in L. L. Pasinetti, *Structural Change and Economic Growth: A Theoretical Essay on the Dynamics of the Wealth of Nations*, Cambridge: Cambridge University Press.

Patarsuk, W. (1989) 'Transnational Corporations in the ASEAN Region: A Case of Thailand', paper prepared for the International Seminar on Industrialization and Development: Focus on ASEAN, Bangkok, Thailand, 21–6 August.

Patel, P. and Pavitt, K. (1991) 'Large Firms in the Production of the World's Technology', *Journal of International Business Studies* 22(1), Winter: 1–21.

Paul, K. and Barbato, R. (1985) 'The Multinational Corporation in the Less Developed Country: The Economic Development Model Versus the North–South Model', *Academy of Management Review* 10(1), January.

Pavitt, K. and Soete, L. L. G. (1982) 'International Differences in Economic Growth and the International Location of Production', in H. Giersch (ed.) *Emerging Technologies: Consequences for Economic Growth, Structural Change and Employment*, Tübingen: J. C. B. Möhr.

Pavitt, K. (1988) 'International Patterns of Technological Accumulation', in N. Hood and J. E. Vahlne (eds) *Strategies in Global Competition*, London: Croom Helm.

Peck, M. and Wilson, R. (1982) 'Innovation, Imitation and Comparative Advantage: the Performance of Japanese Colour Television Set Producers in the US Market', in H. Giersch (ed.) *Emerging Technologies: Consequences for Economic Growth, Structural Change and Employment*, Tübingen: J. C. B. Möhr.

Penrose, E. T. (1959) *The Theory of the Growth of the Firm*, Oxford: Blackwell.

Perez, C. and Soete, L. (1988) 'Catching Up in Technology: Entry Barriers and Windows of Opportunity', in G. Dosi, C. Freeman, R. Nelson, G. Silverberg and L. Soete (eds) *Technical Change and Economic Theory*, London: Pinter Publishers.

Perez, C. (1983) 'Structural Change and the Assimilation of New Technologies in the Economic and Social System', *Futures*, October.

Perez, C. (1988) 'New Technologies and Development', in C. Freeman and B-Å. Lundvall (eds) *Small Countries Facing the Technological Revolution*, London: Pinter Publishers.

Perlmutter, H. V. and Heenan, D. A. (1986) 'Cooperate to Compete Globally', *Harvard Business Review*, March–April.

Perroux, F. (1968) *Multinational Investment in the Economic Development and Integration of Latin America*, Bogotá: Inter-American Development Bank.

Phipatseritham, K. (1982) *The Distribution of Ownership in Thai Big Business*, Bangkok: Thammasat University Press.

Poblador, N. S. (1969) 'Foreign Investment in the Major Non-Financial Corporate Sector of the Philippines, 1964 and 1965', Institute of Economic Development and Research, School of Economics, University of the Philippines, Quezon City, 26 December.

Porter, M. E. (1990) *The Competitive Advantage of Nations*, New York: The Free Press.

Posner, M. V. (1961) 'International Trade and Technical Change', *Oxford Economic Papers* 13, October.

Prado Júnior C. (1955) 'Nacionalismo Brasileiro e Capitais Estrangeiros', *Revista Brasiliense* 2 November–December: 80–93.

Prasad, A. J. (1979) *Technology Policy for Industry*, New Delhi: Allied Publishers.

Prasad, A. J. (1981a) 'Technology Transfer to Developing Countries Through Multinational Corporations', in R. G. Hawkins and A. J. Prasad (eds) *Technology Transfer and Economic Development*, Research in International Business and Finance, vol. 2, Greenwich, Connecticut: JAI Press.

Prasad, A. J. (1981b) 'Licensing As An Alternative to Foreign Investment for Technology Transfer', in R. G. Hawkins and A. J. Prasad (eds) *Technology Transfer and Economic Development*, Research in International Business and Finance, vol. 2, Greenwich, Connecticut: JAI Press.

Prasartset, S. (1988) 'Technological Domination by the Transnational Corporation in Thailand', in W. Teng and N. T. Wang (eds) *Transnational Corporations and China's Open Door Policy*, Lexington, Massachusetts: Lexington Books.

Prebisch, R. (1978) 'Notas Sobre el Desarrollo del capitalismo Periférico', *Estudios Internacionales* 11, July–September: 3–25.

Quijano, A. (1974) 'Imperialism and International Relations in Latin America', in J.

Cotler and R. Fagan (eds) *Latin America and the Changing Political Realities*, Stanford: Stanford University Press.

Quinn, J. B. (1969) 'Technology Transfer by Multinational Companies', *Harvard Business Review*, November–December.

Rana, P. B. (1987) 'Foreign Direct Investment and Economic Growth in the Asian and Pacific Region', *Asian Development Review* 1.

Ranis, G. (1976) 'The Multinational Corporation as An Instrument of Development', in D. E. Apter and L. W. Goodman (eds) *The Multinational Corporation and Social Change*, New York: Praeger.

Ranis, G. (1984) 'Determinants and Consequences of Indigenous Technological Activity', in M. Fransman and K. King (eds) *Technological Capability in the Third World*, New York: St Martin's Press.

Rayner, B. C. P. (1989) 'India: Struggling to Enter the Electronics Age', *Electronic Business* 15(1), 9 January: 130–4.

Reuber, G. L. Crookel, H., Emerson, M. and Gallais-Hamonno, G. (1973) *Private Foreign Investment in Development*, Oxford: Clarendon Press.

Rhee, Y. W. and Westphal, L. E. (1978) 'A Note on the Exports of Technology from the Republic of China and Korea', Mimeo, World Bank, Washington, DC.

Richards, J. B. (1944) Report TC–1 of the Technical Committee to the President of the Philippines, American Philippine Trade Relations, Washington, DC.

Riedel, J., Büttner, V. and Ernst, A. (1988) 'External Debt Alleviation through Foreign Direct Investment: A Real Issue for Third World Countries', *Tokyo Club Papers* 1:67–134.

Riedel, J. (1975) 'The Nature and Determinants of Export-Oriented Direct Foreign Investment in a Developing Country: A Case Study of Taiwan', *Weltwirtschaftliches Archiv* 111.

Robinson, H. (1933) *The Economics of Imperfect Competition*, London: Macmillan.

Robock, S. H. and Simmonds, K. (1983) *International Business and Multinational Enterprises*, Homewood, Illinois: Irwin.

Rodney, W. (1972) *How Europe Underdeveloped Africa*, London: Bogle-l'Ouverture Publications.

Rodrigues, C. A. (1985) 'A Process for Innovators in Developing Countries to Implement New Technology', *Columbia Journal of World Business* 20(3), Autumn.

Rohlwink, A. (1987) 'Formulating Strategies for Industrial Development: An Electronics Example', *Multinational Business* 4, Winter: 47–53.

Root, F. R. and Ahmed, A. A. (1978) 'The Influence of Policy Instruments on Manufacturing Direct Foreign Investment in Developing Countries', *Journal of International Business Studies* 9.

Rosenberg, N. (1976) *Perspectives on Technology*, Cambridge: Cambridge University Press.

Rosenberg, N. (1982) *Inside the Black Box: Technology and Economics*, Cambridge: Cambridge University Press.

Rothschild, K. (1947) 'Price theory and oligopoly', *The Economic Journal* 57.

Rugman, A. M. and McIlveen, J. (1985) *Mega Firms: Strategies for Canada's Multinationals*, Toronto: Methuen.

Rugman, A. M. (1975) 'Motives for Foreign Investment: The Market Imperfections and Risk Diversification Hypothesis', *Journal of World Trade Law* 9.

Rugman, A. M. (1976) 'Risk Reduction by International Diversification', *Journal of International Business Studies* 7.

Rugman, A. M. (1977) 'Risk, Direct Investment and International Diversification', *Weltwirtschaftliches Archiv* 113.

Rugman, A. M. (1979) *International Diversification and the Multinational Enterprise*, Lexington, Massachusetts: Lexington Books.

Rugman, A. M. (1980) 'Internalisation as a General Theory of Foreign Direct Investment: A Reappraisal of the Literature', *Weltwirtschaftliches Archiv* 116.

Rugman, A. M. (1981a) *Inside the Multinationals*, London: Croom Helm.

Rugman, A. M. (1981b) 'Canadian Multinational Enterprises and Developing Countries', Ecole des Hautes Etudes Commerciales Discussion Paper, December.

Rugman, A. M. (1982) *New Theories of the Multinational Enterprise*, London: Croom Helm.

Rugman, A. M. (1985a) 'New Theories of the Multinational Enterprise: An Assessment', Mimeo, Dalhousie University Center for International Business Studies.

Rugman, A. M. (1985b) 'The Marketing Advantages of Canadian Multinationals', Mimeo, Dalhousie University Center for International Business Studies.

Saghafi, M. M. and Davidson, C.-S. (1989) 'The New Age of Global Competition in the Semiconductor Industry: Enter the Dragon', *Columbia Journal of World Business* 24(4), Winter: 60–71.

Sanna Randaccio, F. (1980) 'European Direct Investments in US Manufacturing', M. Litt. thesis, Oxford University.

Schirmer, D. B. and Shalom, S. R. (eds) (1987) *The Philippines Reader: A History of Colonialism, Dictatorship, and Resistance*, Boston: South End Press.

Schirmer, D. B. (1975) 'The Conception and Gestation of a Neocolony', *The Journal of Contemporary Asia* 5(1). Reprinted in D. B. Schirmer and S. R. Shalom (eds) (1987) *The Philippines Reader: A History of Colonialism, Dictatorship, and Resistance*, Boston: South End Press.

Schive, C. and Kuang-Tao, H. (1985) 'Taiwan's investment in ASEAN countries and its competitiveness', Mimeo, April.

Schive, C. (1988) 'Foreign Investment and Technology Transfer in Taiwan: Past Experience and Future Potential', *Industry of Free China*, September: 13–30.

Schumpeter, J. A. (1934) *The Theory of Economic Development*, Cambridge, Massachusetts: Harvard University Press.

Schumpeter, J. A. (1943) *Capitalism, Socialism and Democracy*, New York: Harper & Row.

Seaward, N. (1987) 'Malaysia: A Question of Quantifying Investments', *Far Eastern Economic Review* 138(49), 3 December: 184-5.

Sekiguchi, S. (1979) *Japanese Foreign Direct Investment*, London: Macmillan.

Sercovich, F. (1984) 'Brazil, Special issue: Exports of Technology by Newly-Industrialising Countries', *World Development* 12.

Servan-Schreiber, J. J. (1967) *The American Challenge*, London: Hamish Hamilton.

Seymour, H. (1987) *International Investment in the Construction Industry*, London: Croom Helm.

Shalom, S. R. (1980) 'Philippine Acceptance of the Bell Trade Act of 1946: A Study of Manipulatory Democracy', *Pacific Historical Review* 49(3): 499–517.

Shrivastava, P. (1984) 'Technological Innovation in Developing Countries', *Columbia Journal of World Business* 19(4), Winter.

Simon, R. (1989) 'Taiwan's US Strategy', *Forbes* 143(11), 29 May: 43.

Singer, H. (1984) 'Industrialisation: Where Do We Stand? Where Are We Going?' Mimeo, The Institute of Development Studies, Brighton, Sussex.

Singh, R. K. D. N. (1986) 'Technology Transfer and Technological Co-operation Among Developing Countries', *Development & South-South Cooperation* II(2): 134–55.

Slaybaugh, C. W. (1981) 'Factors in Technology Transfer: A Multinational Firm Perspective', in R. G. Hawkins and A. J. Prasad (eds) *Technology Transfer and Economic Development*, Research in International Business and Finance, vol. 2, Greenwich, Connecticut: JAI Press.

Soete, L. (1981a) 'Technological Dependency: A Critical View', in D. Seers (ed.) *Dependency Theory: A Critical Assessment*, London: Pinter Publishers.

Soete, L. (1981b) 'A General Test of the Technological Gap Trade Theory', *Review of World Economics* 117, 638–66.

Soete, L. (1982) 'Innovation and International Trade: What We Know and What We Do Not Know', Mimeo, Science Policy Research Unit, University of Sussex.

Soete, L. (1985) 'International Diffusion of Technology, Industrial Development and Technological Leapfrogging', *World Development* 13(3).

Soete, L. (1986) 'Long Cycles and the International Diffusion of Technology', in C. Freeman (ed.) *Design, Innovation and Long Cycles in Economic Development*, New York: St Martin's Press.

Staubus, J. (1989) 'As Bunge y Born Goes, So Goes Argentina,' *Multinational Business* 3, Autumn.

Stewart, F. (1977) *Technology and Underdevelopment*, London: Macmillan.

Stewart, F. (1979) 'International Technology Transfer: Issues and Policy Options', World Bank Staff Working Paper, no. 344, Washington, DC.

Stewart, F. (1981) 'Arguments for the Generation of Technology by Less Developed Countries', *The Annals of the American Academy of Political and Social Science* 458, November.

Stewart, F. (1984) 'Facilitating Indigenous Technical Change in Third World Countries', in M. Fransman and K. King (eds) *Technological Capability in the Third World*, New York: St Martin's Press.

Stiglitz, J. E. (1987) 'Learning to Learn, Localized Learning and Technological Progress', in P. Dasgupta and P. Stoneman (eds) *Economic Policy And Technological Performance*, Cambridge: Cambridge University Press.

Stobaugh, R. B., Jr (1968) 'The Product Life Cycle, US Exports and International Investment', DBA dissertation, Harvard Business School, June.

Sunkel, O. (1972) 'Big Business and "Dependencia"', Foreign Affairs 50, April: 517–31.

Svedberg, P. (1981) 'Colonial Enforcement of Foreign Direct Investment', *Manchester School* 50.

Swedenborg, B. (1979) *The Multinational Operations of Swedish Firms*, Stockholm: Industrial Institute for Economic and Social Research.

Swedenborg, B. (1985) 'Sweden', in J. H. Dunning (ed.) *Multinational Enterprises, Economic Structure and International Competitiveness*, London: John Wiley & Sons.

Tai, C. B. (1987) 'The Effectiveness of China's Absorption of Foreign Investment', in R. D. Robinson (ed.) *Foreign Capital and Technology in China*, New York: Praeger.

Taiwan Research Institute (1983) *Taiwan Survey 1983*, Taipei: Taiwan Research Institute.

Tambunlertchai, S. and McGovern, I. (1984) 'An Overview of the Role of MNCs in the Economic Development of Thailand', paper prepared for the Conference on the Role of Multinational Corporations in Thailand organised by Thammasat University, Asia Pattaya Hotel, Cholburi, Thailand, 7–9 July.

Teece, D. J., Pisano G. and Shuen A. (1990) 'Firm Capabilities, Resources, and the Concept of Strategy', Mimeo, University of California at Berkeley, September.

Teece, D. J. (1976) *The Multinational Corporation and the Resource Cost of International Technology Transfer*, Cambridge, Massachusetts: Ballinger Publishing Company.

Teece, D. J. (1977) 'Technology Transfer by Multinational Firms: The Resource of International Technology Transfer', *The Economic Journal*, June.

Teece, D. J. (1981a) 'The Multinational Enterprise: Market Failure and Market Power Considerations', *Sloan Management Review* 22, September.

Teece, D. J. (1981b) 'The Market for Know-How and the Efficient International

Transfer of Technology', *The Annals of the Academy of Political and Social Science* 458.

Teece, D. J. (1981c) 'Technology Transfer and R&D Activities of Multinational Firms: Some Theories and Evidence', in R. G. Hawkins and A. J. Prasad (eds) *Technology Transfer and Economic Development*, Research in International Business and Finance, vol. 2, Greenwich, Connecticut: JAI Press.

Teece, D. J. (1982) 'A Transaction Cost Theory of the Multinational Enterprise', University of Reading Discussion Papers in International Investment and Business Studies, no. 66.

Teece, D. J. (1983) 'Technological and Organisational Factors in the Theory of the Multinational Enterprise', in M. C. Casson (ed.) *The Growth of International Business*, London: George Allen & Unwin.

Teece, D. J. (1985) 'Multinational Enterprise, Internal Governance and Industrial Organisation', *American Economic Review* 75(2), May.

Teece, D. J. (1988) 'Technological Change and the Nature of the Firm', in G. Dosi, C. Freeman, R. Nelson, G. Silverberg and L. Soete (eds) *Technical Change and Economic Theory*, London: Pinter Publishers.

Teece, D. J. (1990) 'Capturing Value from Technological Innovation: Integration, Strategic Partnering and Licensing Decisions', in F. Arcangeli, P. A. David and G. Dosi (eds) *Modern Patterns in Introducing and Adapting Innovations*, Oxford: Oxford University Press.

Teitel, S. and Sercovich, F. C. (1984) 'Latin America in special issue: Exports of Technology by Newly-Industrializing Countries', *World Development* 12.

Tiglao, R. (1990a) 'Wheels within Wheels: People's Car Project Threatens Philippine Cartel', *Far Eastern Economic Review* 147(13), 29 March: 71.

Tiglao, R. (1990b) 'Regional Development: All that Glitters is not in Manila', *Far Eastern Economic Review* 149(28), 12 July: 47–8.

Tiglao, R. (1990c) 'The Philippine Paradox', *Far Eastern Economic Review* 149(28), 12 July: 31–3.

Tiglao, R. (1991) 'Catching the Wind: Philippine Tycoon Rises to Prominence', *Far Eastern Economic Review* 151(7), 14 February: 51–2.

Ting, W. L. and Schive, C. (1981) 'Direct Investment and Technology Transfer from Taiwan', in K. Kumar and M. G. McLeod (eds) *Multinationals from Developing Countries*, Lexington, Massachusetts: D. C. Heath and Company.

Ting, W. L. (1980a) 'A Comparative Analysis of the Management Technology and Performance of Firms in Newly Industrialising Countries', *Columbia Journal of World Business*.

Ting, W. L. (1980b) 'NIC Multinationals and the Transfer of Technology', paper presented at the Economic Seminar in Greater Taipei Area, April.

Tolentino, P. E. E. (1987) 'The Global Shift in International Production: the Growth of Multinational Enterprises from the Developing Countries', Ph.D. dissertation, University of Reading.

Torre de la, J. (1974) 'Foreign Investment and Export Dependency', *Economic Development and Cultural Change* 23.

Trotsky, L. (1959) *The Russian Revolution*, Garden City, New York: Doubleday.

Trotsky, L. (1964) 'The Theory of Permanent Revolution', in Isaac Deutscher (ed.) *The Age of Permanent Revolution: A Trotsky Anthology*, New York: Dell Publishing.

Tsuda, M. *et al.* (1978) 'A Review of Foreign Investment in the Philippines', Mimeo, University of the Philippines Law Center and the UN Asian & Pacific Development Administration, Quezon City, the Philippines.

Tsurumi, Y. (1976) *The Japanese are Coming: The Multinational Spread of Japanese Firms*, Cambridge, Massachusetts: Ballinger.

UK Department of Industry (1974) *Census of Overseas Assets*, Wales, Her Majesty's Stationary Office.

UNCTAD (1984) 'Transnational Corporations and Science and Technology in the NIEO, in P. K. Ghosh (ed.) *Multinational Corporations and Third World Development*, Westport, Connecticut: Greenwood Press.

UNCTAD (1991) 'Transfer and Development of Technology in a Changing World Environment: The Challenges of the 1990s', Report by the UNCTAD Secretariat, TD/B/C6/153, 25 January.

UNCTC (1986) *Transnational Corporations in the International Semiconductor Industry*, New York: United Nations.

UNCTC (1987) *Transnational Corporations and Technology Transfer: Effects and Policy Issues*, New York: United Nations.

UNCTC (1990a) *Transnational Corporations and the Transfer of New and Emerging Technologies to Developing Countries*, New York: United Nations.

UNCTC (1990b) 'Regional Economic Integration and Transnational Corporations in the 1990s: Europe 1992, North America and Developing Countries', Current Studies no. 15, Series A.

UNCTC (1991) 'The Triad in Foreign Direct Investment', World Investment Report, ST/CTC/118.

UNCTC (1992) Transnational Corporation and Management Division, *World Investment Directory*, Asia and Pacific, New York: United Nations.

UNCTC (1993) Transnational Corporation and Management Division, *World Investment Directory, Latin America and Caribbean*, New York: United Nations.

Unger, K. (1988) 'Industrial Structure, Technical Change and Microeconomic Behaviour in LDCs', in G. Dosi, C. Freeman, R. Nelson, G. Silverberg and L. Soete (eds) *Technical Change and Economic Theory*, London: Pinter Publishers.

UNIDO Secretariat (1981) 'The Transnational Corporation as an Agent for Industrial Restructuring', in United Nations *World Industry in 1980*, New York: United Nations.

US Department of Commerce (1960) *US Business Investments in Foreign Countries: A Supplement to the Survey of Current Business*, Washington, DC: US Department of Commerce, Bureau of Economic Analysis.

Usher, A. P. (1929) *A History of Mechanical Inventions*, Cambridge, Massachusetts: Harvard University Press.

Vahlne, J. E. and Nordstrom, K. (1988) 'Choice of Market Channel in a Strategic Perspective', in N. Hood and J.-E. Vahlne (eds) *Strategies in Global Competition*, London: Croom Helm.

Valeriano, L. L. (1991) 'Other Asian Investors Follow Japanese to US Pouring Cash Into High-Technology Companies', *The Asian Wall Street Journal Weekly*, 14 January.

Vaupel, J. W. and Curhan, J. P. (1974) *The World's Multinational Enterprises*, Geneva: Centre for Education in International Management.

Vernon, R. (1966) 'International Investment and International Trade in the Product Cycle', *Quarterly Journal of Economics* 80(2), May.

Vernon, R. (1971) *Sovereignty at Bay*, Harmondsworth: Penguin Books.

Vernon, R. (1974) 'The Location of Economic Activity', in J. H. Dunning (ed.) *Economic Analysis and the Multinational Enterprise*, London: George Allen & Unwin.

Vernon, R. (1977) *Storm Over the Multinationals: The Real Issues*, London: Macmillan.

Vernon, R. (1979) 'The Product Cycle Hypothesis in the New International Environment', *Oxford Bulletin of Economics and Statistics* 41(4), November.

Villela, A. (1983) 'Multinationals from Brazil', in S. Lall (ed.) *The New Multinationals*, Chichester: John Wiley & Sons.

Virata, C. (1972) 'Foreign Investment in Developing Countries: The Philippines', in P. Drysdale (ed.) *Direct Foreign Investment in Asia and the Pacific*, Toronto: University of Toronto Press.

Walker, W. (1979) *Industrial Innovation and International Trading Performance*, Greenwich, Connecticut: JAI Press.

Wallace, C. D. (1989) 'Foreign Direct Investment in the Third World: US Corporations and Government Policy', corporate survey conducted by the Center for Strategic and International Studies, Washington, DC.

Wallerstein, I. (1979) *The Capitalist World Economy: Essays*, New York: Cambridge University Press.

Wang, J.-Y. (1990) 'Growth, Technology Transfer, and the Long-Run Theory of International Capital Movements', *Journal of International Capital Movements* 29: 255–71.

Warren, B. (1973) 'Imperialism and Capitalist Industrialisation', *New Left Review* 81, September–October.

Warren, B. (1980) *Imperialism: Pioneer of Capitalism*, London: Verso New Left Books.

Weinblatt, J. and Lipsey, R. E. (1980) 'A Model of Firms' Decisions to Export or Produce Abroad', Mimeo.

Wells, C. (1988) 'Brazilian Multinationals', *Columbia Journal of World Business*, Winter.

Wells, L. T., Jr (1973) 'Economic Man and Engineering Man', *Public Policy*, Summer.

Wells, L. T., Jr (1977) 'The Internationalization of Firms from Developing Countries', in T. Agmon and C. P. Kindleberger (eds) *Multinationals from Small Countries*, Cambridge, Massachusetts: MIT Press.

Wells, L. T., Jr (1978) 'Foreign Investment from the Third World: The Experience of Chinese Firms from Hong Kong', *Columbia Journal of World Business*, Spring.

Wells, L. T., Jr (1980) 'Multinationals from Latin American and Asian Developing Countries: How They Differ', paper submitted to the Harvard Graduate School of Business Administration, 10 November.

Wells, L. T., Jr (1981) 'Foreign Investors from the Third World', in K. Kumar and M. G. McLeod (eds) *Multinationals from Third World Countries*, Lexington, Massachusetts: D. C. Heath and Company.

Wells, L. T., Jr (1983) *Third World Multinationals*, Cambridge, Massachusetts: MIT Press.

Wells, L. T., Jr (1984) 'Multinationals from Asian Developing Countries', in R. W. Moxon, T. W. Toehl and J. F. Truitt (eds) *International Business Strategies in the Asia-Pacific Region, Research in International Business and Finance*, Vol. 4, part A, Greenwich, Connecticut: JAI Press.

Wells, L. T., Jr (1986) 'New and Old Multinationals: Competitors or Partners?', in K. M. Khan (ed.) *Multinationals of the South*, London: Pinter Publishers.

Wells, L. T., Jr (ed.) (1972) *The Product Life Cycle and International Trade*, Boston: Harvard University Press.

Westphal, L. E., Kim, L. and Dahlman, C. J. (1984) 'Reflections on Korea's Acquisition of Technological Capability', Development Research Department Discussion Paper, no. DRD77, World Bank, Washington, DC.

Westphal, L. E., Rhee, Y. W., Kim, L. and Amsden, A. (1984) 'Exports of Capital Goods and Related Services from the Republic of Korea', World Bank Staff Working Paper, no. 629, Washington, DC.

Westphal, L. E., Rhee, Y. W. and Pursell, G. (1979) 'Foreign Influences on Korean Industrial Development', *Oxford Bulletin of Economics and Statistics* 41(4), November.

Westphal, L. E., Rhee, Y. W. and Pursell, G. (1984) 'Sources of Technological Capability in South Korea', in M. Fransman and K. King (eds) *Technological Capability in the Third World*, New York: St Martin's Press.

White, E. (1981) 'The International Projection of Firms from Latin American Countries', in K. Kumar and M. G. McLeod (eds) *Multinationals from Developing Countries*, Lexington, Massachusetts: D. C. Heath and Company.

Wilkins, M. (1974) 'Multinational Enterprises', in H. Daems and H. Van der Wee (eds) *The Rise of Managerial Capitalism*, Louvain: Louvain University Press.

Williams, J. H. (1929) 'The Theory of International Trade Reconsidered', *The Economic Journal* 39(2).

Williamson, O. E. (1975) *Markets and Hierarchies: Analysis and Anti-Trust Implications*, New York: Free Press.

Williamson, O. E. (1979) 'Transaction Cost Economics: The Governance of Contractual Relations', *Journal of Law and Economics* 22, October.

Williamson, O. E. (1981) 'The Modern Corporation: Origins, Evolution, Attributes', *Journal of Economic Literature* 19, December.

Wilmot, P. D. (1977) 'Technology Transfer in a Multinational Firm', in M. J. Cetron and H. F. Davidson (eds) *Industrial Technology Transfer*, Leiden: Noordhoff.

Wilson, D. (1988) 'European Attraction: East Asian Manufacturers Set Up Factories in Britain', *Far Eastern Economic Review* 141(34), 25 August: 60–1.

Wolf, B. M. (1977) 'Industrial Diversification and Internationalisation: Some Empirical Evidence', *Journal of Industrial Economics* 26, December.

Wong, J. (1989) 'Transnational Corporations and Industrialization in Singapore', paper prepared for the International Seminar on Industrialization and Development: Focus on ASEAN, Bangkok, Thailand, 21–6 August.

Woog, D. (1986) 'South America Isn't Completely Barren', *High-Tech Marketing* 3(3), March.

World Bank (1989) 'Foreign Direct Investment from the Newly Industrialized Countries', Industry and Energy Department Working Paper, no. 22, December.

World Technology Patent Licensing Gazette 22, 1989, October.

Xiaoning, Z. (1989) 'Country Paper: People's Republic of China', paper prepared for ESCAP/UNCTC seminar on TNCs from developing Asian economies, Bangkok, 1–3 February.

Yang, D. J. (1989) 'Taiwan Goes for Broke on Semiconductors', *Business Week* 3119 (Industrial/Technology Edition), 14 August: 90, 92.

Yu, P. I. (1988) 'The Role of Government Policy and the Development of the Computer Industry in Korea', paper presented at the International Symposium on Technology Policy in the Americas, Stanford University, 1–3 December.

Notes

1 INTRODUCTION

1 In principle, the growth of outward investments in the services sector may be incorporated in the dynamic analysis of the industrial development of Third World MNEs in particular as well as in the formulation of the concept of stages of development in international production in general. However, apart from the construction sector where technological advantages play an important role, the growth of outward investments in other service sectors may also be explained by factors outside the framework of technological competence and therefore is excluded here. Readers interested in development patterns of Third World outward investment in general, in other service sectors such as trading and banking and finance, may refer to Annex 2. Some discussion of overseas investments by Philippine firms in the trading and banking and finance sectors is presented in Annex 1.

2 The concept of technological development or capacity adopted here is consistent with that of Stewart (1984) and Ranis (1984) which includes not only the capacity to create new technologies, but also the ability to choose, obtain and then locally develop, assimilate, adapt, improve or modify technology already known elsewhere and to diffuse the best practice, with appropriate modifications, across the economy. Technological capacity is therefore acquired only through human capital formation (Dahlman, 1984). Indigenous technological capacity pertains to the technological capability of the nation defined as an economic unit (Bienefeld, 1984).

5 THE CONCEPT OF AN INVESTMENT DEVELOPMENT CYCLE: SOME ECONOMETRIC TESTING

1 The relative stages of development approach to NOI is consistent with this formulation rather than $NOI_i/POP_i = \alpha + \beta(GNP_i/POP_i) + \gamma(GNP_i/POP_i) + \mu_i$ adopted by Dunning (1981a, 1981b) because in terms of cross-country variations between NOI and GNP, the population component cancels out in the latter relationship.

2 The regression was also undertaken for 1967–75, the period used by Dunning (1981a, 1981b) and the results show the following equation: $NOI_i = 3,760.92 + -8.4367 (GNP_i/POP_i) + .0022 (GNP_i/POP_i)^2$ with respective t-statistics of 0.9250, −2.2655 and 3.0742. The t-tests show that the constant term is not significantly different from zero while β and γ are significantly different from zero at the 5 per cent level. The coefficient γ is also significantly different from zero at the 1 per cent level.

6 THE ROLE OF FOREIGN TECHNOLOGY IN THE DEVELOPMENT OF THIRD WORLD MULTINATIONALS

1 The concept of technological development or capacity adopted here is consistent with that of Stewart (1984) and Ranis (1984) which includes not only the capacity to create new technologies, but also the ability to choose, obtain and then locally develop, assimilate, adapt, improve or modify technology already known elsewhere and to diffuse the best practice with appropriate modifications across the economy. Technological capacity is therefore acquired only through human capital formation (Datilman, 1984). Indigenous technological capacity pertains to the technological capability of the nation defined as an economic unit (Bienefeld, 1984).

2 The concept of a techno-economic paradigm was first advanced by Perez (1983) and is defined as a cluster of interrelated technical, organisational and managerial innovations that lead to the use of a new range of products and processes and significant changes in relative cost structure of all possible inputs to production. A key factor in each new paradigm that represents a particular input or set of inputs may be described by falling relative costs and universal availability. Changes in a techno-economic paradigm are more pervasive than Schumpeter's concept of radical innovations introduced at intervals of forty to sixty years as the effects of change in the technological system exert structural changes throughout the whole economy. The development of a new techno-economic paradigm is associated with a process of *economic* selection from a range of *technically* feasible combinations of innovations that emerge from the perceived limitations on further growth within the existing paradigm in terms of productivity, profitability and markets (Freeman, 1988a, 1988b).

3 The establishment of indigenous technological capacity is also important *inter alia* in order to reduce technological dependence on advanced countries which tends to involve loss of local control over many aspects of production, lessens a country's ability to bargain effectively on terms of technology transfer and exerts adverse effects in terms of self-respect and self-reliance (Stewart, 1984).

4 There are two broad categories of the learning mechanism according to Bell (1984): first, mechanisms where learning is a costless by-product of doing (i.e. learning by operating, learning by changing); and second, mechanisms where learning depends on the allocation of resources (i.e. system learning, and learning by training, by hiring and searching).

5 Katz (1984b) also refers to four major sets of micro and macro variables that affect both the rate and direction of technological research efforts of domestic enterprises in developing countries, namely: (i) strictly microeconomic circumstances or firm-specific factors emerging from the product and production technology originally available to the firm; (ii) signals emerging from the changing competitive atmosphere prevailing in the market(s) catered for by the firm; (iii) changes in macroeconomic parameters affecting firms in general, such as the rate of interest, the exchange rate, tarrifs, etc.; and (iv) new technical information emerging from the international technology frontier. Similarly, Stewart (1984) classifies four theoretical approaches to technical change, namely: (i) empirical case studies at a micro-level; (ii) the neoclassical approach; (iii) the political economy approach; and (iv) the institutional approach. Stewart (1984) stresses that the causality of these various factors on technical change is not unidirectional, but that the nature of the technical change itself influences the environment.

6 However, evidence gathered by Desai (1984) shows that out of thousands of Indian industrial producers, fewer than four thousand imported technology; the rest either bought, borrowed or stole technology within the country.

7 In the period before the Second World War the initial step towards industrial technology development in Japan was initiated by the government. Through the

establishment of government-owned factories in several sectors with the assistance of foreign educators, engineers and other experts, the initiation of compulsory education, investment in infrastructure for technological development and a selective trade and foreign investment regime, the government laid the foundations for economic growth and technology development in the post-war decades (Nagaoka, 1989).

8 In some cases, externalised forms of technology transfer are favoured, especially where host country restrictions and policies make technology transfer through foreign direct investment difficult. Imports of technology may take place through sales of machinery, or the training of entrepreneurs, technicians, workers and technical literature (UNCTC, 1987).

9 The outward direct investment of Singapore firms analysed in this book excludes those of developed country investors based in Singapore. These investments include only domestic companies that are either privately or government owned; firms that have transferred headquarters into Singapore from abroad, mostly owned by Chinese families; and ex-foreign companies (Pang and Komaran, 1985).

10 Singapore remains one of the three largest offshore assembly locations in the world, together with Malaysia and the Philippines. In addition, Singapore has also become an important testing and distribution centre for offshore assembly operations throughout Southeast Asia (UNCTC, 1986).

11 For the textile exporting firms, especially those involving Hong Kong and Japanese interests, the aim of the joint projects, apart from gaining access to complementary skills, may be to blur the national identity of participating textile exporters (Wells, 1978).

12 Mosel, Qusel and Vitelic operate both in Taiwan's science-based industrial park and in Silicon Valley, and each uses Taiwan capital. All of these firms have generated a great number of patents for the design and manufacture of very large-scale integrated circuits (Schive, 1988).

13 Samsung Semiconductor is perhaps the most advanced of the group, having been the first Korean firm to make a 64K DRAM chip without a foreign licence. The firm obtained design capabilities from Tristan Semiconductors, its subsidiary in the US Silicon Valley, and in addition hired engineers from competitors to start up prototype production. Since then, Samsung has accumulated production experience ranging from relatively mature products to more technology-intensive products such as higher memory dynamic random access memory (DRAM) semiconductors as well as microprocessors and certain other memory devices (UNCTC, 1986).

14 However, although South Korea ranks third in the world in the production of semiconductors, after Japan and the United States, it is a distant third, accounting for less than 7 per cent of the world market (Crane, 1990).

15 For example, besides research and development, the Korean Institute of Electronics Technology (KIET) provides contractual services to private industry in the areas of design, wafer fabrication and testing. These services are effected through contract manufacturing or contract testing arrangements which the institute concludes with foreign-based transnational corporations, by means of which the institute gains access to the product and process technologies and computerised testing know-how of foreign-based MNEs in return for a guarantee of a reliable and cost-competitive supply of integrated circuits. Among the foreign firms that have provided assistance to local firms in their high technology industrial park in Gumi are Hewlett-Packard, AMI and Fairchild, which have helped train Korean personnel in process technology, circuit design and layout and hardware and software design respectively. In addition, KIET has arrangements with RCA to provide circuit testing on contract to its semiconductor assembly plant in Taiwan and Malaysia in exchange for testing know-how. Other foreign-based MNEs with which KIET has conducted negotiations to supply various types of semiconductor design, manufacturing and/or testing

expertise include: Siemens, Philips, Western Digital Corporation, Terradyne, Data General, Digital Equipment (UNCTC, 1986).

16 Such recourse to imported know-how rather than efforts by local firms to develop the required technology has been said to weaken the technological capability of local manufacturing companies and hinder the mastery of industrial technologies by Brazilian consulting and design engineering firms. For example, the introduction of government policies to increase the participation of local engineering firms in investment projects in the 1970s actually resulted in their limited participation in the area of detail engineering as the basic technology was provided by foreign suppliers (Guimaraes, 1986).

17 Malaysia has taken nearly twenty-five years to build up a core of workers in the electrical equipment and electronics industry. They numbered approximately 100,000 at the end of the 1980s and are capable of learning, adopting, absorbing and improving on the ever higher level of technology in the industry (UNCTAD, 1991).

18 Apart from a high-powered regulatory and monitoring agency, the effectiveness of active technology strategies is fundamentally dependent on the existence of a dynamic entrepreneurial class, well-trained professionals, and a significant science and technology infrastructure (UNCTC, 1987).

19 National policies also interact with global economic forces in determining the pace of industrialisation and technological performance. Some of these global economic forces are financial and balance-of-payments constraints and the resolution of outstanding international trade policy issues, especially in the area of intellectual property, services and investment measures (UNCTAD, 1991).

7 THE DYNAMIC INTERDEPENDENCE BETWEEN INWARD AND OUTWARD INVESTMENT IN THE PHILIPPINES

1 It may be argued that although the Laurel-Langley Agreement terminated in principle in 1974, as recently as 1983 the provisions of the parity clause accorded to US investors under the previous colonial legislation, in particular the liberty to have a 100 per cent equity stake instead of the 40 per cent equity limit accorded foreign investors, were again re-enacted in a decree signed by the incumbent president, Ferdinand E. Marcos. However, unlike the previous legislation which discriminated against other non-US foreign investors, this decree applied the parity provision to all foreign investors for a period of one year in the hope of luring increased amounts of foreign investment to counteract the mounting external debt crisis.

2 By the late 1980s, the Progressive Car Manufacturing Program had been replaced by the Car Development Program. As with the earlier programme, the new one, composed of such Japanese MNEs as Toyota, Mitsubishi and Nissan, was entrusted with the responsibility of building a no-frills and affordable people's car (Tiglao, 1990a).

8 A PROFILE OF PHILIPPINE MULTINATIONALS

1 The firm is part of Ayala International Holdings which is registered in Liberia and whose overseas investments are also in the management of hotels in the USA and farms in Spain on behalf of Southeast Asian clients (Tiglao, 1991).

2 In principle, the growth of outward investments in the services sector may be incorporated in the dynamic analysis of the industrial development of Third World MNEs in particular as well as in the formulation of the concept of stages of development in international production in general. However, apart from the construction sector, where technological advantages play an important role, the growth of outward investments in other service sectors may also be explained by

other factors outside the framework of technological competence and therefore is excluded here. Readers interested in development patterns of Third World outward investment in general, in other service sectors such as trading and banking and finance, may refer to Annex 2. Some discussion of overseas investments by Philippine firms in the trading and banking and finance sectors is presented in Annex 1.

3 No data available for 1982, 1985 and 1986.

4 The Construction & Development Corporation of the Philippines ceased to exist in the mid-1980s.

5 Other examples of Philippine banks establishing operations in Hong Kong are as follows: The Manila Banking Corporation established Manila & Hong Kong Capital Corporation Limited, a wholly owned subsidiary in Hong Kong. The Bank of the Philippine Islands has a wholly owned deposit-taking financial outfit in Hong Kong, International Finance (HK) Limited, which the bank intends to convert to a full-service commercial bank as part of its planned international expansion. Allied Banking Corporation and Philippine Commercial International Bank also have wholly owned deposit-taking subsidiaries, Allied Capital Resources Limited and PCI Capital Asia Limited, in Hong Kong. Finally, the Equitable Banking Corporation has a 100 per cent interest in Equitable Finance Company (Hong Kong) Limited.

9 THE INDUSTRIAL DEVELOPMENT OF THIRD WORLD OUTWARD INVESTMENT

1 Through their empirical analysis of eleven OECD countries, Patel and Pavitt (1991) confirm earlier studies that home country-specific factors dominate over firm-specific factors in determining national technological performance. These home country-specific factors significantly stimulate the nature and extent of technological change through various inducement mechanisms and positive externalities that influence the effectiveness of local firm response to these stimuli.

2 Some Third World MNEs, such as Singapore and Hong Kong, are unique in Third World outward investment because of their nature as city states with small domestic markets and as a result they have pursued a different path of industrial development. The initial outward investments of firms from Hong Kong were not based on the extraction of overseas natural resources for domestic industrial expansion but rather on maintaining the international competitiveness of their labour-intensive manufactured export products in the face of rising labour costs, declining labour supply in home countries and protectionist pressures in their primary export markets in the developed countries. Hong Kong direct investments started in the late 1960s in the manufacturing sector and gained momentum in the late 1970s. These investments were in the textiles and clothing sector, prompted by the imposition of quotas by developed countries prior to the establishment of the Multi-Fibre Arrangement (Chen 1981, 1983b). Export-oriented firms from Singapore internationalised only in the 1980s since the limited supply of domestic labour in the 1970s was temporarily assuaged by the government's liberal worker policy that allowed the importation of foreign labour. As in Hong Kong, resource-based outward investments to ensure supply security of key natural resources have not been particularly important as Singapore does not have the large heavy industry base that Japan and other Asian NICs such as South Korea and Taiwan have. Rather, the main purpose of outward investments by Singapore firms was to find new markets and investment opportunities and to sell their technological expertise in industries in which they had advanced relative competence, especially in neighbouring countries where there is cultural and geographical proximity. In the case of relocation of labour-intensive industries in lower labour cost developing countries, which became more significant in the 1980s because of the very tight labour market, accelerated wage increases and

growing restrictions on the importation of unskilled foreign workers associated with concern about over-dependence on foreign labour, there is the motive to invest abroad to appropriate returns on accumulated managerial and technical expertise. Many large Singapore-owned firms are in mature industries such as food, beverages, basic metals, electrical and electronic products, plastics, textiles and garments, where opportunities for domestic market expansion are limited by the size of the Singapore and Malaysian markets. Growing protectionist policies in major export markets and the need to diversify, particularly in high technology sectors, are less important motives to engage in direct investment abroad (Pang and Komaran, 1985; Chia, 1989; Pang and Hill, 1991). In addition, although resource-based investments are important for Taiwan outward investors to access essential foreign-based raw materials, their initial outward investments were in a cement plant in Peninsular Malaysia in 1959 and in textile manufacturing in the 1960s (Ting and Schive, 1981). For Indian MNEs, the beginning of direct investment in the post-independence period may be associated with the establishment of a textile mill in Ethiopia in 1955 by the Birla Group. Investments to extract natural resources have not been very important for Indian MNEs even in recent years (Bhatt and Dalal, n. d. ; Morris, 1987). However, the case of South Korea follows the typical example of Japan and other more traditional investors in the concentration of initial outward investments in the resource-based sector. Although Korean trading companies take the lead in overseas investments, for example, the purchase by the Korean Traders Association of a commercial building in New York City for its US branch office in 1959, overseas investments for resource development accounted for the largest share of outward investments in terms of amount (Jo, 1981; Korea Institute of Foreign Investment, 1989). Outward foreign investment by Korean firms in the early stages was limited mainly to forestry development such as that in timber exploitation in Malaysia in 1963. Similarly, P. T. Korea Development Company Limited invested in a timber exploitation project in Indonesia in 1968 for the purpose of providing logs to the Korean plywood industry, then a major export industry. Until 1975, Korea's outward investments were mainly in the forestry and fishery sectors. The first overseas manufacturing investment of Korean firms occurred in 1973, to manufacture food seasonings in Indonesia. After 1974, as protectionism began to grow in industrial countries in the aftermath of the first oil shock, many Korean manufacturing firms and general trading companies established trade agencies in various countries to promote exports to these countries. By 1982, the share of these investments accounted for 13.6 per cent of total outward investments. By the late 1970s, mining began to emerge as another major area of outward investment as resource nationalism strengthened in many countries and as the need for a stable raw material supply became paramount for sustained development of domestic industries. Investments in transportation and warehousing activities and construction also started to emerge in importance (Koo, 1984). Hence, unlike other Asian NICs like Hong Kong and Singapore, Korea's outward investments can be explained mainly by its need to acquire natural resources apart from its need to find an expanded market for their goods (Kwag, 1987). Even in the late 1980s, investments in the mining and manufacturing sectors accounted for about 90 per cent of the total overseas investments of Korean firms. Mining and oil exploration investments alone in 1987 accounted for 50 per cent of total investments, reflecting the ongoing concern with securing a stable supply of major raw materials for the Korean industrial sector.

3 Although some attempts are made to describe differences in the general characteristics among firms and countries belonging to a particular group, for example, the differences in the general characteristics *within* the group of higher-income newly industrialising countries, as opposed to the lower-income developing countries, there is no attempt at this stage to describe more specific or detailed idiosyncratic

characteristics or deviations from the general characteristics of firms and countries belonging to a particular group. For purposes of the present analysis, the emphasis is macroeconomic oriented in as much as comparison is made *between* groups of firms, i.e. between MNEs from the NICs on the one hand and the lower-income developing countries, as opposed to a comparison *within* these groups.

4 This qualification of ultimate beneficial ownership of Third World outward investment is particularly significant for Singapore and Hong Kong which are used as a base for indirect outward investment by firms from both developed and developing countries, especially Hong Kong which is an international financial centre. Sometimes the investments are undertaken by subsidiaries of foreign-based MNEs which operate in these countries. Estimates by Chen (1983a, b) show that foreign direct investment by Hong Kong Chinese firms (including joint ventures with non-Chinese partners) in manufacturing would probably account, by a very rough estimate, for 33 to 44 per cent of Hong Kong's total overseas investment (including investment in both manufacturing and services). For purposes of the analysis, only locally based firms are included. In the case of Singapore, a study by Aggarwal (1987) shows that twelve of seventy-two Singapore-based industrial MNEs listed on the Stock Exchange of Singapore are foreign-owned firms. The studies here cover only outward investment by domestic companies that are either privately or government owned; companies that have transferred headquarters into Singapore from abroad, mostly owned by Chinese families; and ex-foreign companies. In the case of South Korea, the issue of ultimate beneficial ownership is not so significant since the incidence of outward investments by subsidiaries of foreign companies in Korea is considered rare. For example, data from Koo (1984) show that in 1982 there were only thirteen cases of such investment, of which four were localised Korean firms, five were Korean companies that received marginal investment from the International Finance Corporation and therefore not foreign, and two were trade agencies established by the subsidiaries of foreign companies to promote their exports. In the case of Brazil, there is a clear predominance of domestic firms among the sixty-three largest Brazilian outward investors in the manufacturing sector. Only five of these investors are foreign subsidiaries in Brazil, accounting for only 4 per cent of the total amount invested (Guimaraes, 1986).

5 As stated in the last chapter, the growth of outward investments in the services sector, in principle, may be incorporated in the dynamic analysis of the industrial development of Third World MNEs in particular, as well as in the formulation of the concept of stages of development in international production in general. However, apart from the construction sector, where technological advantages play an important role, the growth of outward investments in other service sectors may also be explained by factors outside the framework of technological competence and therefore is excluded here. Readers interested in development patterns of Third World outward investment in other service sectors such as trading and banking and finance may refer to Annexes 1 and 2.

6 However, as shown above, Singapore has outward investments in large-scale, downstream processing in comparison to Hong Kong which has mainly final assembly operations.

7 The growth of the Brazilian capital goods industry is largely attributed to Petrobrás, one of the biggest clients of the industry.

8 It is not thoroughly clear in the literature whether these resource-based investments by Hong Kong firms aim to supply domestic industry with essential raw materials. On the one hand, Chen (1981) states that there is no necessary production relationship between the Hong Kong parent firm and the overseas subsidiaries engaged in timber extraction and processing in Indonesia, while Pang and Komaran (1985) state that Hong Kong investments in plastics and wood in Southeast Asia provide raw materials for production purposes in Hong Kong.

9 Although Indian MNEs in the extractive sector have been very few, the few investments that firms have undertaken in the extraction of natural resources such as magnesite, zinc and lead ore are also downstream processing operations.

10 Unicord, an aggressive, young Thai company, now controls one-fifth of the world market for canned tuna (*The Economist*, 22 December 1990).

11 Although outward investments by the NICs in the developed countries, mainly in the USA and the EC, are initially assembly operations of intermediate or semi-finished products imported from the home country, in recent years the imposition by the developed host countries of anti-dumping duties, not only on finished products but also on semi-finished products and product components, has forced the increase in local content or added value of their production. However, other firms have established or plan to establish plant in Southeast Asian countries or those neighbouring the USA and the EC to maintain competitiveness. For example, Gold Star has established a plant in Mexico to produce colour TV chassis for its Huntsville plant, utilising Mexico's advantage as a duty-free country (Korea Institute for Foreign Investment, 1989).

12 The capital goods production of Brazil and Argentina has enabled local entrepreneurs to produce small batches of goods in relatively skill-intensive industries which are then sold competitively in both local and foreign markets, even against products produced by domestic subsidiaries of large MNEs. The consistent strategy in the development of the local capital goods sector in Brazil and Argentina is reflected in the very high share of capital goods in their technology exports. In addition, the advanced development of the indigenous capital goods sector of India, as a result of the active policies of the Indian government to encourage domestic technology creation by restricting the use of foreign technologies, has enabled Indian firms to have a high degree of embodied and disembodied forms of technological innovation. By contrast, the low capital goods capability of Hong Kong and Mexico is shown in the importation of machinery and equipment by Hong Kong MNEs and, with the exception of those in the glass and cement sectors, by the importation of embodied technology in the form of capital goods for Mexican firms. By contrast, although South Korea has a high ratio of imports of capital goods to gross domestic investment, the ratio of firms' capital goods exports to their capital goods imports is increasing fast, reflecting their rapid development, high degree of specialisation and reliance on foreign technology to acquire additional capabilities and new technologies (Chen, 1983a, b; Katz, 1984a, b; Dahlman and Sercovich, 1984; Lall, 1985a).

13 San Miguel beer is one of the most popular brands in Hong Kong.

14 Similar motives explain the establishment in 1990 of Tilecera Incorporated, a US $45 million joint venture between Siam Cement Group, Thailand's largest industrial conglomerate, and Italian-based Jiacobazzi Stilgres in Tennessee, USA, to produce ceramic tiles. While proximity to the market will enable the firm to discern consumer wants, it will also be able to overcome protectionist barriers (*The Economist*, 22 December 1990).

15 Within the Asian NICs, export-oriented investments have always been most important for MNEs from Hong Kong. The fundamental motives for the outward investments of Hong Kong firms are defensive, prompted by the need to circumvent tariff barriers imposed on their export products and to maintain the competitiveness of these products in their established markets, mainly in the developed countries, through relocation of production in lower-cost host developing countries (Wells, 1978; Chen, 1981).

16 It should be noted that the rate of expansion overseas of Brazilian MNEs accelerated rapidly between 1984 and 1988, when Brazil started accumulating large trade surpluses. A number of fiscal and financial incentives to make exports a stimulus for industrial growth were created in the late 1960s with the aim of increasing exports

of manufactured goods and foreign exchange revenues. From 1970 to 1982, Brazilian exports grew at an annual rate of 16.1 per cent from US $2.7 billion to US $20.2 billion, while manufactured exports grew at an even faster rate of 21 per cent to reach an overall share of total exports of 50 per cent by 1982 (Guimaraes, 1986). However, trade surpluses may have also occurred in response to policies of domestic restraint that curtailed imports through exchange rate and price control policies as well as through discretionary applications of tariffs. In any case, the rather limited outward investments of Brazilian firms in the late 1980s in the manufacturing sector were propelled by exports and have been linked to the sales of aircraft, electricity meters, pencils and iron ore (Guimaraes, 1986; Wells, 1988) unlike the more labour-intensive export products of East Asian NICs that require an outward relocation largely because of rising labour costs or diminishing labour supply in the home country or protectionism in export markets; therefore overseas investments in the manufacturing sector are much more significant. The rather limited manufacturing overseas investments by a few Brazilian firms that are neither the largest nor the most important in the industry are not indicative of an emerging trend, unlike the foreign investment of Brazilian financial institutions and the state-owned oil firm, which present a definite and meaningful trend. Many Brazilian manufacturing firms are in a weak and subordinate position in their domestic markets compared to foreign-based multinational enterprises, which does not favour a massive expansion abroad (Guimaraes, 1986). Nevertheless, despite the differing importance of outward direct investments in the manufacturing sector and concentration in different sectors of manufacturing activity, Brazilian and East Asian MNEs share a common goal, which emerges from the need to preserve existing markets as well as to open new ones based on several years of export experience. Their objectives for export-oriented investments stem from the need to overcome trade barriers by having a springboard for third country exports and, as a result, reduce the volume of exports emerging from the parent company or home country.

17 For example, data provided by Wells (1984) suggest that at least three Argentine firms had a network of subsidiaries abroad before the Second World War, namely Alpargatas which was manufacturing shoes and textiles in Brazil and Uruguay by 1900; Siam di Tella which was manufacturing in Brazil, Uruguay and Chile by 1928, and Bunge y Born which established operations in Brazil in 1929. By contrast, there were few, if any, locally owned and managed companies from Asian developing countries that manufactured outside national borders prior to 1950.

18 The continued access to major export markets is particularly important to the Asian NICs since these countries have lost their preferential, or GSP, status and exhausted their quotas in footwear, apparel and electronic products, which constitute their major export products. Apart from their previous preferential trading status, accorded through GSP, the Asian NICs have no other preferential trading access to the developed countries (Chia, 1989). In addition, the issues of a diminishing labour supply in conjunction with rising labour costs also emerged as important factors behind the growth of offshore, labour-intensive, export-oriented investments of firms from Taiwan and Hong Kong. In the case of Hong Kong, both the limited supply and high cost of labour are associated with GDP growth and inflationary pressures, and with the exodus of skilled and semi-skilled labour from Hong Kong because of future political uncertainty. Emigration of Hong Kong individuals is proceeding at a rate of 62,000 per year, representing 1 per cent of the total population and more than 10 per cent of the middle class. In the case of firms from South Korea, the issues of domestic labour instability, currency fluctuations and market limitations in the home country are more important. Another equally important factors in relation to outward FDI is rising property prices in the home country (Lee, 1989; World Bank, 1989; *The Economist*, 8 December 1990).

19 Government support in respect of FDI in labour-intensive industries is most

comprehensive in South Korea of all the Asian NICs. Since 1986, the South Korean government has, through the provision of financial and tax incentives, supported the restructuring of the domestic industrial base through the transfer of the no longer competitive labour-intensive industries to less developed countries that have a supply of abundant and low-cost labour. In particular, the government has initiated incentives which include deregulation, financial and tax benefits and the provision of information on overseas business activities to Korean investors. Apart from simplifying the procedures for approval and screening, the terms and conditions for overseas investment finance managed by the Export-Import Bank have also been improved since March 1988. In 1989, overseas investment project loans granted by the Export-Import Bank reached $150 million. Approved projects are eligible for Export-Import Bank financing of 80–90 per cent of the total investment value, regardless of the size of the firm, and for Overseas Resources Development Funds and dividend income tax exemptions for investments in petroleum or natural resources. In 1988, a total of US $126 billion was provided to small and medium Korean businesses engaged in labour-intensive industries in addition to US $300 million in restructuring funds to those upgrading their domestic activities towards higher value-added and technologically-intensive industries. Moreover, foreign investors are allowed to deduct corporate taxes paid abroad, and can claim up to 20 per cent of the investment in the event of losses being incurred. An Overseas Investment Insurance Program covers Korean investors abroad against loss of principal, dividends, or interest revenues due to non-commercial risks. In the case of Taiwan, approved investments are eligible for Export-Import Bank credit and insurance and incentives (Korea Institute for Foreign Investment, 1989; World Bank, 1989).

20 However, in certain cases, large firms in Taiwan and South Korea have also located in higher cost centres in their major export markets in the developed countries (mainly in the USA) to ensure security of market access. In most cases, although this involves the transfer of the final assembly production stage to developed countries with high labour costs, the final products remain competitive because the parent companies provide the raw materials and intermediate products at reasonable costs from the home country (Korea Institute of Foreign Investment, 1989). This trend of investments in the developed countries is expected to continue with the establishment of the US–Canada free trade area and the emergence of the integrated European Common Market (Chia, 1989). Given the current restructuring of international economic order towards multilateral systems and regional blocs, developing countries will face increasing protectionism which will force them to invest further in developed countries to overcome such trade barriers.

21 The CBI programme excluded export of textiles, apparel, petroleum and petroleum products, and exports of sugar were subject to quotas (*Business Latin America*, 3 December 1990; *The Economist*, 25 August 1990).

22 The NICs and some lower-income developing countries seek to dynamically change their comparative advantage towards the local design, production and marketing of products that embody more advanced technology. The NICs are interested in a number of emerging technologies. For example, Taiwan has placed special emphasis on investments relating to machine manufacturing, computer and communications industries, electronics and precision instruments. On the other hand, Singapore has chosen to be involved in biotechnology, micro-electronics, robotics, information technology, communications technology, laser-technology and electro-optics. This is partly seen in the establishment of high-tech enclaves or 'technopolises' (Andrianov, 1990) in a number of Asian NICs and developing countries, similar to Silicon Valley in the USA, to foster research and investment in the production of more technologically intensive products. Examples are the Hsin-chu Science-based Industrial Park in Taiwan whose aim is to increase the technological competitiveness

of local firms in the electronics industry, biotechnology and fine chemicals. Another example is Taeduk Research Park, the South Korean counterpart of the Japanese technopolis at Tsukuba, where basic R&D projects are related to the production of high-tech goods, new technologies and materials and which is also engaged in basic research. In Singapore, a technological park was also established in 1981 for the development of information technology. Two technological parks have also been established in Hong Kong, specialising in the development of new generations of electronic hardware. Venture companies have also become the most widespread means by which new high-tech products are being developed in Hong Kong. In Malaysia, the government appropriated 2 million ringgits for the establishment of a technopark at Tamon-Tin-Razak, where main lines of research are on microelectronics, technical rubber, wood working and biotechnology (Andrianov, 1990). The governments and private sectors of South Korea and Singapore have also established science towns to promote technical research. Since Taiwan has few conglomerate firms with the financial capacity to spend much on R&D, the government, in addition, established the Industrial Technology Research Institute in 1973 to develop new ideas, or license them from abroad and to transfer the results to the private sector, primarily to small and medium businesses. The government of Taiwan is also providing start-up capital for small firms as well as exemption from tariff payments on certain equipment. The South Korean government has imposed an import ban on foreign-made electronics since the early 1980s, provided substantial amounts of low interest capital for companies and made heavy investments in electronics R&D with the formation of the Electronics and Telecommunications Research Institute (ETRI). The major projects of ETRI are telecommunications, semiconductors and computer architecture. In addition to ETRI, DACOM was formed by a number of privately owned electronics firms and controlled by the Ministry of Communications to handle initiatives in data communications such as value-added networks and electronic mail. The Korea Telecommunication Authority (KTA) plans to invest 3 trillion won (US $4.5 billion) in R&D, representing 6 per cent of total expected revenues by 2001. R&D allotment for 1990 is 133 billion won or 3 per cent of total expected revenues for 1990, which will gradually increase to 5 per cent in 1996. R&D expenditure by large Korean industrial conglomerates has also grown substantially. In 1989, out of the total amount of capital investments by Samsung of 8.1 trillion won, 800 billion went into R&D; for Hyundai, the equivalent figures are 1.9 trillion and 400 billion; for Lucky Goldstar, 1.5 trillion and 400 billion; for Daewoo, 1.3 trillion and 200 billion; for Sunkyong, 914 billion and 21 billion; for Ssangyong, 400 billion and 40 billion; for Hsoyung, 385 billion and 65 billion; and for Kyo, 525 billion and 45 billion (Andrianov, 1990). Investments by domestic semiconductor companies in new facilities and R&D, largely for megabit DRAM production, reached an estimated 1.4 trillion won in 1989 (Crane, 1990). In Singapore, the government indirectly encourages the development of indigenous technological capacity by encouraging foreign firms to upgrade the technological basis of their operations. Foreign firms are therefore encouraged to play a tutorial role for local firms. In addition, the Singaporean government is also offering tax incentives, venture capital funds and low interest loans to local firms. The emergence and probable future development of the domestic semiconductor industry in South Korea, Taiwan and Hong Kong were spurred by the successful integration of experience by indigenous firms, partly obtained from the presence of offshore assembly plant of US, Japanese and European semiconductor manufacturers in these countries. However, other factors behind the growth of competitive domestic producers of semiconductors are growth in local university training in the scientific and engineering fields and the repatriation of technical talent. In Malaysia, the government established the Technology Park in Selangor where research and development activities receive the highest attention. The rapid growth of Brazil's computer industry in the 1980s was largely due to government limitations on inward foreign direct investment as embodied in the National Informatics Law of 1984. The Law offers Brazilian firms major incentives to develop technological self-reliance (Berney, 1985; Colson, 1985; Davis and Hatano, 1985; Brody, 1986; Cummings, 1987; Kotkin, 1987; ESCAP/UNCTC, 1988; Karlin, 1988; Hock, 1988a; Evans and Tigre, 1989; Henderson, 1989; *Korea Business World*, April 1990).

23 Brazil and Taiwan are among the ten largest computer makers in the world. Taiwan in particular produced 10 per cent of the world's personal computers in 1988, of which 2 per cent is accounted for by Acer Incorporated, Taiwan's top computer manufacturer and the tenth largest producer of IBM-compatible personal computers in the world. The high technological capability of Taiwan in the production of computers is seen in its technology exports in this area. For example, Mitac International (Taiwan) announced in September 1989 that through a licensing agreement it was transferring plant technology for the manufacture of two of its existing personal computer models to Multipolar, an Indonesian firm. Brazil was also one of the earliest countries to commit itself to a fully digital telecommunications system. Electronics products such as satellite receivers and television sets are the second largest export earner of Hong Kong, amounting to HK $22.3 billion in 1986. There are currently seven local firms in the semiconductor industry of Hong Kong. Six of these firms have fabrication capabilities in addition to their assembly and testing functions. South Korea is also a major producer, exporter and foreign direct investor in the area of electrical and electronic products. These products are expected to become the country's most important source of export earnings in the 1990s. Auto parts and components are another area in which South Korea has achieved international competitiveness. Other industrial sectors for which South Korean *chaebols* or conglomerates aim to become fiercer competitors are aerospace, semiconductors, computers and communications (Samsung); office and factory-automation gear or robotics (Daewoo); and memory chips (Lucky-GoldStar Group). Their technological capability in the production of memory chips is no more than one year behind the leading Japanese makers. Overall, it is seen that the technological development of South Korea has been much faster than Japan during an earlier period when it was at a similar stage of development. Korean companies have already caught up in the production of state-of-the-art semiconductor chips like the 256K NMOS DRAM. Their country is currently the third largest producer of semiconductor chips, after Japan and the USA. India is a major exporter of computer software, an area where there is a considerable indigenous technology. The first ever fully electronic private telephone exchange (PABX) was designed, produced and marketed by a small Venezuelan firm (Woog, 1986; Clifford, 1988; Liu, 1988; Moore, 1988; Perez, 1988; Hock, 1988b; Henderson, 1989; Nakarmi and Neff, 1989; Ohmae 1989).

24 Non-Japanese Asian investors, particularly from Singapore, began investments in Silicon Valley in the early 1980s and from the late 1980s capital inflow began to accelerate. While the Koreans are just beginning to invest in start-up companies, the Taiwanese concerns have become the most aggressive of the new investors (Valeriano, 1991).

25 Statement of Lip-bu Tan, general partner of the Walden Group of Venture Capital Funds, San Francisco, USA, as stated by Lourdes Lee Valeriano, 'Other Asian Investors Follow Japanese to US Pouring Cash Into High-Technology Companies', *The Asian Wall Street Journal Weekly*, 14 January 1991.

26 Other motives important for investments by high-technology companies are to seek distribution channels in the USA and to utilise excess manufacturing capacity in the home country through new manufacturing arrangements. For example, the acquisition of Altos Computer Systems by the Acer Group based in Taiwan was motivated as much by the desire to obtain a distribution network in the USA as by

the desire for technology. The purchase by Hualon Microelectronics Corporation, also Taiwanese, of 10 per cent of Seeq Technology, Incorporated, a San Jose based computer chip maker, will enable Hualon to manufacture Seeq chips in Taiwan and therefore help Hualon to use more of its capacity as a foundry for semiconductor chips.

27 The advantages of such joint ventures and other forms of strategic alliance include access to advanced and rapidly changing technology and competitive markets, creation of new products, restructuring of new industries and the opportunity to spread the costs and risks of new product development. These can take several forms such as joint production, joint technology development, production–marketing linkage, R&D production linkage or cross-licensing. In the late 1970s, on the basis of foreign technology licensing agreements with both American and Japanese companies, four semiconductor firms emerged as divisions of the South Korean electronics and industrial conglomerates: Samsung, GoldStar, Hyundai and Daewoo. Recently, the US firm Hewlett-Packard (HP) granted a licence to the South Korean firm Samsung Electronics Company to develop, manufacture and resell chip sets, workstations and other computers using HP's Reduced-Instruction Set Computing (RISC)-based Precision Architecture. The same firm has also been granted the licence to use HP's sub-micron complementary metal oxide semiconductor process technology. The Korean firm Sanchok Industrial is able to produce silicon wafers under a joint venture agreement with the US producer Monsanto. Some other examples of joint ventures and strategic alliances of Korean firms as well as other Third World firms have already been discussed in Chapter 6. By and large, the most common form of foreign technology transfer is supplier contracts with original equipment manufacturers their their (OEM). In the steel industry, the main incentive for the joint venture between the Korean firm Pohang Iron & Steel Company and US Steel Corporation was essentially to overcome import restrictions on Pohang's export volume to the USA as well as to improve the sluggish productivity and solve the labour disputes of US Steel (Lenae, 1985; Juilland, 1986; Perlmutter and Heenan, 1986; UNCTC, 1986; Crawford, 1987; Harrigan, 1987; Contractor and Lorange, 1988; Henderson, 1989; Korea Institute for Foreign Investment, 1989; *World Technology Patent Licensing Gazette*, 1989).

28 A similar view was put forward by Dahlman and Sercovich (1984), Sercovitch (1984), Westphal *et al.* (1984) and Pang and Hill (1991) who suggested that exports of technology and technical services by developing countries reflect the changes in comparative advantage that accompany industrialisation and may also dynamically change it by broadening and deepening technological capability.

29 For example, although the Asian NICs such as Hong Kong and Taiwan are increasingly gaining the capacity to become semiconductor producers, their firms have concentrated on the less technologically intensive memory chips and have not been able to produce the most technologically advanced integrated circuits (Henderson, 1989). Even in the case of South Korea, firms still lack the design capability to compete profitably in the most technology intensive sectors of the semiconductor industry (Crane, 1990).

10 THE GEOGRAPHICAL DEVELOPMENT OF THIRD WORLD OUTWARD INVESTMENT

1 In order to examine the geographical distribution of Third World outward investment, the data examined in this chapter is based on a home country perspective, except for Hong Kong where host country data is used because of the absence of official home country data, associated with its *laissez-faire* policy on outward investment. Both home and host country data must be interpreted with caution and not taken as indicative of the actual value of outward direct investments, for several

reasons. First, there are differences in methods of data collection, accounting conventions, and FDI definitions across countries. For example, it is believed that an overwhelming proportion of outward FDI statistics for Pakistan are in the financial sector and in particular in portfolio holdings in the Middle East. Similarly, Brazilian data on outward direct investment include investments in the banking sector but exclude oil exploration by Petrobrás, the national oil corporation, and other capital invested overseas is registered as expenses. In any case, most of Brazilian outward investments are in the financial sector as a result of the objective of the Brazilian banking system to attract resources in the international capital markets, and their expansion in Latin America is associated with Brazilian export expansion in the region (Guimaraes, 1986). Second, in the case of the Asian city states, an important factor to consider in the interpretation of outward direct investment data, as mentioned in the last chapter, is the issue of ultimate beneficial ownership, with the probable inclusion in the data of reinvestments by foreign affiliates in these countries. Being international financial centres, these countries host large business communities and are used as a base by many foreign firms from both developed and developing countries for their regional or international investments. For example, indirect investments in Hong Kong by the United Kingdom and Taiwan have been particularly notable. The issue of ultimate beneficial ownership is complicated by the fact that some Third World MNEs, especially those from Singapore, are often the primary implementors and managers of nominally non-Singaporean FDI in developing countries. Third, data in some cases may be on government approved outflows and registered investments, and not actual flows. This is especially true in countries such as the Philippines, Taiwan (until 1986) and Peru which maintain controls on outflows of FDI and therefore firms may not report outward investment to escape host country restrictions on foreign exchange and taxes. Chapter 8 has shown that the data on outward investment from the Philippines does not reflect the unofficial outward FDI undertaken by Philippine firms that can be detected in host country data on FDI inflows. Such unofficial outward FDI activities that are not reflected in approved data provided by the government are even more significant in Taiwan as detected by aggregate balance of payments data and host country data on FDI inflows (see World Bank, 1989).

2 A study of a sample of thirty-one MNEs from the Asian NICs shows that the choice of investment location is driven largely by considerations of geographical proximity associated with the availability of information on business conditions and lower costs of production. However, within the Asian NICs, ethnic, religious and cultural ties are also important, primarily to MNEs from Hong Kong, Taiwan and Singapore whose outward investments in developing countries are almost exclusively in Asia because of the presence of an ethnic Chinese network. Many Hong Kong and Taiwan firms consider investment opportunities outside Asia to be much riskier because of the absence of the support system provided by their Chinese connections and the lack of familiarity with business conditions. Hong Kong MNEs dominate in China's Guangdong Province as in other parts of Asia and elsewhere because ethnic and family ties have facilitated FDI negotiations and implementation. Such reliance on ethnic Chinese connections may be a reflection of strong business preference, the low stage of organisational development, the limited capability for information collection, their regional concentration or their small size. By comparison, MNEs from Korea that lack an ethnic network are more concerned with commercial considerations such as relative cost and quality of labour, market potential and political stability, assisted by the information provided by the global trading and investment network of their *chaebol*. Consequently, Korean outward investments are also significant in developing countries outside Asia. In some cases their investments within Asia, such as those in Malaysia and Indonesia, have

circumvented prejudices that offshore Chinese networks have engendered against ethnic Chinese (World Bank, 1989).

3 For example, a field survey of forty-four outward investment projects by Asian NICs undertaken in 1989 showed that 93 per cent and 100 per cent of outward investment projects by Hong Kong and Singapore respectively exported more than 50 per cent of output. By comparison, only 60 per cent and 75 per cent of outward investment projects by Taiwan and Korea respectively exported a majority of their output (World Bank, 1989).

4 In 1989, 65 per cent of the total number of Chinese MNEs were located in developing countries (Xiaoning, 1989).

5 By contrast, Indian subsidiaries abroad, i.e. those firms that are 100 per cent held by firms from India, and are mostly in trade, investments, agency, business, consultancy services or that act as holding companies of other companies, are more evenly distributed between developed and developing countries. In any case, FDI from India in the form of joint ventures did not constitute more than 2 per cent of all FDI into Malaysia, Indonesia, Thailand, Nigeria, Mauritius and Singapore in the 1970s, and even in the case of Kenya, its share was only about 5 per cent. Only in Sri Lanka did Indian direct investments account for approximately 32 per cent of all FDI in the 1970s (Morris, 1987).

6 According to host country data, Singapore is the eighth largest investor in Indonesia, the fourth largest in China and the largest foreign investor in Malaysia where its investments were valued at S \$1.1 billion as at end 1987. These investments in Malaysia are mainly in textiles and clothing, electronics, food manufacturing, plastics, rubber products and fabricated metal products. Singapore is also a major investor in Thailand.

7 Data from the database of the Transnational Corporations and Management Division of the United Nations suggest that Hong Kong has been the most important investor in China since the open door policy was introduced in 1979, with total actual investments of US \$2.1 billion and total employment of approximately 2 million by 1988. Hong Kong is the second largest foreign direct investor in Indonesia after Japan, the third largest investor in Taiwan (after Japan and the USA), Singapore and Thailand, and the fifth largest investor in Malaysia. Hong Kong is also responsible for a considerable proportion of total FDI inflows in Sri Lanka and Bangladesh. The investments in China are predominantly in assembly subcontracting.

8 The presence of an ethnic Chinese network in Mauritius explains the surge of Hong Kong investments in the textile sector in recent years, resulting in the export boom of textiles from Mauritius (World Bank, 1989).

9 For example, although China's political upheaval in the summer of 1989 did not disrupt production in the Chinese factories of the Hong Kong based Kader Industrial Company, the maker of Cabbage Patch dolls, orders from the USA sharply declined with the mere threat of interrupted deliveries. As a result, Kader opened two joint venture companies in Thailand in 1988 to satisfy buyers in the USA (Jiang, 1990).

10 However, when Hong Kong, the largest Third World outward investor, is excluded from the analysis of geographical distribution of Asian outward investment, the share of neighbouring countries or ethnically related territories in the total outward stock of Asia declines to 41 per cent, reflecting no significantly large difference from that of Latin America. This finding reflects the dominant role of Hong Kong as a Third World outward investor and its concentration in neighbouring Asian countries.

11 In some cases, these investments have exceeded Japanese investment and are expanding faster.

12 This is so in the case of India, where outward investments are also significant in ethnically related territories in Africa as well as those in Asia.

13 It should be re-emphasised, however, that the Brazilian data exclude investments by Petrobrás, the state-owned oil company, which are channelled through the United States (Guimaraes, 1986).

14 Although Brazil is far from being export oriented in the East Asian mould, Brazilian outward investment has generally grown concurrently with the export drive. The rate of overseas expansion accelerated rapidly after 1984, when Brazil started accumulating large trade surpluses (Wells, 1988). Some of the distinctive features *vis-à-vis* Asian MNEs, including domestic policies implemented to achieve export growth, have been analysed in the last chapter. See note 16.

15 For example, Nansen, which exports between 10 per cent and 20 per cent of output to Latin America, established a plant in Medellin, Colombia, to serve as a spring-board for Third World country sales. Moreover, many capital goods leaders in Brazil have been forging links with firms in neighbouring Argentina, which also suffers from idle capacity, to gain economies of scale through joint production for the Latin American and African markets. For example, aircraft maker Embraer and the Argentine firm Fama formally agreed in May 1987 jointly to produce a nineteen-seat turboprop commuter liner, the CBA–123. In addition, Petrobrás, the national oil corporation, and Argentina's federally owned oil company, Petroliferos Fiscales, are studying joint oil exploration and pipeline construction in Argentina. Villares, a Brazilian manufacturer of elevators and other heavy equipment, also has protocol agreements with Argentina's Pescarmona jointly to manufacture for export to the rest of Latin America (Wells, 1988).

16 Within the Asian NICs, investments in 'tariff factories' in the developed countries, especially in the USA, are more actively pursued by Taiwan and South Korea than by Hong Kong and Singapore. Nevertheless, Hong Kong FDI in the USA and other developed countries is also considerable and is expected to become more significant with the uncertain future political situation in Hong Kong. Outward investments in the manufacturing sector are largely geared to maintaining access to markets and brand names as well as to upgrading technology. However, service sector invest-ments, particularly in banking and finance, hotels and trading and real estate, are also increasingly significant areas of investment for Hong Kong MNEs in the developed countries. In the case of South Korea, concentration in the developed countries, particularly in the USA and the EC, reflects its need to secure and better serve the largest export markets in the face of ever strengthening protectionism.

17 Data provided by Taiwan Investment Commission, Ministry of Economic Affairs and Bank of Korea.

11 IN CONCLUSION: IMPLICATIONS FOR THEORY AND POLICY

1 Direct investments abroad by the developed countries reached a similar value, $67 billion, in 1960. This represented 99.2 per cent of the estimated global stock of outward foreign direct investment at that time.

2 A similar view was put forward by Dahlman and Sercovich, 1984, Westphal *et al.*, 1984, and Pang and Hill, 1991, who suggested that exports of technology and technical services by developing countries reflect the changes in comparative advan-tage that accompany industrialisation and may also dynamically change it by broadening and deepening technological capability.

3 However, the nature of foreign direct investment in the developing countries at this stage is likely to be different from US FDI in developed countries at an earlier stage. While international production in developing countries is more likely to be of an export-oriented kind that is not demand driven, international production in the more developed countries is more likely to be of an import-substituting kind that is prompted more by demand factors in the host country.

4 Singapore remains one of the three largest offshore assembly locations for semiconductors in the world, together with Malaysia and the Philippines. In addition, Singapore has also become an important testing and distribution centre for offshore assembly operations throughout Southeast Asia (UNCTC, 1986).

5 Malaysia has taken nearly twenty-five years to build up a core of workers in the electrical equipment and electronic industry, numbering approximately 100,000 at the end of the 1980s and capable of learning, adopting, absorbing and improving on the ever higher level of technology in the industry (UNCTAD, 1990).

6 Such recourse to imported know-how rather than efforts by local firms to develop the required technology has been said to weaken the technological capability of local manufacturing companies and hinder the mastery of industrial technologies by Brazilian consulting and design engineering firms. For example, the introduction of government policies to increase the participation of local engineering firms in investment projects in the 1970s actually resulted in their limited participation in the area of detail engineering as the basic technology was provided by foreign suppliers (Guimaraes, 1986).

7 However, years of protectionism have resulted in the emergence of an inefficient, home-grown computer industry. Although computer sales by Brazilian industry have rocketed from the equivalent of $1.8 billion to $7.4 billion since the reserve was created, Brazilian computer models are outmoded and sell for three times the price of their foreign counterparts. As a result, in October 1992 (subject to Senate approval), protection for Brazil's information technology industry ends with the liberalisation of foreign investment and imports in this sector (Lamb, 1991a, b).

8 The concept of a techno-economic paradigm was first advanced by Perez (1983) and is defined as a cluster of interrelated technical, organisational and managerial innovations that lead to the use of a new range of products and processes and significant changes in the relative cost structure of all possible inputs to production. A key factor in each new paradigm that represents a particular input or set of inputs may be described by falling relative costs and universal availability. Changes in a techno-economic paradigm are more pervasive than Schumpeter's concept of radical innovations introduced at intervals of forty to sixty years, as the effects of change in the technological system exert structural changes throughout the whole economy. The development of a new techno-economic paradigm is associated with a process of *economic* selection from a range of *technically* feasible combinations of innovations that emerge from the perceived limitations on further growth within the existing paradigm in terms of productivity, profitability and markets (Freeman, 1988a, b).

9 As a result, Indian foreign investment policies were liberalised in the mid–1980s and since then inward investment as a modality for the transfer of imported foreign technology has taken on a much more significant role. The same liberalising trend is evident in the information technology industry of Brazil. See note 7.

10 Apart from a high-powered regulatory and monitoring agency, the effectiveness of active technology strategies is fundamentally dependent on the existence of a dynamic entrepreneurial class, well-trained professionals, and a significant science and technology infrastructure (UNCTC, 1987).

ANNEX 1

1 The Manila Banking Corporation ceased to exist in the late 1980s.

Index